Aerosol Technology

Properties, Behavior, and Measurement of Airborne Particles

Second Edition

William C. Hinds

Department of Environmental Health Sciences
Center for Occupational and Environmental Health
UCLA School of Public Health
Los Angeles, California

A WILEY-INTERSCIENCE PUBLICATION

JOHN WILEY & SONS, INC.

NEW YORK / CHICHESTER / WEINHEIM / BRISBANE / SINGAPORE / TORONTO

Library of Congress Cataloging-in-Publication Data

Hinds, William C.
    Aerosol technology : properties, behavior, and measurement of
airborne particles / William C. Hinds. —2nd ed.
      p.   cm.
    "A Wiley-Interscience publication."
    Includes bibliographical references and index.
    ISBN 0-471-19410-7 (cloth : alk. paper)
    1. Aerosols.   2. Aerosols—Measurement.   I. Title.
QC882.42.H56   1998   98-23683
648.5′3—dc21

Aerosol Technology

CONTENTS

Preface to the First Edition   xi

Preface to the Second Edition   xiii

List of Principal Symbols   xv

1   Introduction   1

     1.1   Definitions / 3
     1.2   Particle Size, Shape, and Density / 8
     1.3   Aerosol Concentration / 10
     Problems / 12
     References / 13

2   Properties of Gases   15

     2.1   Kinetic Theory of Gases / 15
     2.2   Molecular Velocity / 18
     2.3   Mean Free Path / 21
     2.4   Other Properties / 23
     2.5   Reynolds Number / 27
     2.6   Measurement of Velocity, Flow Rate, and Pressure / 31
     Problems / 39
     References / 41

3   Uniform Particle Motion   42

     3.1   Newton's Resistance Law / 42
     3.2   Stokes's Law / 44
     3.3   Settling Velocity and Mechanical Mobility / 46
     3.4   Slip Correction Factor / 48
     3.5   Nonspherical Particles / 51
     3.6   Aerodynamic Diameter / 53
     3.7   Settling at High Reynolds Numbers / 55
     3.8   Stirred Settling / 62

3.9   Instruments That Rely on Settling Velocity / 65

3.10 Appendix: Derivation of Stokes's Law / 67

Problems / 70

References / 73

4   Particle Size Statistics   **75**

4.1   Properties of Size Distributions / 75

4.2   Moment Averages / 82

4.3   Moment Distributions / 84

4.4   The Lognormal Distribution / 90

4.5   Log-Probability Graphs / 94

4.6   The Hatch–Choate Conversion Equations / 97

4.7   Statistical Accuracy / 100

4.8   Appendix 1: Distributions Applied to Particle Size / 104

4.9   Appendix 2: Theoretical Basis for Aerosol Particle
      Size Distributions / 105

4.10 Appendix 3: Derivation of the Hatch–Choate Equations / 105

Problems / 108

References / 110

5   Straight-Line Acceleration and Curvilinear Particle Motion   **111**

5.1   Relaxation Time / 111

5.2   Straight-Line Particle Acceleration / 112

5.3   Stopping Distance / 117

5.4   Curvilinear Motion and Stokes Number / 119

5.5   Inertial Impaction / 121

5.6   Cascade Impactors / 128

5.7   Virtual Impactors / 134

5.8   Time-of-Flight Instruments / 136

Problems / 138

References / 140

6   Adhesion of Particles   **141**

6.1   Adhesive Forces / 141

6.2   Detachment of Particles / 144

6.3   Resuspension / 145

6.4   Particle Bounce / 146

Problems / 147

References / 148

7   Brownian Motion and Diffusion   **150**

    7.1   Diffusion Coefficient / 150

    7.2   Particle Mean Free Path / 154

    7.3   Brownian Displacement / 156

    7.4   Deposition by Diffusion / 160

    7.5   Diffusion Batteries / 165

    Problems / 168

    References / 169

8   Thermal and Radiometric Forces   **171**

    8.1   Thermophoresis / 171

    8.2   Thermal Precipitators / 176

    8.3   Radiometric and Concentration Gradient Forces / 178

    Problems / 180

    References / 180

9   Filtration   **182**

    9.1   Macroscopic Properties of Filters / 182

    9.2   Single-Fiber Efficiency / 190

    9.3   Deposition Mechanisms / 191

    9.4   Filter Efficiency / 196

    9.5   Pressure Drop / 200

    9.6   Membrane Filters / 202

    Problems / 204

    References / 204

10   Sampling and Measurement of Concentration   **206**

    10.1   Isokinetic Sampling / 206

    10.2   Sampling from Still Air / 213

    10.3   Transport Losses / 216

    10.4   Measurement of Mass Concentration / 217

    10.5   Direct-Reading Instruments / 222

    10.6   Measurement of Number Concentration / 225

    10.7   Sampling Pumps / 228

    Problems / 230

    References / 231

11 Respiratory Deposition   **233**

    11.1   The Respiratory System / 233
    11.2   Deposition / 235
    11.3   Deposition Models / 242
    11.4   Inhalability of Particles / 245
    11.5   Respirable and Other Size-Selective Sampling / 249
    Problems / 257
    References / 258

12 Coagulation   **260**

    12.1   Simple Monodisperse Coagulation / 260
    12.2   Polydisperse Coagulation / 268
    12.3   Kinematic Coagulation / 272
    Problems / 276
    References / 277

13 Condensation and Evaporation   **278**

    13.1   Definitions / 278
    13.2   Kelvin Effect / 281
    13.3   Homogeneous Nucleation / 283
    13.4   Growth by Condensation / 285
    13.5   Nucleated Condensation / 288
    13.6   Condensation Nuclei Counters / 292
    13.7   Evaporation / 294
    Problems / 301
    References / 302

14 Atmospheric Aerosols   **304**

    14.1   Natural Background Aerosol / 304
    14.2   Urban Aerosol / 307
    14.3   Global Effects / 312
    Problems / 314
    References / 315

15 Electrical Properties   **316**

    15.1   Units / 316
    15.2   Electric Fields / 318
    15.3   Electrical Mobility / 320
    15.4   Charging Mechanisms / 323
    15.5   Corona Discharge / 331

15.6   Charge Limits / 333
15.7   Equilibrium Charge Distribution / 335
15.8   Electrostatic Precipitators / 338
15.9   Electrical Measurement of Aerosols / 341
Problems / 346
References / 347

16   Optical Properties   **349**

16.1   Definitions / 350
16.2   Extinction / 352
16.3   Scattering / 358
16.4   Visibility / 364
16.5   Optical Measurement of Aerosols / 370
Problems / 376
References / 377

17   Bulk Motion of Aerosols   **379**

Problems / 385
References / 385

18   Dust Explosions   **386**

Problems / 392
References / 392

19   Bioaerosols   **394**

19.1   Characteristics / 394
19.2   Sampling / 396
Problems / 400
References / 400

20   Microscopic Measurement of Particle Size   **402**

20.1   Equivalent Sizes of Irregular Particles / 402
20.2   Fractal Dimension of Particles / 408
20.3   Optical Microscopy / 413
20.4   Electron Microscopy / 416
20.5   Asbestos Counting / 422
20.6   Automatic Sizing Methods / 424
Problems / 425
References / 426

21 Production of Test Aerosols   **428**

    21.1   Atomization of Liquids / 428

    21.2   Atomization of Monodisperse Particles in Liquid Suspensions / 434

    21.3   Dispersion of Powders / 438

    21.4   Condensation Methods / 443

    Problems / 445

    References / 446

Appendices   **447**

A1. Useful Constants and Conversions Factors / 447

A2. Some Basic Physical Laws / 449

A3. Relative Density of Common Aerosol Materials / 451

A4. Standard Sieve Sizes / 451

A5. Properties of Gases and Vapors at 293 K [20°C] and 101 kPa [1 atm] / 452

A6. Viscosity and Density of Air versus Temperature / 452

A7. Pressure, Temperature, Density, and Mean Free Path of Air versus Altitude / 453

A8. Properties of Water Vapor / 455

A9. Properties of Water / 455

A10. Particle Size Range of Aerosol Properties and Measurement Instruments / 456

A11. Properties of Airborne Particles at Standard Conditions / 458

A12. Slip Correction Factor for Standard and Nonstandard Conditions / 460

A13. Properties of Selected Low-Vapor-Pressure Liquids / 461

A14. Reference Values for Atmospheric Properties at Sea Level and 293 K [20°C] / 462

A15. Greek Symbols Used in This Book / 464

A16. SI Prefixes / 464

Index   **465**

PREFACE TO THE FIRST EDITION

Airborne particles are present throughout our environment. They come in many different forms, such as dust, fume, mist, smoke, smog, or fog. These aerosols affect visibility, climate, and our health and quality of life. This book covers the properties, behavior, and measurement of aerosols.

This is a basic textbook for people engaged in industrial hygiene, air pollution control, radiation protection, or environmental science who must, in the practice of their profession, measure, evaluate, or control airborne particles. It is written at a level suitable for professionals, graduate students, or advanced undergraduates. It assumes that the student has a good background in chemistry and physics and understands the concepts of calculus. Although not written for aerosol scientists, it will be useful to them in their experimental work and will serve as an introduction to the field for students starting such careers. Decisions on what topics to include were based on their relevance to the practical application of aerosol science, which includes an understanding of the physical and chemical principles that underlie the behavior of aerosols and the instruments used to measure them.

Although this book emphasizes physical rather than mathematical analysis, an important aspect of aerosol technology is the quantitative description of aerosol behavior. To this end I have included 150 problems, grouped at the end of each chapter. They are an important tool for learning how to apply the information presented in the book. Because of the practical orientation of the book and the intrinsic variability of aerosol properties and measurements, correction factors and errors of less than 5 percent have generally been ignored and only two or three significant figures presented in the tables.

Aerosol scientists have long been aware of the need for a better basic understanding of the properties and behavior of aerosols among applied professionals. In writing this book, I have attempted to fill this need, as well as the long-standing need for a suitable text for students in these disciplines. The book evolved from class notes prepared during nine years of teaching a required one-semester course on aerosol technology for graduate students in the Department of Environmental Health Sciences at Harvard University School of Public Health.

Chapters are arranged in the order in which they are covered in class, starting with simple mechanics and progressing to more complicated subjects. Particle statistics is delayed until the student has a preliminary understanding of aerosol properties and can appreciate the need for the involved statistical characterization. Applications are discussed in each chapter after the principles have been presented. The more complicated applications, such as filtration and respiratory deposition, are introduced as soon as the underlying principles have been covered. The operating principles of different types of aerosol measuring instruments are given in general terms so that one may correctly interpret data from them and explain the frequent differences in results between instruments. Discussion of specific instruments is limited because they change rapidly and are covered well in *Air Sampling Instruments,* 5th edition, ACGIH, Cincinnati, OH (1978). The latter (or any future edition) makes an excellent companion to this text. Several general references are given at the end of each chapter. Tables and graphs are provided in the appendix for general reference and for help in dealing with the problems at the end of each chapter.

While many people have contributed to this book, I would like to acknowledge particularly Klaus Willeke of the University of Cincinnati, who reviewed the manuscript and made many helpful suggestions; Kenneth Martin, who provided the SEM photos; and Laurie Cassel, who helped prepare and type the manuscript.

WILLIAM C. HINDS

Boston, Massachusetts
February 1982

PREFACE TO THE SECOND EDITION

More than 16 years have passed since the first edition of *Aerosol Technology* was published in 1982. During this time the field of aerosol science and technology has expanded greatly, both in technology and in the number of scientists involved. When the first edition was published there were two national aerosol research associations, now there are 11 with regular national and international meetings. Growth areas include the use of aerosols in high-technology material processing and the administration of therapeutic drugs, and there is an increased awareness of bioaerosols, aerosol contamination in microelectronic manufacturing, and the effect of aerosols on global climate. While the first edition proved to be popular and useful, and became a standard textbook in the field, changes in technology and growth of the field have created the need to update and expand the book.

The objective of the book has remained the same: to provide a clear, understandable, and useful introduction to the science and technology of aerosols for environmental professionals, graduate students, and advanced undergraduates. In keeping with changes in the field, this edition uses dual units, with SI units as the primary units and cgs units as secondary units. Besides updating and revising old material, I have added a new chapter on bioaerosols and new sections on resuspension, transport losses, respiratory deposition models, and fractal characterization of particles. The chapter on atmospheric aerosols has been expanded to include sections on background aerosols, urban aerosols, and global effects. There are 26 new examples and 30 new problems. The latest edition of *Air Sampling Instruments* remains an excellent companion book, as does *Aerosol Measurement*, by Willeke and Baron. Both provide greater depth and detail on measurement methods and instruments.

Of the many people who have helped with this edition. I would like to particularly acknowledge Janet Macher, Robert Phalen, and John Valiulis for reviewing specific chapters; Rachel Kim and Vi Huynh for typing manuscript changes; doctoral student Nani Kadrichu for entering the equations; and finally, my wife Lynda for her continued support during this long process.

WILLIAM C. HINDS

Los Angeles, California

LIST OF PRINCIPAL SYMBOLS

a	acceleration, particle radius
a_c	centrifugal acceleration, Eq. 3.15
A	area, cross-sectional area
A_p	cross-sectional area of a particle
A_s	surface area
b	coefficient for Hatch–Choate equation, Eq. 4.47
B	particle mobility, Eq. 3.16
B_0	luminance of an object, Eq. 16.26
B'	luminance of background, Eq. 16.26
c	molecular velocity; velocity of light
$\bar{c}$	mean molecular velocity, Eq. 2.22; mean thermal velocity of a particle, Eq. 7.10
c_{rms}	root mean square molecular velocity, Eq. 2.18; root mean square thermal velocity of a particle, Eq. 7.9
c_x, c_y, c_z	velocity in the x, y, z directions
C	particle concentration in sampling probe
C_c	Cunningham correction factor, Eq. 3.19; slip correction factor, Eq. 3.20
C_D	drag coefficient, Eq. 3.4
C_m	mass concentration, mass of particles per unit volume of aerosol
CMD	count median diameter
C_N	number concentration, number of particles per unit volume of aerosol
C_R	apparent contrast, reduced contrast, Eqs. 16.27 and 16.33
C_0	true concentration, inherent contrast, Eq. 16.26
CE_R	collection efficiency for respirable precollector, Eq. 11.14
CE_T	collection efficiency for thoracic precollector, Eq. 11.18
d	particle diameter; derivative
$\bar{d}$	arithmetic mean diameter, Eq. 4.11
d^*	Kelvin diameter, Eq. 13.5
d_a	aerodynamic diameter, Eq. 3.26
d_A	specified average diameter, Eq. 4.47
d_c	diameter of cylinder
d_d	droplet diameter
d_e	equivalent volume diameter, Eqs. 3.23 and 19.3
d_f	fiber diameter
d_F	Feret's diameter, Fig. 20.1
d_g	geometric mean diameter, Eq. 4.14

d_i	midpoint diameter of the ith group
d_m	diameter of a gas molecule
$d_{\bar{m}}$	diameter of average mass, Eq. 4.19
d_{mm}	mass mean diameter, Eq. 4.26
d_M	Martin's diameter, Fig. 20.1
d_p	particle diameter
$d_{\bar{p}}$	diameter of average property proportional to d^p, Eq. 4.22
d_{PA}	projected-area diameter, Fig. 20.1
$(d_{qm})_{\bar{p}}$	p moment average of the qth moment distribution, Eq. 4.36
d_s	Stokes diameter, Eq. 3.26
$d_{\bar{s}}$	diameter average surface, Eq. 4.22
d_{sm}	surface mean diameter, Eqs. 4.27 and 4.31
d_t	tube diameter
$d_{\bar{v}}$	diameter of average volume, Eq. 4.22
d_w	wire diameter
d_{50}	particle diameter for 50% collection efficiency, Eqs. 5.28 and 19.1
D	particle diffusion coefficient, Eqs. 7.1 and 7.7
D_{ba}	diffusion coefficient of gas b in air, Eq. 2.35
D_F	fractal dimension, Eq. 20.5
D_j	impactor jet diameter
D_s	sampling probe diameter
D_v	diffusion coefficient of vapor in air
D_0	duct diameter
DF	deposition fraction, total, Eq. 11.5
DF_{AL}	deposition fraction, alveolar, Eq. 11.4
DF_{HA}	deposition fraction, head airways, Eq. 11.1
DF_{TB}	deposition fraction, tracheobronchial, Eq. 11.3
e	charge of an electron; coefficient of restitution, Eq. 6.6; base for natural logarithms
E	efficiency; electrical field strength, Eqs. 15.6 and 15.10
E	overall filter efficiency, Eqs. 9.1 and 9.2
E_D	single-fiber efficiency for diffusion, Eq. 9.27
E_{DR}	single-fiber efficiency for diffusion-interception interaction, Eq. 9.28
E_G	single-fiber efficiency for settling, Eq. 9.30
E_I	impactor efficiency, Eq. 5.27; single-fiber efficiency for impaction, Eq. 9.24
E_L	surface field limit, Eq. 15.28
E_q	single-fiber efficiency for electrostatic attraction, Eq. 9.32
E_R	single-fiber efficiency for interception, Eq. 9.21
E_Σ	total single-fiber efficiency, Eqs. 9.14 and 9.33
f	fraction; frequency; frequency of light, fraction of sites with colonies, Eq. 19.3
f_{ab}	fraction between sizes a and b
$f(d_p)$	frequency function of particle size distribution, Eq. 4.4
f_n	fraction of particles having n charges, Eqs. 15.30 and 15.31

F	force
$F(a)$	cumulative frequency at a, Eq. 4.8
$F(x)$	cumulative fraction at x, Eq. 11.12
F_{adh}	force of adhesion, Eqs. 6.1-6.4
F_D	drag force, Eqs. 3.4 and 3.8
F_E	electrical force, Eq. 15.8
F_f	frictional force on a fluid element, Eq. 2.36
F_G	force of gravity, Eq. 3.11
F_I	inertial force on a fluid element, Eq. 2.39
F_n	form component of Stokes drag, Eq. 3.6
F_{th}	thermal force, Eqs. 8.1 and 8.4
F_v	volume fraction of spheres in liquid, Eq. 21.6
F_τ	frictional component of Stokes drag, Eq. 3.7
g	acceleration of gravity
G	gravitational settling parameter, Eq. 9.29; ratio of cloud velocity to particle velocity, Eqs. 17.6 and 17.7
GSD	geometric standard deviation, σ_g, Eq. 4.40
h	height; velocity head, Eq. 2.43
H	height of chamber; thermophoretic coefficient, Eq. 8.5; latent heat of evaporation of a liquid
i_1	Mie intensity parameter for perpendicular component of scattered light, Eqs. 16.23 and 16.24
i_2	Mie intensity parameter for parallel component of scattered light, Eqs. 16.23 and 16.25
I	number of intervals for grouped size data, Eq. 4.14; light intensity, Eq. 16.7
I_0	incident light intensity, Eq. 16.7
$I_1(\theta)$	intensity of scattered light at angle θ, perpendicular polarization, Eq. 16.24
$I_2(\theta)$	intensity of scattered light at angle θ, parallel polarization, Eq. 16.25
IF	inhalable fraction, Eq. 11.7, 11.8
IF_N	inhalable fraction for nose breathing, Eq. 11.9
J	diffusion flux, Eqs. 2.30 and 7.1
k	Boltzmann's constant
k_v	thermal conductivity of a gas or vapor
K	a constant; corrected coagulation coefficient, Eq. 12.13
K_0	uncorrected coagulation coefficient, Eq. 12.9
$\overline{K}$	effective coagulation coefficient for polydisperse aerosols, Eq. 12.17
K_E	electrostatic constant of proportionality (SI units), Eq. 15.1 and Table 15.1
KE	kinetic energy
Kn	Knudsen number $= 2\lambda/d_p$
Ku	Kuwabara hydrodynamic factor, Eq. 9.22
K_R	Kelvin ratio, Eq. 13.5
K_{st}	Pressure rise index, Eq. 18.1

$K_{1,2}$	coagulation coefficient of particle size 1 with size 2, Eq. 12.16
L	length; length of fluid element, length of chamber, duct, or tube; path length of light beam, Eq. 16.7
L_R	limit of resolution, Eq. 20.9
L_V	visual range, Eq. 16.35
m	mass of molecule; mass of particle; index of refraction, Eq. 16.2
m_r	relative index of refraction, Eq. 16.5
M	molecular weight; total mass
MMD	mass median diameter
n	number of molecules per unit volume; number concentration; number of elementary charges
n_A	number concentration at A
n_c	rate of capture, Eq. 12.20; number of organisms collected, Eq. 19.3
n_i	number of particles in the ith group
n_L	charge limit, Eqs. 15.28 and 15.29
n_m	number of moles
$n(t)$	number of charges at time t, Eqs. 15.24, 15.25, and 15.33
n_z	rate of molecular collisions, Eq. 2.24
n_0	initial number concentration; initial number of charges
N	number of molecules; total number of particles in sample; particle number concentration
N_a	Avogadro's number
NA	numerical aperture, Eq. 20.8
N_i	ion concentration
$N(t)$	particle number concentration at time t, Eq. 12.12
N_0	particle number concentration at time zero
p	pressure; partial pressure
p_A	partial pressure of component A, Eq. 13.1
p_d	partial pressure of vapor at droplet surface, Eq. 13.5
p_s	saturation vapor pressure, Eq. 13.2
p_T	total pressure
p_v	velocity pressure, Eqs. 2.43 and 2.44
p_∞	partial pressure of vapor away from droplet
P	pressure, perimeter
P	penetration, overall filter penetration, Eqs. 9.3 and 9.4
Pe	Peclet number, Eq. 9.26
PF	PM-10 fraction, Eq. 11.19
$P(n)$	probability of n solid spheres in a droplet, Eq. 21.5
q	amount of charge; amount of charge on a particle, Eq. 15.2; weighting parameter for moment distributions
q_F	filter quality, Eq. 9.12
qMD	median of the qth moment distribution, Eq. 4.48
Q	flow rate
Q_a	absorption efficiency, Eq. 16.10
Q_e	extinction efficiency, Eq. 16.8

Q_s	sample flow rate; scattering efficiency, Eq. 16.10
Q_0	duct flow rate
r	radial position
R	gas constant, Eq. 2.1; radius; interception parameter, Eq. 9.20; separation distance of electric charges, Eq. 15.2
Re	Reynolds number, particle or flow, Eq. 2.41
Re_f	fiber Reynolds number, Eq. 9.13
Re_0	initial Reynolds number, Eq. 5.21
RF	respirable fraction, Eq. 11.10
S	stopping distance, Eq. 5.19
S_R	saturation ratio, Eq. 13.3
SMD	surface median diameter
Stk	Stokes number, Eqs. 5.23 and 5.24
Stk_{50}	Stokes number for 50% collection efficiency, Eq. 5.28
t	time; thickness of filter
T	temperature
T_d	temperature at droplet surface
TF	thoracic fraction, Eq. 11.15
T_∞	temperature away from droplet
U	velocity; gas velocity; gas velocity inside filter, Eq. 9.6; gas velocity in sampling probe
$\overline{U}$	average velocity in duct
U_0	face velocity of filter; free-stream velocity
v	gas volume
v_d	droplet volume
v_p	particle volume
v_m	volume of a molecule, Eq. 13.9
v_1, v_2	volume of gas or vapor at state 1 or 2
V	velocity of particle; relative velocity between particle and gas
V_c	critical velocity for bounce, Eq. 6.5; cloud velocity, Eq. 17.4
V_{dep}	deposition velocity, Eq. 7.27
V_f	final velocity
VMD	volume median diameter
V_r	gas velocity in the r direction, Eq. 3.41
$V(t)$	particle velocity at time t, Eq. 5.15
V_{th}	thermophoretic velocity, Eqs. 8.2 and 8.6
V_T	tangential velocity, Eq. 3.15
V_{TC}	terminal centrifugal velocity, Eq. 3.14
V_{TE}	terminal electrical velocity, Eq. 15.15
V_{TF}	terminal velocity for constant external force F, Eq. 5.5
V_{TS}	terminal settling velocity, Eqs. 3.13 and 3.21
$\overline{V}_x$	average velocity in the x-direction, Eq. 3.37
V_0	initial velocity; velocity at time zero
V_∞	gas velocity far away from particle or fiber
V_θ	gas velocity in the θ direction, Eq. 3.42
W	width of slot; voltage

x	separation distance; distance from wall
$\bar{x}$	average number of spheres per droplet, Eq. 21.6
$\bar{x}_{MMD}$	average number of spheres in an MMD-sized droplet, Eq. 21.7
x_{rms}	rms displacement of particle, Eq. 7.18;
$x(t)$	position of particle at time t, Eq. 5.18
y	vertical distance
z	number of molecular collisions per unit area, Eq. 2.15
Z	electrical mobility, Eq. 15.21
Z_i	ion mobility
α	volume fraction of fibers in a filter, solidity, Eq. 9.7; size parameter for light scattering, Eq. 16.6
α_v	volume shape factor, Eq. 20.2
β	correction factor for coagulation coefficient, Eq. 12.13
γ	surface tension; fraction captured per unit thickness of filter, Eqs. 9.11 and 9.19
Γ	velocity gradient
δ	diffusion boundary-layer thickness, Eq. 7.30
∂	partial derivative
Δd	diameter interval
Δp	pressure drop, pressure differential, Eqs. 2.47, 2.52, and 9.36
∇T	temperature gradient
ε	relative permittivity (dielectric constant); threshold of brightness contrast, Eq. 16.34
ϵ_0	permittivity of vacuum, Eq. 15.2
η	viscosity, Eq. 2.26
Θ	angle between flow direction and sampling probe
θ	scattering angle
λ	gas mean free path, Eq. 2.25; wavelength of light; step size, Eq. 20.5
λ_p	particle mean free path, Eq. 7.11
μ	deposition parameter for diffusion loss in tubes, Eqs. 7.28 and 7.33
ρ	density of gas; density of particle
ρ_b	density of bulk material
ρ_c	density of cloud, Eq. 17.2
ρ_g	density of gas
ρ_L	density of liquid
ρ_p	density of particle
ρ_0	standard density, 1000 kg/m^3 [1.0 g/cm^3]
σ	standard deviation, Eq. 4.38
σ_a	absorption coefficient, Eq. 16.11
σ_e	extinction coefficient, Eq. 16.7
σ_g	geometric standard deviation, GSD, Eq. 4.40
σ_s	scattering coefficient, Eq. 16.11
τ	relaxation time, Eq. 5.3
ϕ	bend angle, Eq. 10.17; Fuchs-effect correction factor, Eq. 13.16
χ	dynamic shape factor, Eq. 3.23
ω	angular frequency, rotational velocity

1 Introduction

The microscopic particles that float in the air are of many kinds: resuspended soil particles, smoke from power generation, photochemically formed particles, salt particles formed from ocean spray, and atmospheric clouds of water droplets or ice particles. They vary greatly in their ability to affect not only visibility and climate, but also our health and quality of life. These airborne particles are all examples of *aerosols*. An aerosol is defined in its simplest form as a collection of solid or liquid particles suspended in a gas. Aerosols are two-phase systems, consisting of the particles *and* the gas in which they are suspended. They include a wide range of phenomena such as dust, fume, smoke, mist, fog, haze, clouds, and smog. The word *aerosol* was coined about 1920 as an analog to the term *hydrosol*, a stable liquid suspension of solid particles. Aerosols are also referred to as suspended particulate matter, aerocolloidal systems, and disperse systems. Although the word *aerosol* is popularly used to refer to pressurized spray-can products, it is the universally accepted scientific term for particulate suspensions in a gaseous medium and is used in that sense in this book.

Aerosols are but one of the several types of particulate suspensions listed in Table 1.1. All are two-component systems having special properties that depend on size of the particles and their concentration in the suspending medium. All have varying degrees of stability that also depend on particle size and concentration.

An understanding of the properties of aerosols is of great practical importance. It enables us to comprehend the process of cloud formation in the atmosphere, a key link in the hydrological cycle. Aerosol properties influence the production, transport, and ultimate fate of atmospheric particulate pollutants. Measurement and control of particulate pollutants in the occupational and general environments require the application of this knowledge. Aerosol technology has commercial application in the manufacture of spray-dried products, fiber optics, and carbon black; the production of pigments; and the application of pesticides. Because the toxicity of inhaled particles depends on their physical as well as their chemical properties, an understanding of the properties of aerosols is required to evaluate airborne particulate hazards. The same knowledge is used in the administration of therapeutic aerosols for the treatment of respiratory and other diseases.

Aerosol technology is the study of the properties, behavior, and physical principles of aerosols and the application of this knowledge to their measurement and control. The particulate phase of an aerosol represents only a very small fraction of its total mass and volume, less than 0.0001%. Bulk properties of aerosols, such as viscosity and density, differ imperceptibly from those of pure air. Consequently, to study the properties of aerosols, one must adopt a *microscopic point of view*. This

1

TABLE 1.1   Types of Particulate Suspensions

Suspending Medium	Type of Suspended Particles		
	Gas	Liquid	Solid
Gas	—	Fog, mist, spray	Fume, dust
Liquid	Foam	Emulsion	Colloid, suspension, slurry
Solid	Sponge	Gel	Alloy

reduces the problem of understanding the complex properties of aerosols to that of understanding the properties of individual particles. The microscopic approach considers one particle at a time and deals with questions about the forces on that particle, its motion, and its interaction with the suspending gas, with electromagnetic radiation, and with other particles.

At the beginning of the 20th century, the study of aerosols was at the forefront of physical science because aerosols represented the smallest observable division of matter. Aerosol science contributed to the early understanding of Brownian motion and diffusion, Millikan's measurement of the charge on the electron, and Wilson's cloud chamber experiments for the study of ionizing radiation. This classical period of aerosol science research continued through the first half of the century, concluding with the publication of *The Mechanics of Aerosols* by Fuchs in 1955. Following World War II, and particularly during the 1970s and 1980s, aerosol technology grew in importance because of an increased environmental awareness and a concern for the health effects arising from air pollution in community and occupational environments. The field expanded rapidly in the 1980s to include the use of aerosols in high-technology production processes and a concern for aerosol contamination in the semiconductor industry (clean technology). The decade of the1990s has seen increased research on the properties of ultrafine particles (<0.1 μm) and on the effect of aerosols on global climate. Aerosol technology has become an important tool in understanding the effect we have on our environment and the impact of that environment on us.

Any subject that touches upon such diverse phenomena as sunsets, silicosis, rain, cascade impactors, global climate change, cross pollination, electrostatic precipitation, and rainbows is not a simple one. Aerosol technology draws on physics, chemistry, physical chemistry, and engineering. It uses some tools, concepts, and terminology of powder technology. It is used in the fields of occupational hygiene, air pollution control, inhalation toxicology, atmospheric physics and chemistry, and radiological health.

A dual system of units is used in this book, with the primary system being the the International System of Units (SI units, or meter-kilogram-second units). Because of a tradition of using cgs (centimeter-gram-second) units in this field, especially in the United States, cgs units are included in square brackets, and some equations and most examples are presented both ways.

Figures 1.1–1.5 show sources of aerosols paired with microscope photographs of the particles produced. They illustrate the range of aerosol-producing activities and the complex nature of the resulting particles.

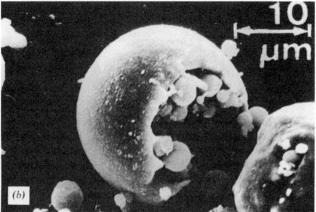

FIGURE 1.1 (*a*) Coal-burning power plant. (*b*) Scanning electron microscope (SEM) photograph of coal fly ash particles.

1.1 DEFINITIONS

Aerosols can be subdivided according to the physical form of the particles and their method of generation. There is no strict scientific classification of aerosols. The following definitions correspond roughly to common usage and are precise enough for most scientific description.

Aerosol A suspension of solid or liquid particles in a gas. Aerosols are usually stable for at least a few seconds and in some cases may last a year or more. The

FIGURE 1.2 (*a*) Granite cutting. (*b*) SEM photograph of quartz particles. Magnification 2650×.

term *aerosol* includes both the particles and the suspending gas, which is usually air. Particle size ranges from about 0.002 to more than 100 μm.

Bioaerosol An aerosol of biological origin. Bioaerosols include viruses, viable organisms, such as bacteria and fungi, and products of organisms, such as fungal spores and pollen.

Cloud A visible aerosol with defined boundaries.

Dust A solid-particle aerosol formed by mechanical disintegration of a parent material, such as by crushing or grinding. Particles range in size from submicrometer to more than 100 μm and are usually irregular.

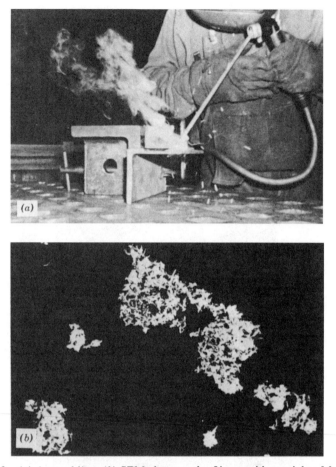

FIGURE 1.3   (*a*) Arc welding. (*b*) SEM photograph of iron-oxide particles. Magnification 2300×.

Fume   *A* solid-particle aerosol produced by the condensation of vapors or gaseous combustion products. These submicrometer particles are often clusters or chains of primary particles. The latter are usually less than 0.05 μm. Note that this definition differs from the popular use of the term to refer to any noxious contaminant in the atmosphere.

Haze   An atmospheric aerosol that affects visibility.

Mist and Fog   Liquid-particle aerosols formed by condensation or atomization. Particles are spherical with sizes ranging from submicrometer to about 200 μm.

Smog   1. A general term for visible atmospheric pollution in certain areas. The term was originally derived from the words *smoke* and *fog*. 2. *Photochemical smog* is a more precise term referring to an aerosol formed in the atmosphere by the action of sunlight on hydrocarbons and oxides of nitrogen. Particles are generally less than 1 or 2 μm.

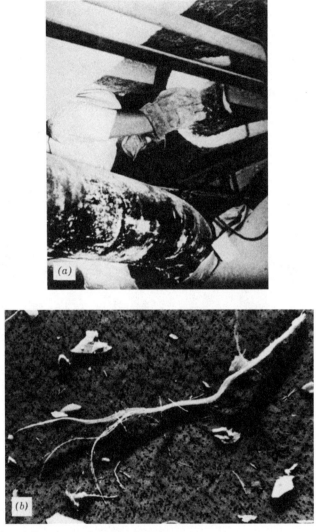

FIGURE 1.4 (*a*) Removal of asbestos pipe covering. (*b*) SEM photograph of asbestos fibers. Magnification 1250×.

Smoke   A visible aerosol resulting from incomplete combustion. Particles may be solid or liquid, are usually less than 1 μm in diameter, and may be agglomerated like fume particles.

Spray   A droplet aerosol formed by the mechanical breakup of a liquid. Particles are larger than a few micrometers.

In this book the preceding distinctions are usually not necessary, and the general term *aerosol* is used. Liquid particles are referred to as *droplets*. The term *par-*

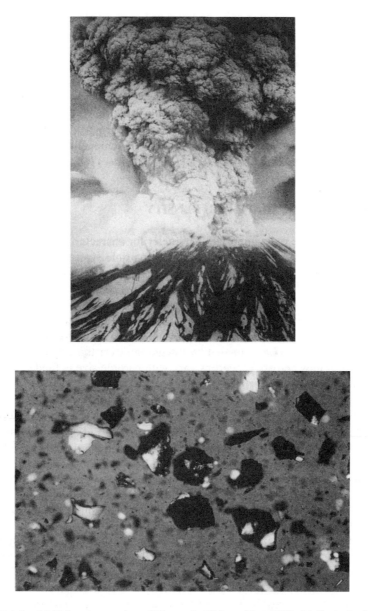

FIGURE 1.5 (a) Volcanic eruption of Mount St. Helens, May 1980. (b) Optical microscope photograph of volcanic ash. Magnification 125×. USGS photograph by Austin Post. Reprinted from *Mount St. Helens: Five Years Later*. Courtesy of Eastern Washington University Press and W. C. McCrone and J. G. Delly, *The Particle Atlas*. Reprinted by permission from McCrone Research Institute.

ticulate matter refers to either solid particles or liquid droplets. A *primary aerosol* has particles that are introduced directly into the atmosphere, whereas a *secondary aerosol* has particles that are formed in the atmosphere by chemical reactions of gaseous components (gas-to-particle conversion). A *homogenous aerosol* is an aerosol in which all particles are chemically identical. *Monodisperse aerosols* have particles that are all the same size and can be produced in the laboratory for use as test aerosols. Most aerosols are *polydisperse*, with a wide range of particle sizes, and statistical measures should be used to characterize their particle size.

In this text, *standard conditions* are defined as a temperature of 293 K [20°C] and an atmospheric pressure of 101 kPa ($1Pa = 1N/m^2$) [760 mm Hg].

1.2 PARTICLE SIZE, SHAPE, AND DENSITY

Particle size is the most important parameter for characterizing the behavior of aerosols. All properties of aerosols depend on particle size, some very strongly. Furthermore, most aerosols cover a wide range of sizes; a hundredfold range between the smallest and largest particles of an aerosol is common. Not only do aerosol properties depend on particle size, but the nature of the laws governing these properties may change with particle size. This emphasizes the need to adopt a microscopic approach and characterize properties on an individual particle basis. Average properties can then be estimated by integrating over the size distribution. An appreciation of how aerosol properties vary with particle size is fundamental to an understanding of their properties.

The "yardstick" for particle size is the micrometer (μm) or its older equivalent, the micron (μ), which is 10^{-6} m, 10^{-4} cm, or 10^{-3} mm. The micron is no longer acceptable as an SI unit. Particle size can refer to particle radius, but in this book it refers to particle diameter. For consistency it is expressed in micrometers for all particle sizes, even though nm is more appropriate for particles less than 0.1 μm. Particle diameter is given the symbol d or, where confusion with other symbols might arise, the symbol d_p. It is customary to refer to particle size in micrometers, but calculations require converting micrometers to meters (SI units) by multiplying by 10^{-6} or to centimeters (cgs units) by multiplying by 10^{-4}.

Figure 1.6 shows size ranges for aerosols and other phenomena. A major dividing line is 1 μm, which marks the upper limit of the submicrometer range (less than 1.0 μm) and the lower limit of the micrometer size range (1–10 nm). Figure 1.6 covers a size range of 10^7, from gas molecules to millimeter-sized particles. The particle sizes of the aerosols shown in the figure range from 0.01 to 100 μm, the size range addressed in this book. In general, dusts, ground material, and pollen are in the micrometer range or larger, and fumes and smokes are submicrometer. The smallest aerosol particles approach the size of large gas molecules and have many of their properties. Ultrafine particles cover the range from large gas molecules to about 100 nm (0.001 to 0.1 μm). Particles less than 50 nm are called nanometer particles or nanoparticles. Particles greater than 10 μm have limited stability in the atmosphere, but still can be an important source of occupational exposure because

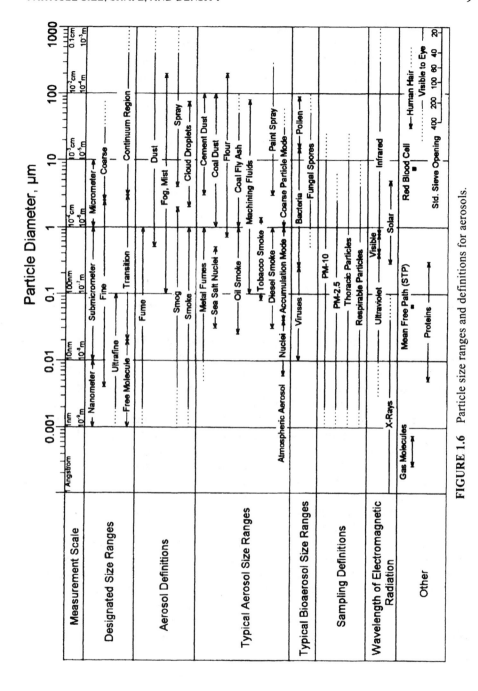

FIGURE 1.6 Particle size ranges and definitions for aerosols.

of a worker's proximity to the source. The largest aerosol particles are visible grains that have properties described by the familiar Newtonian physics of baseballs and automobiles. The dot over the letter *i* has a diameter of about 400 μm, and the smallest grains of flour that one can see under normal conditions are 50–100 μm. The finest wire mesh sieves have openings of about 20 μm. The wavelength of visible light is in the submicrometer size range, about 0.5 μm.

EXAMPLE

What is the ratio of the volume of a 10-μm spherical particle to that of a 0.1-μm particle?

$$\text{Volume} = \frac{\pi d^3}{6}$$

$$\text{Ratio} = \frac{(\pi/6)\, d_{10}^3}{(\pi/6)\, d_{0.1}^3} = \left(\frac{d_{10}}{d_{0.1}}\right)^3 = \left(\frac{10}{0.1}\right)^3 = 10^6$$

Liquid aerosol particles are nearly always spherical. Solid aerosol particles usually have complex shapes, as shown in Figs. 1.1–1.5. In the development of the theory of aerosol properties, it is usually necessary to assume that the particles are spherical. Correction factors and the use of equivalent diameters enable these theories to be applied to nonspherical particles. An *equivalent diameter* is the diameter of the sphere that has the same value of a particular physical property as that of an irregular particle. For approximate analysis, shape can usually be ignored, as it seldom produces more than a twofold change in any property. Particles with extreme shapes, such as long, thin fibers, are treated as simplified nonspherical shapes in different orientations. The complex shape of some fume and smoke particles can be characterized by their fractal dimension. (See Section 20.2.)

The remaining physical property of interest is *particle density,* usually expressed in kg/m^3 [g/cm^3]. Particle density refers to the mass per unit volume of the particle itself, not of the aerosol (the "density" of which is called concentration, as described in the next section). Liquid particles and crushed or ground solid particles have a density equal to that of their parent material. Smoke and fume particles may have apparent densities significantly less than that predicted from their chemical composition. This is a result of the large amount of void space in their highly agglomerated structure, which may resemble a cluster of grapes. In this book particles are assumed to have *standard density* ρ_0—that is, the density of water, 1000 kg/m^3 [1.0 g/cm^3]—unless specified otherwise.

1.3 AEROSOL CONCENTRATION

The most commonly measured aerosol property, and the most important one for health and environmental effects, is the *mass concentration,* the mass of particu-

TABLE 1.2 Examples of Mass Concentration Expressed in Parts per Million[a]

	Mass Concentration, Mass/Volume (mg/m³)	Parts per Million, Volume/Volume (ppm)	Parts per Million, Mass/Mass (ppm)
U.S. PM-10, annual average	0.05	5×10^{-5}	0.04
Threshold limit value for nuisance dusts			
(Particulates not otherwise classified)	10	0.01	8
Uncontrolled stack effluent (typical)	10,000	10	8,000

[a]Standard-density spheres.

late matter in a unit volume of aerosol. Common units are g/m^3, mg/m^3, and $\mu g/m^3$. The mass concentration is equivalent to the density of the ensemble of aerosol particles in air; however, the latter term is not used because of possible confusion with particle density.

Another common measure of concentration is *number concentration,* the number of particles per unit volume of aerosol, commonly expressed as number/cm³ or number/m³. An older unit is mppcf (million particles per cubic foot). Concentrations of bioaerosols and fibers are expressed in terms of number concentration.

Unlike the situation with gaseous contaminants, volume ratio or mass ratio in parts per million (ppm) is not used for aerosols, because two phases are involved and aerosol concentrations are numerically very low when expressed in this way. It is informative, however, to make such calculations for some standard concentrations, as shown in Table 1.2. Note that, on a volume basis, a dense combustion plume is 99.999% pure air.

Figure 1.7 shows the extremely wide range (from 10^{-13} to 10^3 g/m³) of aerosol concentrations that one encounters in practice.

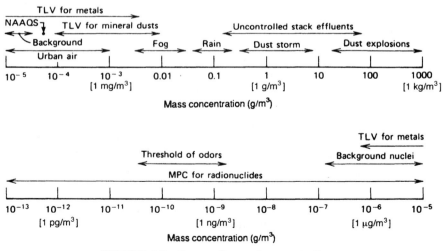

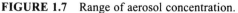

FIGURE 1.7   Range of aerosol concentration.

If you have not already done so, now is a good time to read the preface, which contains important information about this book and how to use it.

PROBLEMS

1.1 How many 1.0-μm-diameter particles are required per cubic centimeter of aerosol for the mass concentration to be 10 mg/m^3? Assume that the particle density is 1000 kg/m^3 [1 g/cm^3].
ANSWER: 19,100/cm^3.

1.2 In smoking one nonfilter cigarette, a person inhales 350 mL of aerosol containing 20 mg of tobacco smoke particles. If these particles are standard-density spheres 0.4 μm in diameter, how many particles does the smoker inhale from one cigarette? What is the mass concentration of the smoke? Make a ratio comparison of this smoke concentration to the U.S. PM-10 standard (annual average) of 50 μg/m^3.
ANSWER: 6.0×10^{11}, 0.057 kg/m^3 [57 g/m^3], 1.1×10^6.

1.3 How many molecules are in a 0.1-μm diameter water droplet?
ANSWER: 1.8×10^7.

1.4 By what factor does the total surface area of a 5-cm-diameter sphere of coal increase on being dispersed into 0.1-μm-diameter spheres?
ANSWER: 5×10^5.

1.5 A person inhales approximately 20 mg of tobacco smoke particles from one cigarette. If smoke particles are standard-density spheres 0.4 μm in diameter, what is the surface area of this amount of smoke?
ANSWER: 0.30 m^2 [3000 cm^2].

1.6 How many particles would be present in a cubic meter of aerosol at the concentration of the U.S. PM-10 standard of 50 μg/m^3 if the particles are (a) 0.1 μm, (b) 1.0 μm, and (c) 10 μm in diameter? Assume standard density.
ANSWER: (a) 9.5×10^{10}, (b) 9.5×10^7, (c) 9.5×10^4.

1.7 If aerosol particles were considered to be extremely large gas molecules, what would be the gram molecular weight of a "gas" of 1.0-μm particles having a density of 1000 kg/m^3?
ANSWER: 3.2×10^{11} g/mole.

1.8 Derive an expression for the surface area per kilogram of material as a function of particle size. Assume that the material is divided into spheres, each having a diameter d and a density ρ = 1000 kg/m^3 [1.0 g/cm^3]. What is the surface area of 1 g of 0.1-μm-diameter particles?

ANSWER: 60 m^2.

1.9 Determine the ratio of the surface area of a sphere to that of a fiber with the same volume. Assume that the fiber is a cylinder with a diameter equal to 20% of the diameter of the sphere. Assume that the sphere and the fiber have standard density.

ANSWER: 0.3.

REFERENCES

Listed below are some general references on aerosol science and technology. A comprehensive list of books and journals is given in *Aerosol Science and Technology*, **14**, 1–4 (1990).

Cadle, R. D., *The Measurement of Airborne Particles*, Wiley, New York, 1975. (Emphasizes microscopic measurement.)

Clift, R., Grace, J. R., and Weber, M. E., *Bubbles, Drops, and Particles*, Academic Press, New York, 1978. (Reference textbook on fluid dynamics, heat transfer, and mass transfer of single bubbles, drops, and particles.)

Davies, C. N. (Ed.), *Aerosol Science*, Academic Press, New York, 1966. (Thorough coverage on selected topics.)

Friedlander, S. K., *Smoke, Dust and Haze*, Wiley, New York, 1977. (Good coverage; oriented toward air pollution.)

Fuchs, N. A., *The Mechanics of Aerosols*, Pergamon, Oxford, UK., 1964. (This outstanding classic has been republished in paperback by Dover Publications, New York, 1989).

Hidy, G. M., and Brock, J. R., *The Dynamics of Aerocolloidal Systems*, Pergamon Press, New York, 1971. (A mathematical approach to the theory of aerosol science.)

Hinds, W. C., *Aerosol Technology: Properties, Behavior and Measurements of Airborne Particles*, Wiley, New York, 1982. (The predecessor to this edition.)

Mercer, T. T., *Aerosol Technology in Hazard Evaluation*, Academic Press, New York, 1973. (Good coverage; oriented toward occupational hygiene.)

Reist, P. C., *Aerosol Science and Technology*, McGraw-Hill, New York, 1993. (An introductory textbook on aerosol science.)

Shaw, D. T. (Ed.), *Fundamentals of Aerosol Science*, Wiley, New York, 1978. (Detailed coverage of selected topics.)

Vincent, J. H., *Aerosol Science for Industrial Hygienists*, Pergamon, Oxford, UK., 1995. (A textbook on the application of aerosol science and technology to the field of occupational hygiene.)

Vincent, J. H., *Aerosol Sampling: Science and Practice*, Wiley, New York, 1989. (Comprehensive coverage of aerosol sampling, especially with blunt samplers.)

Willeke, K., and Baron, P., *Aerosol Measurement: Principles, Techniques, and Applications*, Van Nostrand Reinhold, New York, 1993. (Comprehensive coverage of aerosol measurement.)

Williams, M. M. R., and Loyalka, S. K., *Aerosol Science: Theory and Practice*, Pergamon, Oxford, 1991. (Strong theoretical coverage.)

Two international journals are devoted to aerosol science and technology:

Journal of Aerosol Science, Pergamon, Elsevier, Exeter, U.K.

Aerosol Science and Technology, Taylor & Francis, Philadelphia.

2 Properties of Gases

Aerosols consist of two phases: the solid or liquid particles and the gas in which they are suspended. Most of this book deals with the properties and behavior of the aerosol particles, but the suspending gas interacts directly with the particles and has a great effect on their behavior. Particle motion is resisted by the gas, and the nature of this resistance changes with particle size, especially as particle size approaches the spacing between the gas molecules. Aerosol particles share energy with gas molecules and exhibit Brownian motion. Temperature gradients in a gas create a force on an aerosol particle called the thermophoretic force. Because of the varied interactions between a particle and the surrounding gas, we review the properties of gases—particularly, the kinetic theory of gases—before dealing with the properties of aerosols.

2.1 KINETIC THEORY OF GASES

To understand the interactions between particles and the suspending gas, it is necessary to consider certain aspects of the kinetic theory of gases. These interactions are described well by the classical kinetic theory of gases, and it is not necessary to consider the more advanced kinetic theory and quantum mechanics to get a physical picture of the interactions that is adequate for the study of aerosol properties.

According to kinetic theory, the temperature, pressure, mean free path, and viscosity of a gas are manifestations of the motion of the gas molecules. As will be shown, temperature is a measure of the kinetic energy of the molecules, pressure comes from the force of molecular impacts on the container walls, viscosity represents the transfer of momentum by molecular motion, and diffusion is the transfer of molecular mass.

The basic assumptions of the kinetic theory are that (1) gases contain a large number of molecules, (2) the molecules are small compared with the distances between them, and (3) the molecules are rigid spheres traveling in straight lines between elastic collisions. Molecules lead a simple, but happy, life in which collisions with other entities are their only means of expression. Kinetic theory uses this "billiard ball" model of gas molecules to describe properties such as temperature, pressure, viscosity, mean free path, diffusion, and thermal conductivity in terms of the properties of the billiard balls—that is, the number per unit volume n, mass m, diameter d_m, and velocity c.

15

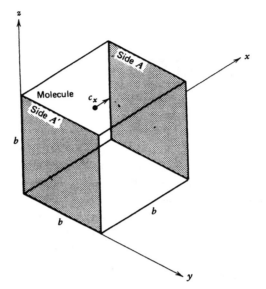

FIGURE 2.1   Cubical box for molecular derivation of Boyle's law.

The familiar ideal-gas law relates the absolute pressure P, the absolute temperature T, the volume v, and the number of moles, n_m, of a gas,

$$Pv = n_m RT \qquad\qquad (2.1)$$

where R is the gas constant, with a value that depends on the units used for the other quantities. For pressure in Pa (newtons per square meter) and volume in cubic meters, $R = 8.31$ J/K · mol (Pa · m³/K · mol); for pressure in atmospheres and volume in cubic centimeters, $R = 82$ atm · cm³/K · mol. Other values are given in Appendix A1. The ideal-gas law is valid for most gases, including air, at pressures less than a few atmospheres.

The ideal gas law is a combination of Boyle's law, Charles's law, and Avogadro's principle. Boyle's law can be written directly from Eq. 2.1 as

$$Pv = \text{constant} \quad \text{for constant } n_m \text{ and } T \qquad\qquad (2.2)$$

It can also be derived from the kinetic theory of gases and Newton's laws of motion. Consider a cubical box with side length b, shown in Fig. 2.1, and containing N molecules, each having a mass m. One of these molecules has a velocity component c_x parallel to the x-axis. Assume that the y and z components of velocity do not affect the component in the x direction and that there are no collisions with other molecules. Then the original molecule will bounce back and forth between side A and side A' at a velocity c_x. The time between successive collisions with either A or A' is

$$\Delta t = \frac{b}{c_x} \tag{2.3}$$

On each collision the momentum of the molecule changes from mc_x to $-mc_x$, for a net change in momentum of

$$\Delta m V = mc_x - (-mc_x) = 2mc_x \tag{2.4}$$

Over a long time, the rate of change of momentum is

$$\frac{\Delta m V}{\Delta t} = \frac{2mc_x}{b/c_x} = \frac{2mc_x^2}{b} \tag{2.5}$$

Newton's second law (see Appendix A2) says that the force exerted by the molecule must be equal to the rate of change of its momentum that is,

$$F = \frac{d(mV)}{dt} = \frac{2mc_x^2}{b} \tag{2.6}$$

The resulting pressure, or force per unit area, due to the molecule is

$$P = \frac{F}{A + A'} = \frac{F}{2b^2} = \frac{2mc_x^2}{2b^3} = \frac{mc_x^2}{v} \tag{2.7}$$

where $b^3 = v$, the volume of the box.

Each of the other molecules will also exert a pressure described by Eq. 2.7, so the total pressure is

$$P = \sum \left(\frac{mc_x^2}{v} \right)_i = \frac{m}{v} \sum (c_x^2)_i \tag{2.8}$$

where the summation is over all the molecules. We can express Eq. 2.8 in terms of the mean square average velocity $\overline{c_x^2}$ by

$$N\overline{c_x^2} = \Sigma (c_x^2)_i \tag{2.9}$$

so that

$$P = \frac{mN\overline{c_x^2}}{v} \tag{2.10}$$

where N is the number of molecules in the box. The speed c of each molecule is given by

$$c^2 = c_x^2 + c_y^2 + c_z^2 \tag{2.11}$$

where c_x, c_y, and c_z are the components of the molecule's velocity in the x, y, and z directions and are all equivalent. Therefore,

$$\overline{c_x^2} = \overline{c_y^2} = \overline{c_z^2} \tag{2.12}$$

and it follows that

$$\overline{c^2} = 3\overline{c_x^2} \tag{2.13}$$

Combining Eqs. 2.10 and 2.13 gives

$$Pv = \frac{mN\overline{c^2}}{3} \tag{2.14}$$

which is a statement of Boyle's law, Eq. 2.2. As required by Boyle's law, the quantities on the right-hand side of Eq. 2.14 are constant for a fixed amount of gas at a constant temperature.

A more detailed analysis shows that collisions with other molecules do not affect the results just derived. Further analysis gives the number of molecular collisions per unit area of a stationary surface each second z as

$$z = \frac{n\overline{c}}{4} \tag{2.15}$$

where $\overline{c}$ is the mean molecular velocity and n is the concentration of molecules— the number per unit volume. Equation 2.15 can be used to make a rough estimate of the collision frequency between air molecules and an aerosol particle. The molecular concentration n is obtained by dividing Avogadro's number by the molar volume of an ideal gas. For air at standard conditions, n is $2.5 \times 10^{25}/m^3$ [$2.5 \times 10^{19}/cm^3$] and $\overline{c}$ is 460 m/s [46,000 cm/s]. Equation 2.15 gives the rate of molecular collisions with any surface as $2.9 \times 10^{27}/m^2$ [$2.9 \times 10^{23}/cm^2$]. A 0.1-μm diameter particle has a surface area of 3.1×10^{-14} m^2 and will thus experience 10^{14} collisions per second! This calculation assumes that the 0.1-μm particle is stationary and that its surface is flat. It will be shown that these are both reasonable assumptions for the purposes of this calculation.

2.2 MOLECULAR VELOCITY

Equations 2.14 and 2.1 can be combined to give an expression for the square root of the mean square molecular velocity, the rms velocity. For one mole of gas,

$$Pv = RT = \frac{mN_a\overline{c^2}}{3} \tag{2.16}$$

$$c_{rms} = (\overline{c^2})^{1/2} = \left(\frac{3RT}{mN_a}\right)^{1/2} \tag{2.17}$$

where N_a is Avogadro's number. The quantity mN_a is the molecular weight of the gas M so

$$c_{\text{rms}} = \left(\frac{3RT}{M}\right)^{1/2} \tag{2.18}$$

Equation 2.18 illustrates the power of the kinetic theory in that it permits a direct calculation of the rms molecular velocity of a known gas by the measurement of one easily observable macroscopic quantity, the temperature of the gas. Molecular velocity increases with an increase in temperature, but rather slowly, being proportional to the square root of temperature.

Equation 2.16 can be rewritten to give the translational kinetic energy KE (KE = ½mV²) of one mole of gas:

$$\text{KE} = \frac{N_a m \overline{c^2}}{2} = \frac{3RT}{2} \tag{2.19}$$

For an ideal gas the kinetic energy is independent of pressure, volume, and molecular weight and thus depends only on temperature. Table 2.1 gives the molecular velocities of several gases and vapors at 293 K [20°C]. It is apparent that the lighter molecules must have greater velocities to maintain the same kinetic energy as the heavier ones.

Although the rms molecular velocity is fixed for a gas at a given temperature, there exists in the gas a wide range or distribution of molecular velocities described by the Maxwell–Boltzmann distribution. (The reader may wish to review the properties of frequency distributions in Section 4.1.) The distribution of the component of molecular velocity along any axis is given by

$$f(c_x)dc_x = \left(\frac{m}{2\pi kT}\right)^{1/2} \exp\left(\frac{-mc_x^2}{2kT}\right) dc_x \tag{2.20}$$

where $f(c_x)dc_x$ is the fraction of velocities between c_x and $c_x + dc_x$ and k is Boltzmann's constant. The latter is the gas constant per molecule, $k = R/N_a$. As shown in Fig. 2.2, velocities in the positive or negative x direction are equally likely, and the arithmetic mean will be zero, which means that the gas as a whole is sta-

TABLE 2.1   Molecular Velocities of Gases and Vapors at 293 K [20°C]

Gas or Vapor	Molecular Weight (g/mol)	rms Velocity (m/s)
H_2	2	1910
H_2O	18	637
Air	29	503
CO_2	44	407
Hg	201	191

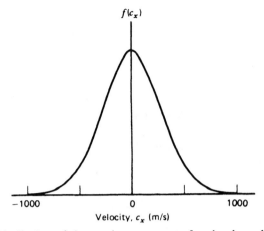

FIGURE 2.2   Distribution of the *x*-axis component of molecular velocity, N_2 at 273 K [0°C].

tionary. The corresponding equation for the distribution of the magnitudes of the molecular velocity in any direction (the molecular speed) is

$$f(c)dc = 4\pi c^2 \left(\frac{m}{2\pi kT}\right)^{3/2} \exp\left(\frac{-mc^2}{2kT}\right) dc \qquad (2.21)$$

As shown in Fig. 2.3, the probability of a molecular speed equal to zero—that is, the probability that a molecule is stopped—is zero. Because we are dealing with a

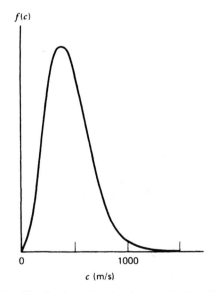

FIGURE 2.3   Distribution of molecular speed, N_2 at 273 K [0°C].

distributed quantity, different ways of computing the average velocity will give different values. The rms average molecular velocity has already been discussed. The other common type of average is the arithmetic mean, which is obtained by dividing the sum of all the molecular velocities at a given instant by the total number of molecules. For the continuous distribution given by Eq. 2.21, the mean is obtained by integrating the product of velocity and its probability over all velocities

$$\bar{c} = \int_0^\infty cf(c)\, dc = \left(\frac{8kT}{\pi m}\right)^{1/2} = \left(\frac{8RT}{\pi M}\right)^{1/2} \tag{2.22}$$

2.3 MEAN FREE PATH

In many situations in aerosol technology, there is an interaction between the particles and the gas, and one must take into account the discontinuous nature of the gas. That is, the gas cannot be treated as a continuous fluid, but must be considered an ensemble of rapidly moving molecules colliding randomly with the particles. The criterion for using such an approach depends on the particle size relative to the spacing between the gas molecules. Instead of the average spacing between molecules, a more useful concept is the *mean free path*, which is defined as the average distance traveled by a molecule between successive collisions. The mean free path λ of a gas can be determined from the average number of collisions a particular molecule undergoes in one second, n_z, and the average distance it has traveled in that second.

$$\lambda = \frac{\bar{c}}{n_z} \tag{2.23}$$

This is analogous to driving a car at 60 mph and having three (elastic) collisions every hour. The average distance between collisions is 20 miles. The quantity n_z is given by an expression related to Eq. 2.15 that takes into account the small dimensions and moving-target character of gas molecules, namely,

$$n_z = \sqrt{2}n\pi d_m^2 \bar{c} \tag{2.24}$$

where d_m is the *collision diameter* of the molecule, defined as the distance between the centers of two molecules at the instant of collision. For air, the collision diameter is approximately 3.7×10^{-10} m [3.7×10^{-8} cm].
  Combining Eqs. 2.23 and 2.24 gives

$$\lambda = \frac{1}{\sqrt{2}n\pi d_m^2} \tag{2.25}$$

The mean free path for air at 101 kPa [1 atm] and 293 K [20°C] is 0.066 μm.
  For a given gas, that is, a fixed d_m, the mean free path depends only on the gas density, which is directly proportional to n. Thus for ambient conditions the mean

free path increases with increasing temperature or decreasing pressure. In the atmosphere, the mean free path increases with altitude.

EXAMPLE

What is the mean free path at the top of Mount Whitney, altitude 4421 m [14,494 ft.]? Assume that the temperature and pressure are 20°C and 0.7 atm.

$$\text{Eq. 2.25: } \lambda \propto \frac{1}{n}, \text{ where } n \propto P$$

$$\text{Thus, } \frac{\lambda_1}{\lambda_0} = \frac{n_0}{n_1} = \frac{P_0}{P_1}$$

$$\lambda_1 = \lambda_0 \frac{P_0}{P_1} = 0.066\left(\frac{1}{0.7}\right) = 0.094 \text{ } \mu m$$

Figure 2.4 and Table 2.2 compare the relative size and spacing of air molecules with a 0.1-μm-diameter particle (only a portion of its edge is shown) at standard conditions. The figure is a two-dimensional representation of the gas molecules, with the three-dimensional spacing of the molecules shown in the plane of the figure. In the figure, a 0.1-μm–diameter particle would have a diameter of 0.28 m [28 cm]. On average, a molecule striking a particle will have traveled a distance equal to the mean free path (0.066 μm) since its last collision with another gas molecule.

In scientific work on the interaction of gases and particles, it is common to use the Knudsen number, a dimensionless number equal to the ratio of the mean free

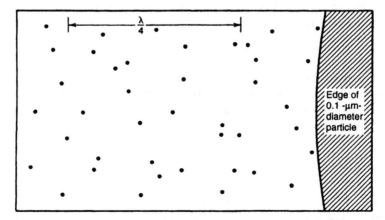

FIGURE 2.4   Relative size and spacing of air molecules at standard conditions.

TABLE 2.2   Properties of Air Molecules at Standard Conditions

Distance	Length (μm)	Ratio to Molecular Diameter
Molecular diameter	0.00037	1
Approximate molecular spacing	0.0034	9
Mean free path	0.066	180
0.1-μm particle	0.1	260

path to the particle radius, Kn = 2λ/d. The Knudsen number is a confusing quantity that gets larger as the particle size gets smaller and is not used in this book.

2.4   OTHER PROPERTIES

The kinetic theory, particularly in its more advanced form, can be used to describe many properties of gases besides temperature and pressure, such as thermal conductivity, viscosity, and diffusivity. Because of the importance of viscosity and diffusion to the study of particle motion, we briefly review the molecular basis for these gas properties.

First we state Newton's law of viscosity, which gives the frictional force between fluid layers moving at different velocities. Consider two parallel plates of area A separated by a distance y that is small compared with the dimensions of the plates (Fig. 2.5). One plate moves with a constant velocity U, and the other is stationary. The gas (or liquid) between the plates resists the motion, so that a force F has to be continuously applied to maintain a constant velocity. This force is proportional to the area of the plates and to the relative velocity of one plate with respect to the other and is inversely proportional to the distance between the plates, so that

$$F = \frac{\eta A U}{y} \tag{2.26}$$

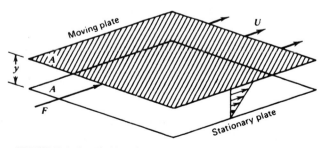

FIGURE 2.5   Fluid resistance between two parallel plates.

where η is the proportionality constant called the *coefficient of dynamic viscosity* or, simply, the *viscosity*. In the SI system, viscosity has units of N · s/m², kg/m · s, or Pa · s, where 1 Pa (pascal) = 1 N/m². [In the cgs system, the unit of viscosity is the poise (pronounced pwaz), which has dimensions of dyn · s/cm²]. The viscosity of air at 20°C is 1.81×10^{-5} Pa · s [1.81×10^{-4} poise (P)]. A fluid with a viscosity of 1 Pa · s requires a force of 1 N to maintain a difference in velocity of 1 m/s between two 1-m² plates separated by 1 m of fluid. Viscosity can be thought of as an internal coefficient of friction for a fluid.

At the molecular level, the viscosity of a gas represents a transfer of molecular momentum from a faster moving layer to a slower moving layer. This transfer is accomplished by the random thermal motion of molecules traveling between the layers.

An appealing physical analogy for this process is that of two trains of flatcars coasting along at slightly different speeds on parallel adjacent tracks. The passengers, with mass m', on the flatcars amuse themselves by jumping back and forth between the trains. Each jump from the faster train to the slower train imparts a momentum of $m'\Delta V$ to speed up the slower train. Similarly, jumps the other way slow the faster train. The greater the difference in velocity and the more frequent the jumping, the greater the effect. For a jumping rate of N' per second, each train receives a force of $N'm'\Delta V$, accelerating the slower train and decelerating the faster train.

Because the transfer of molecular momentum takes place in the velocity gradient between the plates, U/y, the difference in fluid velocity at the ends of the average transit (the mean free path) is $2\lambda U/3y$, averaged for all orientations of molecular motion. A molecule of mass m transfers an amount of momentum equal to $2m\lambda U/3y$ during the average transit. The number of molecules crossing (both ways) an area A parallel to the plates is given by Eq. 2.15 as $2(n\bar{c}/4)$. The total momentum transfer per second is

$$\frac{nm\bar{c}\lambda A U}{3y} \tag{2.27}$$

This rate of transfer of momentum is equal to the frictional resisting force (Eq. 2.26). Combining Eqs. 2.26 and 2.27 gives

$$\eta = \frac{1}{3}nm\bar{c}\lambda = \frac{1}{3}\rho_g\bar{c}\lambda \tag{2.28}$$

Combining Eq. 2.22 for $\bar{c}$ and Eq. 2.25 for λ gives

$$\eta = \frac{2(mkT)^{1/2}}{3\pi^{3/2}d_m^2} \tag{2.29}$$

Viscosity is independent of pressure and depends only on molecular constants and temperature. It increases with increasing temperature. The temperature dependence given by Eq. 2.29 is only approximately correct, and the true temperature dependence is greater than $T^{1/2}$ for most gases, because the rigid-sphere assumption is not met. For example, over the temperature range from 223–773 K [−70 to 500°C], the viscosity of air is proportional to its absolute temperature raised to the 0.74 power. The viscosity of air is shown as a function of temperature in Fig. A6.

That the viscosity of a gas is independent of pressure is a surprising result, but one that has been verified experimentally over a pressure range of 0.001–100 atm. That the viscosity of a gas should increase with temperature also contradicts our intuition, which is based on liquids such as oil or honey that get less viscous as they are heated. The viscosity of liquids, unlike that of gases, is governed primarily by cohesive forces between the closely spaced liquid molecules. These forces decrease rapidly with an increase in temperature, resulting in a decrease in viscosity. Gas molecules are too far apart most of the time for these cohesive forces to be significant.

EXAMPLE

What is the viscosity of air at 100°C?

$$\eta \propto T^{0.74} \qquad \eta_2 = \eta_1 \left(\frac{T_2}{T_1} \right)^{0.74}$$

$$\eta_{100°C} = 1.81 \times 10^{-5} \left(\frac{373}{293} \right)^{0.74} = 2.16 \times 10^{-5} \ \text{Pa} \cdot \text{s}$$

$$\left[\eta_{100°C} = 1.81 \times 10^{-4} \left(\frac{373}{293} \right)^{0.74} = 2.16 \times 10^{-4} \text{dyn} \cdot \text{s/cm}^2 \right]$$

A slightly more accurate determination can be made by using the Sutherland equation:

$$\eta_T = \frac{1.458 \times 10^{-6} \, T^{1.5}}{T + 110.4} \ \text{Pa} \cdot \text{s}$$

$$\eta_{100°C} = \frac{1.458 \times 10^{-6} \times 373^{1.5}}{373 + 110.4} = 2.17 \times 10^{-5} \ \text{Pa} \cdot \text{s}$$

The Sutherland equation is accurate over the range 100–1800 K.

The transfer of mass of one gas through another in the absence of fluid flow is called *diffusion*. It is the result of the motion of the gas molecules in a concentration gradient. For gases diffusing through air, the flux J, the quantity transferred

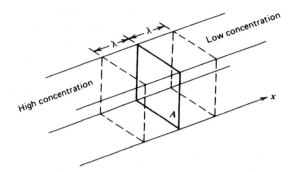

FIGURE 2.6 Diagram for derivation of diffusion coefficient.

per unit time through a unit area perpendicular to the direction of diffusion (molecules/m$^2 \cdot$ s) under the influence of a concentration gradient dC/dx, is given by Fick's first law of diffusion:

$$J = -D_{ba} \frac{dC}{dx} \tag{2.30}$$

The quantity D_{ba} is the constant of proportionality called the *diffusion coefficient*, or *diffusivity*, of gas b in air and has units of m^2/s [cm^2/s]. The negative sign is required because the mass transfer is always toward the region of lower concentration, or "down the gradient." At the molecular level, the diffusion coefficient can be described in terms of the mean free path and the mean molecular velocity $\bar{c}$. Consider only motion along the positive x-axis in a concentration gradient dn/dx, as shown in Fig. 2.6. The number of molecules of gas b crossing a unit area of plane A from left to right in unit time is

$$J^+ = (n^+)\bar{c} \tag{2.31}$$

where n^+ *is* the number of molecules of gas b having motion in the positive x direction. Assuming that one-third of the molecules have motion in the x direction, one-third in the y direction, and one-third in the z direction, then one-sixth of the molecules will have motion in the positive x direction. The number concentration n^+ will be then one-sixth of the concentration at a point one mean free path away from plane A, because this is the region where the molecules experienced their most recent collisions before passing through A. Similarly, the concentration in the negative x direction, n^-, is based on the concentration one mean free path on the other side of A. Thus,

$$n^+ = \frac{1}{6}\left(n_A - \lambda \frac{dn}{dx}\right)$$

$$\tag{2.32}$$

$$n^- = \frac{1}{6}\left(n_A + \lambda \frac{dn}{dx}\right)$$

where n_A is the concentration at plane A. The net number of molecules passing through a unit area of A is

$$J = J^+ - J^- = -\frac{2\bar{c}\lambda}{6}\frac{dn}{dx} = -\frac{1}{3}\bar{c}\lambda\frac{dn}{dx} \qquad (2.33)$$

Comparing Eq. 2.33 with Eq. 2.30, we find

$$D_{ba} = \frac{1}{3}\bar{c}\lambda \qquad (2.34)$$

Substituting for $\bar{c}$ and λ (Eqs. 2.22 and 2.25) yields

$$D_{ba} = \left(\frac{2}{3\pi^{3/2}}\right)\frac{1}{nd_m^2}\left(\frac{RT}{M}\right)^{1/2} \qquad (2.35)$$

A more rigorous theoretical analysis gives the first factor in Eq. 2.35 as $3\sqrt{2}\pi/64$. Calculation of the diffusion coefficient for air at standard conditions by Eq. 2.35 gives 1.8×10^{-5} m^2/s, about 10% less than the correct value of 2.0×10^{-5} m^2/s [0.20 cm^2/s]. A similar derivation can be applied to the transfer of kinetic energy to obtain the thermal conductivity of a gas.

2.5 REYNOLDS NUMBER

A key to understanding the aerodynamic properties of aerosol particles is the Reynolds number, a dimensionless number that characterizes fluid flow through a pipe or around an obstacle such as an aerosol particle. The Reynolds number has the following properties:

1 It is an index of the flow regime; that is, it provides a benchmark to determine whether the fluid flow is laminar or turbulent.
2 It is proportional to the ratio of inertial forces to frictional forces acting on each element of the fluid. This ratio is the key to determining which flow resistance equation is correct in a given situation.
3 Equality of Reynolds numbers is required for geometrically similar flow to occur around geometrically similar objects. This similarity means that the pattern of the streamlines will be the same for flow around different-sized objects or flow of different fluids. A streamline is the path traced by a tiny element of fluid as it flows around an obstacle.

The Reynolds number can be derived by evaluating the ratio of inertial to frictional force acting on an element of fluid in a steady-flow system. The *frictional force*, defined by Eq. 2.26, is

$$F_f = \eta A \frac{dU}{dy} \propto \eta L^2 \frac{dU}{dL} \qquad (2.36)$$

where L is some characteristic length of the fluid element and dU/dy is the velocity gradient in the region of the fluid element. The *inertial force is* equal to the rate of change of momentum of the fluid element,

$$F_I = ma = m \frac{d'U}{dt} \tag{2.37}$$

where $d'U/dt$ is the total acceleration of the fluid element. The latter has two components, an acceleration due to a change in the total flow of the system, dU/dt, and the acceleration that results when the fluid element moves into a region of higher (or lower) velocity, such as around an obstacle or in a pipe of varying diameter. The second acceleration, $U\, dU/dx$, is proportional to the fluid velocity and the rate of change of velocity with position. Thus, the total acceleration is

$$\frac{d'U}{dt} = \frac{dU}{dt} + U \frac{dU}{dx} \tag{2.38}$$

where the first term on the right is zero for the steady flow conditions considered here. Combining Eqs. 2.37 and 2.38 gives

$$F_I = \rho L^3 U \frac{dU}{dL} \tag{2.39}$$

Combining Eqs. 2.36 and 2.39 gives the Reynolds number Re, the ratio of inertial to frictional force,

$$\text{Re} = \frac{F_I}{F_f} = \frac{\rho L^3 U}{\eta L^2} = \frac{\rho U L}{\eta} \tag{2.40}$$

where ρ is the density of the fluid. The relative velocity V between the fluid and an object such as a particle is proportional to U, and the characteristic linear dimension, such as the diameter d of a pipe or particle, *is* proportional to L for a geometrically similar flow. The Reynolds number expressed in terms of these more useful quantities is

$$\text{Re} = \frac{\rho V d}{\eta} \tag{2.41}$$

The Reynolds number is dimensionless in any consistent system of units. The quantities associated with inertia are in the numerator and viscosity is in the denominator, as expected from the derivation. Care should be taken to note that the density is that of the gas, not the aerosol particle, an occasional source of confusion in the application of the Reynolds number to aerosol particles.

The properties of air given in Table 2.3 provide a simpler form of the Reynolds number for use with SI or cgs units at standard conditions:

$$\begin{aligned} \text{SI Units:} \quad &\text{Re} = 66{,}000 Vd \quad \text{for } V \text{ in m/s and } d \text{ in m} \\ [\text{cgs units:} \quad &\text{Re} = 6.6 Vd \quad \text{for } V \text{ in cm/s and } d \text{ in cm}] \end{aligned} \tag{2.42}$$

TABLE 2.3   Properties of Air at Standard Conditions: 293K, 101 kPa [20°C, 760 mmHg]

Property	SI Units	cgs Units
Viscosity	1.81×10^{-5} Pa · s (N · s/m^2)	1.81×10^{-4} dyn · s/cm^2
Density	1.20 kg/m^3	1.20×10^{-3} g/cm^3
Diffusion Coefficient	2.0×10^{-5} m^2/s	0.20 cm^2/s
Mean Free Path	0.066 μm	0.066 μm

Equation 2.42 applies equally well to flow in a pipe or particle motion; the latter is called the *particle Reynolds number*.

The Reynolds number depends only on the relative velocity between an object such as an aerosol particle and the surrounding gas. It is aerodynamically equivalent for air to flow past a stationary particle or for the particle to move through stationary air at the same velocity. *Laminar flow* around a particle occurs at low Reynolds numbers (Re < 1), for which viscous forces are much greater than inertial forces. This kind of flow is characterized by a smooth pattern of streamlines that are symmetrical on the upstream and downstream sides of the particle, as shown in Fig. 2.7(a). As Reynolds number increases above 1.0, eddies form downstream

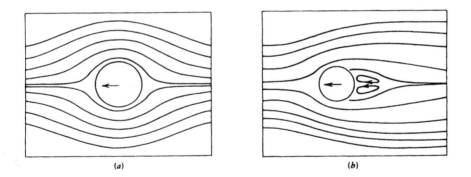

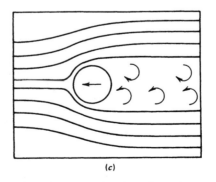

FIGURE 2.7   Flow around a sphere. (*a*) Laminar flow, Re = 0.1. (*b*) Turbulent flow, Re ≅ 2. (c) Turbulent flow, Re ≅ 250.

of the particle, gradually becoming more numerous and vigorous as shown in Figs. 2.7(b) and (c). Flow through pipes is laminar for Re < 2000 and turbulent for Re > 4000. The reason for the different upper limits of the laminar-flow region for particles and pipes is the enhanced importance of inertial forces for a fluid flowing around a particle compared with the straight-line flow along a pipe.

EXAMPLE

A sample of oil mist is taken at a flow rate of 1.2 L/min through a horizontal tube 10 mm in diameter. The aerosol consists of 2-μm diameter oil droplets in air at standard conditions. The particles are moving through the tube with the air, but are settling at 1×10^{-4} m/s [0.01 cm/s]. What is (a) the flow Reynolds number and (b) the particle Reynolds number due to the settling of the particles?

$$\text{Flow Re} = \frac{\rho V d}{\eta} = 66{,}000 \ Vd = [6.6 \ Vd] \quad \text{(Eqs. 2.41 and 2.42)}$$

where V is the flow velocity—the velocity of the air relative to the tube—and d is the diameter of the tube.

$$V = \frac{Q}{A}$$

where

$$Q = \frac{1.2 \ \text{L/min} \times 0.001 \ \text{m}^3/\text{L}}{60 \ \text{s/min}} = 2.0 \times 10^{-5} \ \text{m}^3/\text{s} \left[\frac{1.2 \times 1000}{60} = 20 \ \text{cm}^3/\text{s} \right]$$

and A is the cross-sectional area of the tube,

$$A = \frac{\pi d^2}{4} = \frac{\pi (0.01)^2}{4} = 7.9 \times 10^{-5} \ \text{m}^2 \left[\frac{\pi 1^2}{4} = 0.79 \ \text{cm}^2 \right]$$

$$V = \frac{2.0 \times 10^{-5} \ \text{m}^3/\text{s}}{7.9 \times 10^{-5} \ \text{m}^2} = 0.25 \ \text{m/s} \left[\frac{20}{0.79} = 25 \ \text{cm/s} \right]$$

substituting gives

$$\text{Flow Re} = 66{,}000 \ Vd = 66{,}000 \times 0.25 \times 0.01 = 170$$

$$= [6.6 \times 25 \times 1 = 170]$$

Since 170 is less than 2000, flow in the tube is laminar.

$$\text{Particle Re} = \frac{\rho V d}{\eta} = 66{,}000 \ Vd = [6.6 \ Vd] \quad \text{(Eqs. 2.41 and 2.42)}$$

where V is the settling velocity of the particle through the gas and d is the diameter of the particle. Substituting gives

$$\text{Particle Re} = 66,000 \ Vd = 66,000(1 \times 10^{-4})(2 \times 10^{-6}) = 1.3 \times 10^{-5}$$

$$= [6.6 \times 0.01 \times (2 \times 10^{-4}) = 1.3 \times 10^{-5}]$$

Since 1.3×10^{-5} is less than 1.0, particle motion is in the laminar region.

2.6 MEASUREMENT OF VELOCITY, FLOW RATE, AND PRESSURE

Table 2.4 gives comparative information on representative types of instruments for the measurement of local velocity, volume flow rate, integrated volume, and pressure of gases. These devices find application in the sampling of aerosols and the calibration of sampling equipment. Measurement of the local velocity of air in a duct is required for proper isokinetic sampling (Section 10.1) and can be used to determine flow rate and to calibrate flow-measuring devices. The *pitot tube* (Fig. 2.8) measures directly the velocity pressure in a moving stream of gas. The device consists of two concentric tubes connected to a pressure-measuring device, such as a manometer. The inner tube is aligned with, and open to, the oncoming flow of gas to measure the total pressure (velocity pressure plus static pressure). The outer tube has holes normal to the gas flow direction to measure the static pressure. The dif-

TABLE 2.4 Instruments for the Measurement of Velocity, Flow Rate, Volume, and Pressure of Gases

Quantity Measured	Instrument	Range	Figure
Velocity	Pitot tube	>5 m/s	2.8
	Hot wire anemometer	50 mm/s to 40 m/s	
Flow rate	Venturi meter	1 L/s to 100 m³/s	2.9
	Orifice meter	1 cm³/s to 100 m³/s	2.10
	Rotameter	0.01 cm³/s to 50 L/s	2.11
	Mass flowmeter	0.1 cm³/s to 2 L/s	
	Laminar-flow element	0.1 cm³/s to 20 L/s	
Volume	Spirometer	1–1000 L	2.12
	Soap bubble spirometer[a]	1–1000 cm³	2.13
	Piston meter[a]	1–1000 cm³	
	Wet test meter	Unlimited	2.14
	Dry gas meter	Unlimited	
Pressure	Manometer	0–200 kPa [0–2 atm]	2.15
	Micromanometer	0–0.5 kPa [0–0.005 atm]	
	Aneroid pressure gauge	0–30 kPa [0–0.3 atm]	
	Bourdon tube gauge	>20 kPa [>0.2 atm]	

[a]Automated version available to measure flow rate.

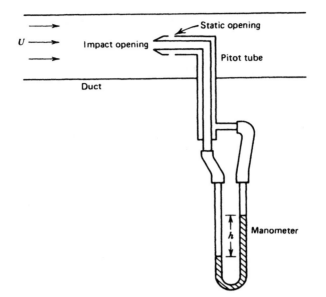

FIGURE 2.8   Pitot tube.

ference of the two pressures is the velocity pressure. The general relationship between velocity and velocity pressure is

$$U = \sqrt{2gh} \tag{2.43}$$

where U is velocity, g is the acceleration due to gravity, and h is the velocity pressure, or head, expressed as the height of the equivalent column of air. By converting the velocity pressure p_v to pascals, Eq. 2.43 can be written in a more useful form for air at standard conditions,

$$U = 0.35\sqrt{p_v} \tag{2.44}$$

where U is in m/s. Often the pressure difference is measured with a water manometer as shown in Fig. 2.8, and the equation

$$U = 4.04\sqrt{p_v} \tag{2.45}$$

where U is in m/s and p_v is in millimeters of water, is more useful.

The pitot tube is a standard air velocity meter for velocities greater than 5 m/s [1000 ft/min]. If made according to standard design, it is considered accurate and requires no calibration.

A *hot wire anemometer* measures air velocity by sensing the cooling effect of the flow of air past a heated wire. The greater the air velocity, the greater the cooling effect. This effect is sensed electronically and converted to a digital or meter display of velocity. These devices and others using heated films can measure velocities as low as 0.05 m/s (10 ft/min) and as high as 40 m/s (8000 ft/min). They

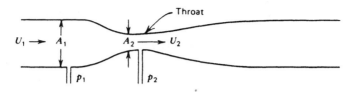

FIGURE 2.9 Venturi meter.

must be temperature compensated for the air stream temperature and must be calibrated periodically. A constant mass flow can be maintained with this device.

The total volumetric flow rate in a duct Q can be determined with a local velocity-measuring instrument by integrating local velocity over the duct cross-sectional area A. This approach is used to calibrate venturi and orifice meters for ventilation systems. For an average velocity $\overline{U}$, the flow rate is given by

$$Q = \overline{U}A \qquad (2.46)$$

Two types of meters provide a direct measurement of flow rate: variable-head meters, such as venturi and orifice meters, and variable-area meters, called rotameters. The *venturi meter* and the *orifice meter* measure the average flow rate by measuring the pressure differential across a calibrated resistance in the flow stream. For a *venturi meter*, this resistance is a streamlined constriction in the duct designed to minimize losses. As shown in Fig. 2.9, pressure is measured upstream and at the venturi throat. The volumetric flow rate for turbulent flow is given by

$$Q = kA_2 \left(\frac{2(\Delta p)}{\rho_g (1 - (A_2/A_1)^2)} \right)^{1/2} \qquad (2.47)$$

where $k = 0.98$ for standard conditions, Δp is the pressure differential $(p_1 - p_2)$, A_1 and A_2 are the cross-sectional areas upstream and at the throat, respectively, and ρ_g is the gas density.

A simpler, less expensive, and more common variable-head meter is the *orifice meter*, a thin plate with a sharp-edged circular orifice at its center inserted into the flow stream. As shown in Fig. 2.10, the orifice causes the flow lines to converge to a shape similar to that for the venturi. The narrowest part of the flow stream is

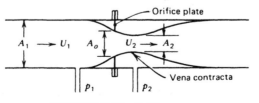

FIGURE 2.10 Orifice meter.

called the *vena contracta*. Because of the difficulty of measuring the cross-sectional area of the stream at the vena contracta A_2, it is customary to replace A_2 with the orifice area A_o in Eq. 2.47. If pressure taps are placed upstream and at the vena contracta, a value of $k = 0.62$ is used. The value of k depends on the location of the pressure taps and other factors, so an experimental calibration is usually required. For both types of variable-head meters, differential pressure is measured with a manometer or an aneroid pressure gauge.

A *critical nozzle* (an orifice with a streamline inlet) will maintain constant flow when upstream conditions are constant and the downstream absolute pressure *is less than* 0.53 of the upstream pressure. Under these conditions the velocity in the throat is the speed of sound, and a further reduction in downstream pressure does not increase the velocity through the throat. A *critical orifice* maintains a nearly constant flow under such conditions. For a critical nozzle or orifice, an approximate equation relating the orifice area A_o in mm^2 to the flow rate Q in L/min at standard conditions is

$$Q = 11.7kA_o, \quad \text{for} \quad 1 < Q < 20 \text{ L/min} \tag{2.48}$$

where k is 0.98 for a critical nozzle and 0.62 for a critical orifice. For conditions other than standard,

$$Q_{\text{STP}} \propto \frac{p_1}{\sqrt{T_1}} \tag{2.49}$$

where Q_{STP} is the volumetric flow rate expressed in terms of air at standard conditions (standard temperature and pressure). These devices are useful for taking constant flow rate samples with a vacuum pump.

A *flow controller* combines a pressure regulator with a metering (needle) valve to maintain a constant flow rate under varying conditions. The regulator maintains a constant pressure drop across the metering valve.

Unlike the variable-head meter, which measures a pressure drop that varies with flow rate, the variable-area meter varies the orifice area with the flow rate to maintain a nearly constant pressure drop. The most common type of variable-area meter is the *rotameter* (Fig. 2.11), consisting of a float free to move up and down in a vertical, tapered tube through which the fluid to be measured passes.

The float rises in the tapered tube until its weight balances the upward drag force due to the fluid flowing up through the tube. The area between the float and the tube wall increases as the float rises, reducing the velocity and drag force of the fluid. For a float of mass m_f and cross-sectional area A_f, the equation for flow is

$$Q = C_r A_o \left(\frac{2gm_f}{\rho_g A_f} \right)^{1/2} \tag{2.50}$$

where C_r is a rotameter coefficient (usually 0.6–0.8) and A_o is the open area of the tube at the float position. Float position is usually calibrated by marks on the tube

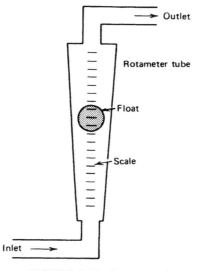

FIGURE 2.11   Rotameter.

in terms of volumetric flow rate at standard atmospheric pressure. For operation at a density or pressure other than standard, Q_{STP}, the true flow rate, expressed in terms of air at standard conditions, is given by

$$Q_{STP} = (\text{indicated } Q)\left(\frac{\rho_r}{\rho_{STP}}\right)^{1/2} = (\text{indicated } Q)\left(\frac{p_r}{p_{STP}}\right)^{1/2} \qquad (2.51)$$

where the subscript r refers to the actual condition in the rotameter. The effect of a change in viscosity depends on the design of the float and its position in the tube. Generally an increase in viscosity increases the reading for a given flow rate.

The flow rate is read on the tube scale at the widest point of the float, unless the rotameter has been calibrated for another reading point. Some rotameters have interchangeable floats of different densities to provide different flow rate ranges with the same tube. Rotameter flow rates range from 0.01 cm³/s to 50 L/s. The accuracy of mass-produced rotameters is ±2 to ±10% of the full-scale reading, so these devices may be quite inaccurate when the float is near the bottom of the scale. The maximum flow rate for a particular rotameter is about 10 times the minimum flow rate.

A *mass flowmeter* produces an electrical signal proportional to the mass flow rate of gas passing through a sensing tube. A constant output heater is located midway along the tube. The gas temperature is measured upstream and downstream of the heater. The resulting temperature difference is proportional to the mass flow rate. The output is usually displayed as volumetric flow rate at standard conditions. These devices can be coupled to electronically controlled valves to maintain a constant mass flow rate.

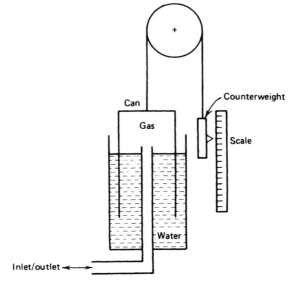

FIGURE 2.12   Spirometer

A typical *laminar-flow element* consists of a tube containing parallel narrow channels through which the gas flows. Flow in the channels is laminar. Differential pressure across the element is directly proportional to the volumetric flow rate and viscosity. This is in contrast to the venturi and orifice meters, for which differential pressure is proportional to Q^2. (See Eq. 2.47.)

The ability to measure integrated gas volume over a period of time is used to calibrate flow rate meters. There are two types, those that use an expandable chamber and positive-displacement meters. The *spirometer* is a primary standard for measuring integrated gas volume. As shown in Fig. 2.12, an inverted cylindrical can, open at the bottom, is free to move up and down in a water seal. In operation, the can is raised a distance that is directly proportional to the volume of the gas introduced into the interior of the can. The capacity of a spirometer is limited by the vertical travel of the can. These devices are primary standards because their calibration, volume per unit displacement, can be determined by direct physical measurement. They range in capacity from 0.001 to 1 m³ [1 to 1000 L].

Small volumes of less than 500 cm³ can best be measured by a simple device called a *soap bubble spirometer* or bubble flowmeter (Fig. 2.13). A soap bubble across the mouth of a burette acts as a piston and is displaced along the burette a distance proportional to the volume removed from the burette. Volume is read directly from the burette scale. A compact, automated version electronically senses the bubble's transit and displays the flow rate. These devices can measure flow rates from 1 cm³/min to 25 L/min.

A *piston meter* uses the displacement of a near-frictionless piston to measure volume. Electronic timing of the piston's movement enables the flow rate to be cal-

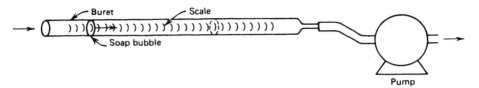

FIGURE 2.13   Soap bubble spirometer.

culated and displayed. These devices can measure flow rates from 1 cm³/min to 50 L/min.

The *wet test meter* (Fig. 2.14*)* uses a rotating system of chambers connected to a revolution counter. As gas is introduced, it fills one chamber at a time, causing it to rotate. The water level seals the chamber and acts as a valve to direct the flow to the proper chamber. There is no limit on the total gas volume that can be measured.

The *dry gas meter* uses a pair of bellows that are alternately filled and emptied by the metered gas. Movement of the bellows controls valves that direct the flow and operate a system of dial counters that indicate the total volume of gas passing through the meter. Several revolutions of the indicator dial are necessary to eliminate the effect of nonlinear strokes. These devices resemble household gas meters.

The simplest device for measuring low differential pressures, and one that requires no calibration, is the *U-tube manometer*. As shown in Fig. 2.15*a*, this device displays the pressure difference $p_1 - p_2$ as a difference in the height of the liquid column, Δh. The pressure is measured in mm, cm, or inches of the liquid in the

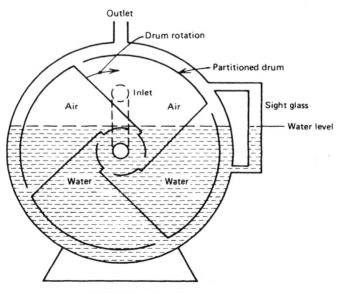

FIGURE 2.14   Wet test meter.

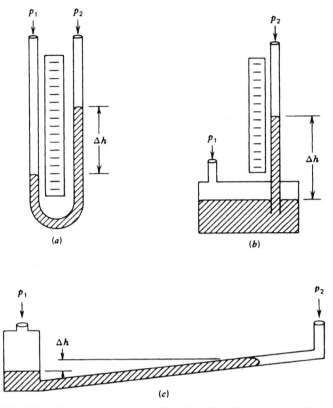

FIGURE 2.15   Manometers. (*a*) U-tube, (*b*) well type, (*c*) inclined.

manometer, which could be colored water, mercury, or another liquid whose density is known. For any liquid, the pressure can be determined by the relationship

$$\Delta p = \rho_L g \Delta h \qquad (2.52)$$

where ρ_L is the density of the liquid. One should always admit pressure (or vacuum) to a manometer cautiously to avoid blowing (or sucking) the liquid out of the manometer tube.

A common modification is the *well-type manometer* (Fig. 2.15*b*), in which the readout scale is compensated to account for the small change in level in the well, so that only a single column-height reading is required. As with all types of manometers, the pressure is always the difference in heights of the liquid columns.

The *inclined manometer* (Fig. 2.15*c*), is a more sensitive device capable of measuring pressure differences as small as 0.1 mm of water. The pressure is still the difference in heights of the liquid columns, but because of one or both legs of the manometer are inclined, the scale is greatly expanded and can be read more accurately. A 1:10 ratio of vertical rise to distance along the scale is common. For such an inclined manometer, a pressure corresponding to 1 unit in the vertical direction is displayed as a 10-unit displacement along the inclined face. Some commercially available manometers have combined inclined and vertical tubes, giving the sensitivity at low pressure of an inclined manometer with the range of a vertical manometer.

A *micromanometer* relies on the manual adjustment of a micrometer and electronic sensing of surface contact to determine the precise position of the liquid surface in a U-tube manometer. The accuracy of the instrument is ±0.005 mm of water.

The aneroid type of pressure gauge contains a diaphragm that is mechanically deformed by pressure. A low-friction linkage system displays movement of the diaphragm as a pressure reading on the dial. By using a large diaphragm, these gauges can be made sensitive to very small pressures. Portable and easy to read, they operate in any orientation and need no fluid; however, being mechanical devices, their calibration should be checked occasionally.

The common Bourdon-tube gauge is used for pressures greater than 20 kPa (3 psi). The gauge responds to the movement of the tube, which tends to straighten out as the pressure inside it is increased.

PROBLEMS

2.1 Estimate the viscosity of CO_2 gas at 20°C if its molecular collision diameter is 3.8×10^{-10} m.

ANSWER: 1.43×10^{-5} Pa · s [1.43×10^{-4} P].

2.2 Determine the number of molecular impacts per second on the surface of a 0.5-μm diameter particle in air at a temperature of 300°C.

ANSWER: 1.6×10^{15}/s.

2.3 What is the magnification of Fig. 2.4, based on the mean free path shown?

ANSWER: ~ 2.8×10^{6}.

2.4 What is the mean free path of air at an altitude of 20 km? At this altitude, the pressure is 5.5 kPa and the temperature is 217 K.

ANSWER: 0.90 μm.

2.5 A hyperbaric chamber is operated at 1.5 atm. What is the mean free path of the air in the chamber? The temperature is 20°C.

ANSWER: 0.044 μm.

2.6 The viscosity of CO_2 (MW = 44) at standard conditions has been found to be 1.48×10^{-5} Pa · s. Based on the kinetic theory of gases presented here, determine the collision diameter of a CO_2 molecule and the number of collisions it experiences in 1 s.

ANSWER: 3.7×10^{-10} m , 5.7×10^9/s.

2.7   What change occurs in the mean free path of a gas in a sealed container when the temperature of the gas is raised from 0 to 500°C?

ANSWER: No change.

2.8   At what pressure will air at 20°C have a mean free path of 1 μm?

ANSWER: 6.6 kPa.

2.9   The velocity of sound in a gas V_s is given by $V_s = (\kappa RT/M)^{\frac{1}{2}}$, where κ is the ratio of the specific heat at constant pressure to that at constant temperature ($\kappa = 1.40$ for air), T is the absolute temperature, R is the gas constant, and M is the molecular weight of the gas. How does V_s vary with mean molecular velocity? Calculate V_s for air at 20°C. What does this proportionality say about the mechanism of sound transmission in air?

ANSWER: 340 m/s [3.4×10^4 cm/s].

2.10 For a given velocity in a duct, how does the Reynolds number vary with absolute temperature?

ANSWER: Re $\propto T^{-1.74}$.

2.11 What must the air temperature be inside a hot-air balloon 10 m in diameter to just lift a 100-kg payload? The pressure inside and outside the balloon is 101 kPa, and the outside temperature is 20°C. [*Hint*: Buoyant force = Vol $(\rho_{out} - \rho_{in})g$.]

ANSWER: 348 K [75°C].

2.12 A rotameter scale is calibrated for standard conditions. What is the volumetric flow rate (in standard L/min) if the device reads 8 L/min at a pressure of 160 kPa? What is the actual volumetric flow rate at 160 kPa?

ANSWER: 10.1 L/min, 6.4 L/min.

2.13 A hole is to be drilled in a thin disk to make a critical orifice for sampling at 5 L/min at standard conditions. What should the diameter of the hole be?

ANSWER: 0.98 mm.

2.14 A rotameter is used downstream of a filter to measure sample flow rate. What absolute pressure will result in a flow error of 5% if uncorrected? How many psi and mm Hg below atmospheric pressure is this pressure?

ANSWER: 92 kPa, −1.4 psi, −71 mm Hg.

REFERENCES

Bird, B. R., Stewart, W. E., and Lightfoot, E. N., *Transport Phenomena*, Wiley, New York, 1960.

Kauzmann, W., *Kinetic Theory of Gases*, W. A. Benjamin, New York, 1966.

Lippmann, M., "Airflow Calibration," in Cohen, B., and Hering, S. (Eds.), *Air Sampling Instruments*, 8th ed., Ch. 7, ACGIH, Cincinnati, 1995.

Moore, W. J., *Physical Chemistry*, Prentice-Hall, Englewood Cliffs, NJ, 1962.

National Oceanic and Atmospheric Administration, National Aeronautics and Space Administration, and U.S. Air Force, *U.S. Standard Atmosphere*, 1976, NOAA, NASA, USAF, Washington, DC, 1976. Values in this standard are identical to ICAO Standard (1964) and ISO Standard (1973) up to an altitude of 32 km.

Prandtl, L., and Tietjens, O., *Fundamentals of Hydro- and Aeromechanics* and *Applied Hydro- and Aeromechanics*, 2 vols., reprinted by Dover, 1957.

3 Uniform Particle Motion

The most common and perhaps the most important type of particle motion is steady, straight-line motion. This uniform motion is typically the result of the action of two forces, a constant external force such as gravity or electrical force and the resistance of the gas to particle motion. Analysis of uniform particle motion is especially useful for the study of aerosols, because in most situations aerosol particles come to a constant velocity almost instantly. The acceleration of aerosol particles is covered in Chapter 5. The resisting force of the gas depends on the relative velocity between the particle and the gas and is the same whether the particle moves through the gas or the gas flows past the particle. Both situations are used in the analysis of particle motion that follows.

3.1 NEWTON'S RESISTANCE LAW

The general equation for the force resisting the motion of a sphere passing through a gas was derived by Newton as part of a ballistic evaluation of cannonballs. Newton's resistance equation is valid for a wide range of particle motion but is useful primarily for Reynolds numbers greater than 1000, a range applicable to cannonballs, but not to aerosol particles. As is discussed in Section 3.2, Stokes's law, although derived from first principles, represents a special case of Newton's resistance law. Newton's resistance equation is used in Section 3.7 for high Reynolds number settling of aerosol particles.

Newton reasoned that the resistance experienced by a cannonball traveling through air is a result of the acceleration of the air that has to be pushed aside to allow the sphere to pass through. In one second, a sphere of diameter d will push aside a volume of gas equal to the projected area of the sphere times its velocity V. The mass of this amount of gas is

$$\dot{m} = \rho_g \frac{\pi}{4} d^2 V \qquad (3.1)$$

The acceleration of the displaced gas is proportional to the relative velocity between the sphere and the gas, and thus the change of momentum per unit time is

$$\frac{\text{change of momentum}}{\text{unit time}} \propto \dot{m}V = \rho_g \frac{\pi}{4} d^2 V^2 \qquad (3.2)$$

By definition, this rate of change of momentum is equal to the force required to move the sphere through the gas, the *gas resistance force* or *drag force* F_D, and given by

$$F_D = K\rho_g \frac{\pi}{4} d^2 V^2 \tag{3.3}$$

where K is a constant of proportionality. Newton originally thought that K was independent of velocity for a given shape. This is true, however, only when the Reynolds number for particle motion is greater than 1000. Equation 3.3 with a constant value of K is the restricted form of Newton's equation. A more general relationship is obtained by replacing K in Eq. 3.3 with the coefficient of drag C_D, giving

$$F_D = C_D \frac{\pi}{8} \rho_g d^2 V^2 \tag{3.4}$$

This is the general form of Newton's resistance equation, valid for all subsonic particle motion. The dimensionless coefficient of drag is constant for spheres having Re > 1000, but changes value for Re < 1000, as shown in Fig. 3.1. The curve in the figure is for spheres; other shapes have similar curves. For example, at Re > 1000, C_D for a delivery van is about 1.0, for a sedan about 0.5, for a sports car 0.2–0.3, and for an airplane 0.04.

The derivation of Eq. 3.3 is based on the inertia of the gas, and it follows that Eq. 3.3 is valid only for motion at high Reynolds numbers, where inertial forces are much larger than viscous forces. For particle motion with Reynolds numbers from 1000 to 2×10^5, C_D has a nearly constant value of 0.44, and Eq. 3.3 can be

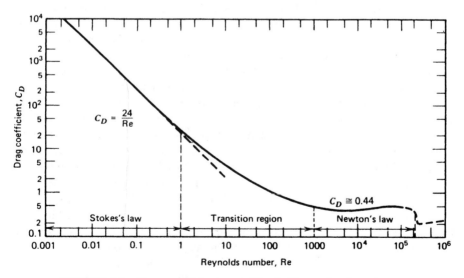

FIGURE 3.1 Drag coefficient versus Reynolds number for spheres.

used. With Reynolds numbers less than 1000, one needs to know the particle Reynolds number to get the correct value of C_D from Fig. 3.1 in order to use Eq. 3.4. Since most aerosol motion problems have either diameter or velocity as one of the unknowns, the Reynolds number cannot be calculated until the problem is solved. A way around this dilemma is covered in Section 3.7.

At the left side of Fig. 3.1, the straight-line segment on the log–log plot implies that the functional relationship between the variables is constant, but different from that given by Eq. 3.3. This is the Stokes region, which is covered in the next section. Between these straight-line regions, the relationship between C_D and Re can be described by

$$C_D = \frac{24}{\text{Re}}(1 + 0.15\,\text{Re}^{0.687})\qquad\qquad(3.5)$$

which agrees with the experimentally based correlation given by Clift et al. (1978) within 4% for Re < 800 and within 7% for Re < 1000.

3.2 STOKES'S LAW

Newton's drag (Re > 1000) applies to particle motion for which the viscous effects of the gas can be neglected compared with the inertial effects. In 1851 Stokes derived an expression for drag at the other extreme, when inertial forces are negligible compared with viscous forces. As discussed in Section 2.5, the Reynolds number is a ratio of inertial to viscous forces; consequently, a condition of negligible inertial forces, compared with viscous forces, implies a low Reynolds number and laminar flow. Because of the low velocities and small particle sizes involved, most aerosol motion occurs at low Reynolds numbers. Stokes's law thus has wide application to the study of aerosols, and it is worthwhile reviewing briefly the derivation of the law and its assumptions and implications.

Stokes's law is a solution of the generally unsolvable Navier–Stokes equations. These equations are the general differential equations describing fluid motion. They are derived from the application of Newton's second law to a fluid element on which the forces include body forces, pressure, and viscous forces. The resulting equations are very difficult to solve, because they are nonlinear partial differential equations. In general, some simplifying assumptions must be made before they can be solved. Stokes's solution does this by assuming that the inertial forces are negligibly small compared with the viscous forces. This assumption eliminates the higher order terms in the Navier–Stokes equations and yields linear equations that can be solved. Further assumptions of Stokes's derivation are that the fluid is incompressible, there are no walls or other particles nearby, the motion of the particle is constant, the particle is a rigid sphere, and the fluid velocity at the particle's surface is zero. A derivation of Stokes's law is given in the appendix at the end of this chapter.

Stokes solved the Navier–Stokes equations with the assumptions just stated to obtain equations for the forces acting at any point in the fluid surrounding a spherical particle. The net force acting on the particle is obtained by integrating the normal and tangential forces over the surface of the particle. The two resulting forces acting in the direction opposite to particle motion are the *form component*,

$$F_n = \pi \eta V d \qquad (3.6)$$

and the *frictional component*,

$$F_\tau = 2\pi \eta V d \qquad (3.7)$$

These components are combined to give the total resisting force on a spherical particle moving with a velocity V through a fluid:

$$F_D = 3\pi \eta V d \qquad (3.8)$$

This is *Stokes's law*. When the resisting force experienced by a particle is described by Eq. 3.8, the particles motion is said to be in the Stokes region.

As a particle moves through a fluid, it deforms the fluid, causing layers of the fluid in the region around the particle to slide over one another. The resisting force is the result of the friction of these layers sliding over one another. The energy spent overcoming this resistance is dissipated as heat throughout the fluid. In practice, the use of Stokes's law is restricted to situations in which the particle Reynolds number is less than 1.0. At this Reynolds number, the error in the drag force as calculated by Stokes's law is 12%; at a Reynolds number of 0.3, it is 5%.

It is instructive to compare the drag force given by Stokes's law with that given by Newton's law:

$$F_D = 3\pi \eta V d = C_D \frac{\pi}{8} \rho_g V^2 d^2 \quad \text{for Re} < 1 \qquad (3.9)$$

Stokes's law contains viscosity, but not factors associated with inertia, such as ρ_g; Newton's law contains ρ_g, but not viscosity. Solving Eq. 3.9 for the coefficient of drag gives

$$C_D = \frac{24\eta}{\rho_g V d} = \frac{24}{\text{Re}} \qquad (3.10)$$

This is the equation for the straight-line portion at the left of Fig. 3.1. Because Eq. 3.10 includes V and d, it changes the functional relationship of Newton's equation from an equation having a drag force proportional to V^2 and d^2 to one with a drag force that is proportional to V and d, Stokes's law. For conditions between the range of application of Newton's and Stokes's laws, the functional dependence of drag gradually changes from V^2 to V and from d^2 to d. This is the curved portion of Fig. 3.1. The frictional component, Eq. 3.7, represents two-thirds of Stokes's drag, and the expression equivalent to Eq. 3.10 for flow in tubes where there is no form component gives $C_D = 16/\text{Re}$, or two-thirds of Eq. 3.10.

How do the other assumptions of Stokes's law limit its application to aerosol particles? The assumption of an incompressible fluid implies not that air is incompressible, but rather that it does not compress significantly near the particle as the particle moves through it. This is equivalent to assuming that the relative velocity is much less than the speed of sound, which is nearly always the case for aerosol particles.

The presence of a wall within 10 diameters of a particle will modify the drag force on the particle. Because of the small size of aerosol particles, only a tiny fraction of the aerosol particles will be within 10 particle diameters of a wall in any real container or tube.

The correction to Stokes's law for nonrigid spheres such as water droplets is generally insignificant. Water droplets settle 0.6% faster than predicted by Stokes's law because of circulations that develop within the droplet caused by the resisting force at the droplet surface.

The effect of nonzero velocity at the particle surface is covered in Section 3.4 and the effect of nonspherical particles in Section 3.5.

3.3   SETTLING VELOCITY AND MECHANICAL MOBILITY

One important application of Stokes's law is the determination of the velocity of an aerosol particle undergoing gravitational settling in still air. When a particle is released in air, it quickly reaches its terminal settling velocity, a condition of constant velocity wherein the drag force of the air on the particle, F_D, is exactly equal and opposite to the force of gravity F_G. Under this condition,

$$F_D = F_G = mg \qquad (3.11)$$

$$3\pi\eta Vd = \frac{(\rho_p - \rho_g)\pi d^3 g}{6} \qquad (3.12)$$

where g is the acceleration of gravity, ρ_p is the density of the particle, and ρ_g is the density of the gas. The latter is included to account for the buoyancy effect, but this can usually be neglected because ρ_p is much greater than ρ_g. For example, a water droplet settling in air has a density ratio $\rho_p/\rho_g = 800$, and neglecting buoyancy introduces an error of only 0.1%. Solving Eq. 3.12 for the terminal settling velocity V_{TS} gives

$$V_{TS} = \frac{\rho_p d^2 g}{18\eta}, \quad \text{for } d > 1 \ \mu\text{m and Re} < 1.0 \qquad (3.13)$$

Terminal settling velocity increases rapidly with particle size, being proportional to the square of the particle diameter. As expected from the derivation, the settling velocity in the Stokes region is inversely proportional to viscosity and does not depend on the density of the gas. As will be shown in Section 5.2, aerosol particles adjust to their terminal settling velocity almost instantly, and V_{TS} is appropriate for characterizing particle motion in most real situations. Equation 3.13 cannot be used for particles smaller than 1.0 μm unless the slip correction factor, covered in the next section, is applied. A simpler version of Eq. 3.13, valid only for spheres with standard density (1000 kg/m³ [1.0 g/cm³]) at standard conditions, is

$$V_{TS} \cong 3 \times 10^{-5} d^2 \text{ m/s} \quad \text{for } 1 < d < 100 \text{ μm}$$

$$[V_{TS} \cong 0.003 d^2 \text{ cm/s} \quad \text{for } 1 < d < 100 \text{ μm}]$$

where d is in μm.

The terminal velocity for other kinds of external forces, such as centrifugal force, can be obtained by derivations similar to that given for gravity. In a centrifugal force field, the terminal velocity is

$$V_{TC} = \frac{\rho_p d^2 a_c}{18\eta} \tag{3.14}$$

where a_c is the centrifugal acceleration at the location of the particle. For a tangential velocity V_T and radius of motion R,

$$a_c = \frac{V_T^2}{R} \tag{3.15}$$

The Stokes settling equation, Eq. 3.13, is of fundamental importance to aerosol studies. However, in the form given, it is accurate ($\pm 10\%$) only for determining the settling velocity of standard-density particles having diameters of 1.5–75 μm. When slip correction is included (see Section 3.4), it is accurate for particles as small as 0.001 μm.

In Stokes's law, Eq. 3.8, the resistance force is directly proportional to velocity. From this relationship, we can define the particle mobility, B, a measure of the relative ease of producing steady motion for an aerosol particle, as

$$B = \frac{V}{F_D} = \frac{1}{3\pi\eta d} \quad \text{for } d > 1 \text{ μm} \tag{3.16}$$

Mobility is the ratio of the terminal velocity of a particle to the steady force producing that velocity. It has units of m/N · s [cm/dyn · s] and is often called mechanical mobility to distinguish it from electrical mobility. The terminal velocity of an aerosol particle is simply the force times mobility; for example,

$$V_{TS} = F_G B \tag{3.17}$$

For particles less than 1 μm, B must be multiplied by the slip correction factor. (See Section 3.4.)

EXAMPLE

What are the terminal settling velocity, drag force, and mobility of a 2.5-μm-diameter iron-oxide sphere settling in still air? The density of iron oxide is 5200 kg/m³ [5.2 g/cm³].

Substituting into Eq. 3.13, we obtain

$$V_{TS} = \frac{\rho_p d_p^2 g}{18\eta} = \frac{5200 \times (2.5 \times 10^{-6})^2 \times 9.81}{18(1.81 \times 10^{-5})} = 9.8 \times 10^{-4} \text{ m/s}$$

$$= \left[\frac{5.2 \times (2.5 \times 10^{-4})^2 \times 981}{18(1.81 \times 10^{-4})} = 0.098 \text{ cm/s} \right]$$

We calculate the particle Reynolds number to ensure that motion is in the Stokes region:

$$\text{Re} = 66,000 Vd = 66,000(9.8 \times 10^{-4})(2.5 \times 10^{-6}) = 1.62 \times 10^{-4}$$

$$= [6.6 Vd = 6.6(0.098)(2.5 \times 10^{-4}) = 1.62 \times 10^{-4}]$$

The motion is well within the Stokes region. The Stokes drag force is

$$F_D = 3\pi\eta Vd = 3\pi(1.81 \times 10^{-5})(9.8 \times 10^{-4})(2.5 \times 10^{-6}) = 4.18 \times 10^{-11} \text{ N}$$

$$= [3\pi\eta Vd = 3\pi(1.81 \times 10^{-4})(0.098)(2.5 \times 10^{-4}) = 4.18 \times 10^{-8} \text{ dyn}]$$

The same result can be obtained by calculating the force of gravity, since, at the terminal settling velocity, $F_D = F_G$.

$$F_G = ma = \rho_p \frac{\pi d^3}{6} g = \frac{5200\pi(2.5 \times 10^{-6})^3 9.81}{6} = 4.17 \times 10^{-13} \text{ N}$$

$$= \left[\frac{5.2\pi(2.5 \times 10^{-4})^3 981}{6} = 4.17 \times 10^{-8} \text{ dyn} \right]$$

The particle mobility is

$$B = \frac{V_{TS}}{F_G} = \frac{9.8 \times 10^{-4}}{4.17 \times 10^{-13}} = 2.35 \times 10^9 \text{ m/N} \cdot \text{s}$$

$$= \left[\frac{0.098}{4.17 \times 10^{-8}} = 2.35 \times 10^6 \text{ cm/s} \cdot \text{dyn} \right]$$

3.4   SLIP CORRECTION FACTOR

An important assumption of Stokes's law is that the relative velocity of the gas right at the surface of the sphere is zero. This assumption is not met for small particles whose size approaches the mean free path of the gas. Such particles settle faster than predicted by Stokes's law, because there is "slip" at the surface of the particle. At standard conditions, this error becomes significant for particles less than 1 μm in diameter. In 1910, Cunningham derived a correction factor for Stokes's law to account for the effect of slip. The factor, called the *Cunningham correction factor C_c*, is always greater than one and reduces the Stokes drag force by

$$F_D = \frac{3\pi\eta Vd}{C_c} \qquad (3.18)$$

where

$$C_c = 1 + \frac{2.52\lambda}{d} \qquad (3.19)$$

and λ is the mean free path, given by Eq. 2.25.

Use of the Cunningham correction factor, Eq. 3.19, extends the range of application of Stokes's law to particles of 0.1 μm in diameter. This range can be extended to still smaller particles by empirical equations based on experimental measurements of slip. The following equation agrees with the correlation (adjusted for mean free path) developed by Allen and Raabe for oil droplets (1982) *and* for solid particles (1985) within 2.1% for all particle sizes:

$$C_c = 1 + \frac{\lambda}{d}\left[2.34 + 1.05\exp\left(-0.39\frac{d}{\lambda}\right)\right] \qquad (3.20)$$

This factor is called the *slip correction factor* and must be used in the form of Eq. 3.20 for particles less than 0.1 μm in diameter. The slip-corrected form of the terminal settling velocity becomes

$$V_{TS} = \frac{\rho_p d^2 g C_c}{18\eta} \quad \text{for Re} < 1.0 \qquad (3.21)$$

Equation 3.21 is valid for all particle sizes when Re < 1.0 and Eq. 3.20 is used for C_c.

The slip correction factor for a 1.0-μm particle at standard conditions is 1.15; that is, the particle settles 15% faster than predicted by Eq. 3.13, which is based on the uncorrected form of Stokes's law. For particles less than 1 μm, slip increases rapidly as size decreases (see Figs. 3.2 and A12), and the Cunningham or slip correction factor must be used. For accurate work, it should be used for particles less than 5 or 10 μm. It is commonly stated that slip correction is necessary for particles as their size approaches the mean free path because "the particles are so small that they slip between the molecules." This is an incorrect, but useful, way of remembering how to apply the slip correction factor.

Slip correction increases as pressure decreases, because the mean free path increases. A useful form of Eq. 3.20 for air at pressures other than standard pressure is

$$C_c = 1 + \frac{1}{Pd}[15.60 + 7.00 \ \exp(-0.059Pd)] \tag{3.22}$$

where P is the absolute pressure in kPa and d is the particle diameter in μm.

Equation 3.22 underscores the dependence of slip correction on pressure and suggests a simple way to determine slip correction factors at pressures other than standard. Pressure appears in Eq. 3.22 only as a product of P and d, so any combination of P and d having the same product will give the same slip correction factor. Dividing Pd by 101 kPa gives the particle size that has the same slip correction factor at standard pressure as does a particle of size d at pressure P. This factor can be determined directly from Fig. 3.2 or Table A11. For example, the slip correction factor for a 1-μm particle at 202 kPa pressure is the same as that for a 2-μm particle at 101 kPa ($C_c = 1.08$), which can be obtained directly from Fig. 3.2 or Table A11.

Although pressure does not appear directly in Stokes's law, pressure affects the settling velocity of small particles because it influences the mean free path, which in turn affects slip correction. As shown in Table 3.1, increasing the gas density slows the settling of particles small enough to have a large slip correction ($d < 0.1$ μm) and particles large enough ($d > 100$ μm) to have a significant inertial component to their drag force. Particles between these sizes—for example, 10-μm particles—show little effect of pressure on settling velocity.

Motion of particles when $d \ll \lambda$ ($d < 0.02$ μm at standard pressure) is said to be in the free-molecule or molecular kinetic region. In this region the air resists the particle's motion, not as a continuous fluid, but as a series of discrete impacts oc-

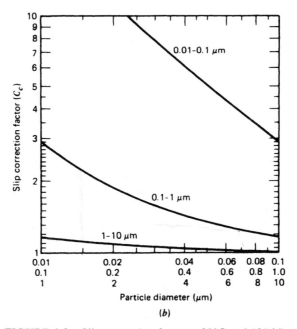

FIGURE 3.2 Slip correction factor at 20°C and 101 kPa.

TABLE 3.1 Effect of Pressure on Terminal Settling Velocity of Standard Density Spheres at 293 K [20°C].

Particle Diameter (μm)	V_{TS} at the Indicated Pressure (m/s)		
	$P = 0.1$ atm	$P = 1.0$ atm	$P = 10$ atm
0.001	6.9×10^{-8}	6.9×10^{-9}	6.9×10^{-10}
0.01	6.9×10^{-7}	7.0×10^{-8}	8.7×10^{-9}
0.1	7.0×10^{-6}	8.8×10^{-7}	3.5×10^{-7}
1	8.8×10^{-5}	3.5×10^{-5}	3.1×10^{-5}
10	0.0035	0.0031	0.0029
100	0.29	0.25	0.17

curring more frequently on the front of the particle than on the back. When $d > 3$ μm, particle motion is in the continuum region ($C_c \approx 1$), and the uncorrected form of Stokes's law, Eq. 3.8, can be used. The region between the molecular kinetic and the continuum region is called the transition region. The slip correction factor in the form of Eq. 3.20 corrects Stokes's law so that it can be used for particle motion in the free molecule and transition (as well as continuum) regions.

3.5 NONSPHERICAL PARTICLES

The equations for drag and settling velocity covered thus far are based on spherical particles. Liquid droplets (less than 1 mm) and some solid particles formed by condensation are spherical; most other types of particles are nonspherical. Some have regular geometric shapes, such as cubic (sea salt particles), cylindrical (bacteria and fibers), single crystals, or clusters of spheres. Others, such as agglomerated particles or crushed materials, have irregular shapes. The shape of a particle affects its drag force and settling velocity.

A correction factor called the *dynamic shape factor* is applied to Stokes's law to account for the effect of shape on particle motion. The dynamic shape factor is defined as the ratio of the actual resistance force of the nonspherical particle to the resistance force of a sphere having the same volume and velocity as the nonspherical particle. The dynamic shape factor χ is given by

$$\chi = \frac{F_D}{3\pi\eta V d_e} \tag{3.23}$$

where d_e, called the *equivalent volume diameter*, is the diameter of a sphere having the same volume as that of the irregular particle. This sphere is called the *equivalent volume sphere*. The equivalent volume diameter can be thought of as the diameter of the sphere that would result if the irregular particle melted to form a droplet. The calculation of d_e from microscopic measurements of particles is covered in Section 20.1. Stokes's law for irregular particles becomes

$$F_D = 3\pi\eta V d_e \chi \tag{3.24}$$

and the terminal settling velocity becomes

$$V_{TS} = \frac{\rho_p d_e^2 g}{18\eta\chi} \tag{3.25}$$

Dynamic shape factors for particles of various shapes are given in Table 3.2. Values for the geometric shapes were determined by measuring the settling velocity of geometric models in liquids. For irregular particles, settling velocities were measured indirectly using the elutriation devices described in Section 3.9. Values given in the table are averaged over all orientations, which is the usual situation for aerosol particle motion (Re < 0.1) because of the Brownian motion of the particles. Except for certain streamlined shapes, the dynamic shape factor is greater than 1.0. This means that nonspherical particles settle more slowly than their equivalent volume spheres.

The calculation of the slip correction factor for irregular particles is complicated, and the approximate factor calculated for the equivalent volume sphere is adequate

TABLE 3.2   Dynamic Shape Factors

| | | Dynamic Shape Factor,[a] χ | | |
| | | Axial Ratio | | |
Shape		2	5	10
Geometric Shapes				
Sphere	1.00			
Cube[b]	1.08			
Cylinder[b]				
Vertical axis		1.01	1.06	1.20
Horizontal axis		1.14	1.34	1.58
Orientation averaged		1.09	1.23	1.43
Straight chain[c]		1.10	1.35	1.68
Compact cluster				
Three spheres	1.15			
Four spheres	1.17			
Dusts				
Bituminous coal[d]	1.05–1.11			
Quartz[d]	1.36			
Sand[d]	1.57			
Talc[e]	1.88			

[a]Averaged over all orientations except where noted.
[b]Calculated from Johnson et al. (1987).
[c]Dahneke (1982).
[d]Davies (1979).
[e]Cheng et al. (1988).

for most irregular particles. The slip factor for randomly oriented fibers ($L/d < 20$) is 0–12% greater than that for the equivalent volume sphere. [See Dahneke (1973).]

Elongated particles such as fibers tend to line up with streamlines while flowing in tubes. These particles are oriented perpendicular to the direction of settling when settling occurs with a Reynolds number greater than about 10. For $0.1 < \mathrm{Re} < 10$ there is partial alignment, and below a Reynolds number of 0.1 there is no alignment.

3.6 AERODYNAMIC DIAMETER

An equivalent diameter that finds wide application in aerosol technology is the *aerodynamic diameter* d_a. This is defined, for a particular particle, as the diameter of the spherical particle with a density of 1000 kg/m³ [1 g/cm³] (the density of a water droplet) that has the same settling velocity as the particle. The aerodynamic diameter standardizes for shape (a sphere) and density (1000 kg/m³ [1 g/cm³]). A related, but less common, equivalent diameter is the *Stokes diameter* d_s, the diameter of the sphere that has the same density and settling velocity as the particle.

Equation 3.25 can be written in terms of these diameters, neglecting slip correction, as

$$V_{\mathrm{TS}} = \frac{\rho_p d_e^2 g}{18\eta\chi} = \frac{\rho_0 d_a^2 g}{18\eta} = \frac{\rho_b d_s^2 g}{18\eta} \tag{3.26}$$

where ρ_0 is the standard particle density, 1000 kg/m³ [1.0 g/cm³]. The aerodynamic diameter can be thought of as the diameter of a water droplet having the same aerodynamic properties as the particle. If a particle has an aerodynamic diameter of 1 μm, it behaves in an aerodynamic sense like a 1-μm water droplet regardless of its shape, density, or physical size. Furthermore, it is aerodynamically indistinguishable from other particles of different size, shape, and density having aerodynamic diameters of 1 μm. The Stokes diameter is usually defined in terms of the normal density of the bulk material of the particle, ρ_b. This definition avoids the problem of defining the true density of the particle, which may be less than ρ_b due to porosity, occlusions, or an agglomerated structure.

An irregular particle and its aerodynamic and Stoke equivalent spheres are compared in Fig. 3.3. Each has the same settling velocity, but a different shape or density. Both aerodynamic and Stokes diameters are defined in terms of their aerodynamic behavior rather than their geometric properties. Settling velocity is a surrogate for most types of aerodynamic behavior. The aerodynamic diameter is the key particle property for characterizing filtration, respiratory deposition, and the performance of many types of air cleaners. In many situations, it is not necessary to know the true size, shape factor, and density of a particle if its aerodynamic diameter is known. Instruments such as elutriators (see Section 3.9) and cascade impactors (see Section 5.6) use aerodynamic separation to measure aerodynamic particle size.

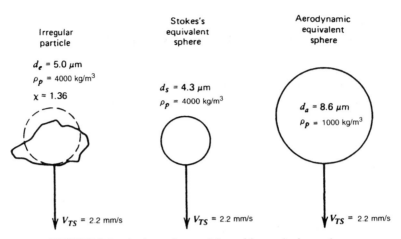

FIGURE 3.3   An irregular particle and its equivalent spheres.

Rearranging Eq. 3.26 gives

$$d_a = d_e \left(\frac{\rho_p}{\rho_0 \chi}\right)^{1/2} = d_s \left(\frac{\rho_b}{\rho_0}\right)^{1/2} \tag{3.27}$$

and, for spheres,

$$d_a = d_p \left(\frac{\rho_p}{\rho_0}\right)^{1/2} \tag{3.28}$$

For the usual case of a sphere with a density greater than 1000 kg/m³ [1.0 g/cm³], d_a is always greater than d_p.

For small particles for which slip correction must be applied, the foregoing definitions still hold, but are less convenient because the slip correction must be applied to the aerodynamic and Stokes diameters. The slip-corrected form of Eq. 3.28 is

$$d_a = d_p \left(\frac{C_c(d_p)}{C_c(d_a)}\right)^{1/2} \left(\frac{\rho_p}{\rho_0}\right)^{1/2}$$

where $C_c(d_p)$ and $C_c(d_a)$ are the respective slip correction factors for d_p and d_a. Likewise, when a particle's Reynolds numbers is greater than 1.0, its motion is outside the Stokes region, and although the preceding definitions hold, Eqs. 3.26–3.28 are not valid.

EXAMPLE

Calculate the aerodynamic diameter for a quartz particle with $d_e = 20$ μm and $\rho_p = 2700$ kg/m^3.

$$d_a = d_e \left(\frac{\rho_p}{\rho_o \chi} \right)^{1/2} \quad \text{where } \chi = 1.36, \text{ from Table 3.2}$$

$$d_a = 20 \left(\frac{2700}{1000 \times 1.36} \right)^{1/2} = 20(1.41) = 28.2 \text{ μm}$$

3.7 SETTLING AT HIGH REYNOLDS NUMBERS

For particle motion in the Stokes region, the settling velocity can be determined explicitly if the particle diameter and density are known; the diameter can be found if the velocity is known. For particle motion with Re > 1.0, this is not the case. Equating Eq. 3.4 to the force of gravity and solving for the settling velocity gives

$$V_{TS} = \left(\frac{4\rho_p d_p g}{3 C_D \rho_g} \right)^{1/2} \tag{3.29}$$

In order to calculate V_{TS} by Eq. 3.29, we must determine the correct value of C_D. To obtain C_D from either Fig. 3.1 or Eq. 3.5, however, we need the particle Reynolds number ($\rho V d / \eta$), which requires knowing V_{TS}. This dilemma arises because the functional dependence of V_{TS} on d_p gradually changes from $V_{TS} \propto d^2$ in the Stokes region to $V_{TS} \propto d^{1/2}$ in the Newton's drag region. This changing relationship manifests itself in a changing value of C_D in Eq. 3.29 for the transition region, Reynolds number of 1–1000.

The first step is to determine whether the particle's motion is outside the Stokes region. By assuming that Stokes's law holds, Eq. 3.13 can be used to calculate V_{TS}, which then allows the calculation of the Reynolds number. If Re is greater than 1.0, the motion of the settling particle is outside the Stokes region and the calculated value of V_{TS} is wrong. In that case, the procedures outlined next can be used to determine the correct value of V_{TS}.

One way around the dilemma is an iterative (trial-and-error solution) obtained by substituting Eq. 3.5 into Eq. 3.29 and trying different values of velocity until Eq. 3.29 holds to within some desired accuracy. Table 3.3 gives V_{TS} for standard-density spheres at standard conditions computed using this process. The values in the table cannot be adjusted for different conditions because any change in ρ_p, ρ_g, or η will change the relationship between C_D and V_{TS}.

TABLE 3.3 Terminal Settling Velocity for 100- to 2000-Micrometer Spheres at Standard Conditions[a]

Diameter (μm)	Re	V_{TS} (m/s)	Diameter (μm)	Re	V_{TS} (m/s)
100	1.65	0.248	360	34.5	1.44
105	1.87	0.268	370	36.5	1.48
110	2.12	0.289	380	38.6	1.52
115	2.38	0.312	390	40.7	1.57
120	2.67	0.335	400	42.9	1.61
125	2.98	0.358	410	45.1	1.65
130	3.31	0.382	420	47.4	1.69
135	3.61	0.402	430	49.8	1.74
140	3.96	0.425	440	52.2	1.78
145	4.33	0.448	450	54.6	1.82
150	4.71	0.472	460	57.1	1.86
155	5.11	0.495	470	59.6	1.90
160	5.52	0.518	480	62.2	1.94
165	5.95	0.542	490	64.8	1.98
170	6.40	0.566	500	67.5	2.02
175	6.87	0.589	550	81.6	2.22
180	7.35	0.613	600	96.8	2.42
185	7.84	0.637	650	113	2.61
190	8.35	0.660	700	130	2.80
195	8.88	0.684	750	149	2.98
200	9.42	0.708	800	168	3.16
210	10.5	0.755	850	189	3.34
220	11.7	0.802	900	210	3.52
230	13.0	0.849	950	233	3.69
240	14.3	0.896	1000	257	3.86
250	15.7	0.943	1100	307	4.19
260	17.1	0.989	1200	361	4.52
270	18.6	1.03	1300	419	4.84
280	20.1	1.08	1400	480	5.15
290	21.7	1.12	1500	546	5.46
300	23.4	1.17	1600	614	5.77
310	25.1	1.21	1700	686	6.06
320	26.9	1.26	1800	762	6.36
330	28.7	1.30	1900	841	6.65
340	30.6	1.35	2000	924	6.94
350	32.5	1.39			

[a]Calculated using the intermediate-region approximation $C_D = (24/Re)(1 + 0.15Re^{0.687})$. Particle density $= 1000$ kg/m^3 [1.0 g/cm^3].

Alternative approaches require the calculation of $C_D(\text{Re})^2$. Rearranging Eq. 3.29 gives

$$C_D = \frac{4\rho_p d_p g}{3\rho_g V^2} \tag{3.30}$$

Multiplying both sides of Eq. 3.30 by $(\text{Re})^2$ yields

$$C_D(\text{Re})^2 = \frac{4d^3 \rho_p \rho_g g}{3\eta^2} \tag{3.31}$$

The quantity $C_D(\text{Re})^2$ can be determined because the right-hand side of Eq. 3.31 *does not include velocity.* $C_D(\text{Re})^2$ can be used to determine the settling velocity *graphically* because it is a unique function of the Reynolds number that can be calculated without knowing the velocity. For a given particle Eq. 3.31 can be written

$$C_D(\text{Re})^2 = K$$

$$C_D = \frac{K}{(\text{Re})^2} \tag{3.32}$$

This is one equation relating C_D to Re that must be valid for the particle at its terminal settling velocity. The other is Eq. 3.5 for the transition region of Fig. 3.1. The two curves intersect at the correct value of Re and C_D. (See Fig. 3.4.) The settling velocity is determined from the Reynolds number of this point.

The settling velocity can be calculated directly using $C_D(\text{Re})^2$ by the empirical equation

$$V_{TS} = \left(\frac{\eta}{\rho_g d_p}\right) \exp(-3.070 + 0.9935J - 0.0178J^2) \tag{3.33}$$

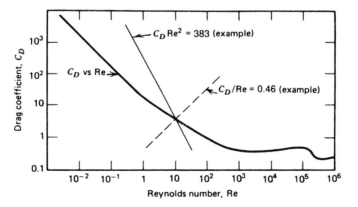

FIGURE 3.4 Drag coefficient versus Reynolds number, showing example lines for $C_D(\text{Re})^2 = 383$ and $C_D/\text{Re} = 0.46$.

where

$$J = \ln[C_D(\text{Re})^2] = \ln\left(\frac{4\rho_p\rho_g d^3 g}{3\eta^2}\right)$$

Equation (3.33) agrees with iterative calculations using the transition region approximation, Eq. 3.5, within 3% for $1 < \text{Re} < 600$ and within 7% for $0.5 < \text{Re} < 1000$. If the Reynold number is less than 0.5, it is more accurate to use Eq. 3.13 or 3.21.

A slightly faster procedure is to use the values of Re tabulated for their corresponding values of $C_D(\text{Re})^2$ given in Table 3.4. The value of $C_D(\text{Re})^2$ is calculated first, and the value of Re associated with this calculated value is obtained from the table. V_{TS} is then calculated from this Reynolds number.

EXAMPLE

What is the gravitational settling velocity at standard conditions of a 100-μm-diameter sphere having a density of 8000 kg/m^3?

$$C_D(\text{Re})^2 = \frac{4d^3\rho_p\rho_g g}{3\eta^2} = \frac{4 \times (100 \times 10^{-6})^3 \times 8000 \times 1.2 \times 9.81}{3 \times (1.81 \times 10^{-5})^2} = 383$$

$$\left[C_D(\text{Re})^2 = \frac{4 \times (10^{-2})^3 \times 8 \times 1.20 \times 10^{-3} \times 981}{3(1.81 \times 10^{-4})^2} = 383\right]$$

The Reynolds number obtained by interpolation from Table 3.4 for a value of $C_D(\text{Re})^2$ equal to 383 is 9.40, giving

$$V_{TS} = \frac{\text{Re}}{66,000d} = \frac{9.40}{66,000 \times 100 \times 10^{-6}} = 1.42 \text{ m/s} \quad \left[\frac{9.40}{6.6 \times 100 \times 10^{-4}} = 142 \text{ cm/s}\right]$$

Alternatively, substituting $J = \ln(383) = 5.95$ into Eq. 3.33 yields

$$V_{TS} = \left(\frac{1.81 \times 10^{-5}}{1.20 \times 10^{-4}}\right) \exp[-3.07 + 0.9935(5.95) - 0.0178(5.95)^2]$$

$$= 0.151 \exp(2.21) = 1.38 \text{ m/s}$$

$$= \left[\left(\frac{1.81 \times 10^{-4}}{1.20 \times 10^{-3} \times 10^{-2}}\right) \exp(2.21) = 138 \text{ cm/s}\right]$$

In this example, if Stokes's law had been unwittingly applied, the calculated velocity would have been

TABLE 3.4 Reynolds Numbers for Settling Spheres Having $C_D(Re)^2$ from 10 to 10^5 [a]

$C_D(Re)^2$	0	1	2	3	4	5	6	7	8	9
10	0.40	0.43	0.47	0.51	0.54	0.58	0.61	0.65	0.69	0.72
20	0.75	0.79	0.82	0.86	0.89	0.92	0.96	0.99	1.02	1.06
30	1.09	1.12	1.15	1.18	1.22	1.25	1.28	1.31	1.34	1.37
40	1.40	1.43	1.47	1.50	1.53	1.56	1.59	1.62	1.65	1.68
50	1.71	1.74	1.77	1.80	1.83	1.86	1.89	1.92	1.95	1.98
60	2.01	2.04	2.07	2.10	2.13	2.16	2.19	2.22	2.25	2.27
70	2.30	2.33	2.36	2.39	2.42	2.45	2.47	2.50	2.53	2.56
80	2.59	2.62	2.64	2.67	2.70	2.73	2.75	2.78	2.81	2.84
90	2.86	2.89	2.92	2.95	2.97	3.00	3.03	3.05	3.08	3.11

$C_D(Re)^2$	0	10	20	30	40	50	60	70	80	90
100	3.14	3.40	3.66	3.92	4.17	4.41	4.66	4.90	5.13	5.36
200	5.59	5.82	6.05	6.27	6.49	6.70	6.92	7.13	7.34	7.55
300	7.75	7.96	8.16	8.36	8.56	8.75	8.95	9.14	9.34	9.53
400	9.72	9.90	10.1	10.3	10.5	10.6	10.8	11.0	11.2	11.4
500	11.5	11.7	11.9	12.1	12.2	12.4	12.6	12.8	12.9	13.1
600	13.3	13.4	13.6	13.8	13.9	14.1	14.2	14.4	14.6	14.7
700	14.9	15.0	15.2	15.4	15.5	15.7	15.8	16.0	16.1	16.3
800	16.4	16.6	16.7	16.9	17.1	17.2	17.3	17.5	17.6	17.8
900	17.9	18.1	18.2	18.4	18.5	18.7	18.8	19.0	19.1	19.2

$C_D(Re)^2$	0	100	200	300	400	500	600	700	800	900
1000	19.4	20.8	22.1	23.4	24.7	26.0	27.2	28.4	29.6	30.7
2000	31.8	33.0	34.0	35.1	36.2	37.2	38.2	39.2	40.2	41.2
3000	42.2	43.2	44.1	45.1	46.0	46.9	47.8	48.7	49.6	50.5
4000	51.4	52.2	53.1	53.9	54.8	55.6	56.4	57.3	58.1	58.9
5000	59.7	60.5	61.3	62.1	62.9	63.6	64.4	65.2	65.9	66.7
6000	67.4	68.2	68.9	69.7	70.4	71.1	71.8	72.6	73.3	74.0
7000	74.7	75.4	76.1	76.8	77.5	78.2	78.9	79.6	80.2	80.9
8000	81.6	82.3	82.9	83.6	84.2	84.9	85.6	86.2	86.9	87.5
9000	88.1	88.8	89.4	90.1	90.7	91.3	92.0	92.6	93.2	93.8

$C_D(Re)^2$	0	1000	2000	3000	4000	5000	6000	7000	8000	9000
10000	94.4	100	106	112	117	123	128	133	138	143
20000	148	152	157	162	166	170	175	179	183	187
30000	191	195	199	203	207	211	214	218	222	226
40000	229	233	236	240	243	247	250	254	257	260
50000	264	267	270	273	277	280	283	286	289	292
60000	295	298	301	304	307	310	313	316	319	322
70000	325	328	331	334	336	339	342	345	348	350
80000	353	356	358	361	364	367	369	372	374	377
90000	380	382	385	388	390	393	395	398	400	403

[a]The value of $C_D(Re)^2$ equals the sum of the row and column headings.

$$V_{TS} = \frac{\rho_p d^2 g}{18\eta} = \frac{8000 \times (10^{-4})^2 \times 9.81}{18 \times 1.81 \times 10^{-5}} = 2.41 \text{ m/s}$$

an overestimate of 70%.

The problem of determining a particle's diameter when its settling velocity is known is attacked in a similar way. The quantity C_D/Re, obtained by combining Eqs. 3.30 and 2.41, can be calculated without knowing the particle diameter:

$$\frac{C_D}{\text{Re}} = \frac{4\rho_p \eta g}{3\rho_g^2 V^3} \tag{3.34}$$

Equation 3.34 defines another relationship between C_D and Re that must be met for a settling particle. The particle diameter can be obtained by iterative, graphical or tabular procedures analogous to those for the settling velocity just described. Values of Reynolds numbers for C_D/Re from 0.01 to 100 are given in Table 3.5. The diameter of a settling particle whose velocity is known can be calculated directly using C_D/Re by the empirical equation.

$$d = \left(\frac{\eta}{\rho_g V_{TS}}\right) \exp(1.787 - 0.577H + 0.0109H^2) \tag{3.35}$$

where

$$H = \ln(C_D/\text{Re}) = \ln\left(\frac{4\rho_p \eta g}{3\rho_g^2 V^3}\right)$$

Equation 3.35 agrees with iterative calculations using the transition region approximation, Eq. 3.5, within 3% for $0.5 < \text{Re} < 1000$.

EXAMPLE

To work the preceding example backwards, assume that the particle diameter is not known, but its settling velocity at standard conditions has been determined to be 1.42 m/s [142 cm/s] and $\rho_p = 8000$ kg/m^3 [8 g/cm^3]. Then

$$\frac{C_D}{\text{Re}} = \frac{4\rho_p \eta g}{3\rho_g^2 V^3} = \frac{4 \times 8000 \times 1.81 \times 10^{-5} \times 9.81}{3(1.20)^2 (1.42)^3} = 0.46$$

$$\left[\frac{C_D}{\text{Re}} = \frac{4 \times 8 \times 1.81 \times 10^{-4} \times 981}{3(1.2 \times 10^{-3})^2 (142)^3} = 0.46\right]$$

TABLE 3.5 Reynolds Numbers for Settling Spheres Having C_D/Re from 0.01 to 100[a]

C_D/Re	0	0.001	0.002	0.003	0.004	0.005	0.006	0.007	0.008	0.009
0.01	106	99.5	93.8	88.8	84.4	80.6	77.2	74.1	71.3	68.8
0.02	66.5	64.3	62.4	60.6	58.9	57.3	55.8	54.5	53.2	52.0
0.03	50.8	49.7	48.7	47.8	46.8	46.0	45.1	44.3	43.6	42.8
0.04	42.1	41.5	40.8	40.2	39.6	39.1	38.5	38.0	37.5	37.0
0.05	36.5	36.0	35.6	35.2	34.7	34.3	34.0	33.6	33.2	32.8
0.06	32.5	32.2	31.8	31.5	31.2	30.9	30.6	30.3	30.0	29.7
0.07	29.5	29.2	29.0	28.7	28.5	28.2	28.0	27.8	27.5	27.3
0.08	27.1	26.9	26.7	26.5	26.3	26.1	25.9	25.7	25.5	25.3
0.09	25.2	25.0	24.8	24.7	24.5	24.3	24.2	24.0	23.9	23.7

C_D/Re	0	0.01	0.02	0.03	0.04	0.05	0.06	0.07	0.08	0.09
0.1	23.6	22.2	21.1	20.0	19.2	18.4	17.7	17.0	16.4	15.9
0.2	15.4	15.0	14.6	14.2	13.8	13.5	13.2	12.9	12.6	12.4
0.3	12.1	11.9	11.7	11.4	11.2	11.1	10.9	10.7	10.5	10.4
0.4	10.2	10.1	9.93	9.80	9.67	9.54	9.42	9.30	9.19	9.08
0.5	8.97	8.87	8.77	8.67	8.58	8.49	8.40	8.32	8.23	8.15
0.6	8.07	8.00	7.92	7.85	7.78	7.71	7.64	7.58	7.51	7.45
0.7	7.39	7.33	7.27	7.21	7.16	7.10	7.05	7.00	6.95	6.90
0.8	6.85	6.8	6.75	6.70	6.66	6.61	6.57	6.53	6.48	6.44
0.9	6.40	6.36	6.32	6.28	6.24	6.20	6.16	6.12	6.09	6.06

C_D/Re	0	0.1	0.2	0.3	0.4	0.5	0.6	0.7	0.8	0.9
1	6.03	5.71	5.44	5.20	4.99	4.8	4.63	4.48	4.34	4.21
2	4.09	3.98	3.88	3.79	3.70	3.62	3.54	3.47	3.40	3.33
3	3.27	3.21	3.16	3.11	3.06	3.01	2.96	2.92	2.88	2.84
4	2.80	2.76	2.72	2.69	2.66	2.62	2.59	2.56	2.53	2.51
5	2.48	2.45	2.43	2.40	2.38	2.35	2.33	2.31	2.29	2.27
6	2.25	2.23	2.21	2.19	2.17	2.15	113	2.12	2.10	2.08
7	2.07	2.05	2.04	2.02	2.01	2.00	1.98	1.97	1.96	1.94
8	1.93	1.92	1.90	1.89	1.88	1.87	1.86	1.84	1.83	1.82
9	1.81	1.80	1.79	1.78	1.77	1.76	1.75	1.74	1.73	1.72

C_D/Re	0	1	2	3	4	5	6	7	8	9
10	1.71	1.63	1.55	1.49	1.43	1.38	1.33	1.29	1.25	1.21
20	1.18	1.15	1.12	1.10	1.07	1.05	1.03	1.01	0.99	0.97
30	0.95	0.93	0.92	0.90	0.89	0.88	0.86	0.85	0.84	0.83
40	0.82	0.81	0.80	0.79	0.78	0.77	0.76	0.75	0.74	0.73
50	0.73	0.72	0.71	0.70	0.70	0.69	0.68	0.68	0.67	0.67
60	0.66	0.65	0.65	0.64	0.64	0.63	0.63	0.62	0.62	0.61
70	0.61	0.61	0.60	0.60	0.59	0.59	0.58	0.58	0.58	0.57
80	0.57	0.57	0.56	0.56	0.55	0.55	0.55	0.54	0.54	0.54
90	0.54	0.53	0.53	0.53	0.52	0.52	0.52	0.51	0.51	0.51

[a]The value of C_D/Re equals the sum of the row and column headings.

The Reynolds number obtained from Table 3.5 for $C_D/\text{Re} = 0.46$ is 9.42, which allows calculation of particle diameter,

$$d = \frac{9.42}{66000 \times V_{TS}} = \frac{9.42}{66000 \times 1.42} = 101 \ \mu\text{m} \quad \left[\frac{9.42}{6.6 \times 142} = 101 \ \mu\text{m} \right]$$

Alternatively, we can use Eq. 3.35, substituting $H = \ln(0.46) = -0.777$ to obtain

$$d = \left(\frac{1.81 \times 10^{-5}}{1.2 \times 1.38} \right) \exp[1.787 - 0.577(-0.777) + 0.0109(-0.777)^2]$$

$$d = (1.06 \times 10^{-5}) \exp(2.242) = 100 \ \mu\text{m}$$

$$\left[d = \left(\frac{1.81 \times 10^{-4}}{1.20 \times 10^{-3} \times 142} \right) \exp(2.242) = 100 \ \mu\text{m} \right]$$

the same as the original diameter.

For nonspherical particles with motion outside the Stokes region, the right-hand sides of Eqs. 3.31 and 3.34 are multiplied by $1/\chi$, where χ is the dynamic shape factor, and the equivalent volume diameter is used throughout. The method for determining the velocity or diameter is the same as that given for spheres. This approach is valid for compact particles with Reynolds numbers less than 100 or for irregular particles with sharp edges and Reynolds numbers less than 10.

The calculations of settling velocity for spheres can be summarized as follows. For the smallest particles, <0.1 μm, use Eq. 3.21 with Eq. 3.20 for slip correction. For particles from 0.1 to 1 μm, use either Eq. 3.19 or 3.20 to calculate the slip correction factor. For particles >1.0 μm, Eq. 3.13 can be used, provided that the particle Reynolds number is less than 1.0. For particle motion with Reynolds number from 0.5 to 1.0, Eq. 3.13, Eq. 3.33, or the tabular procedure can be used. For large and heavy particles for which Re >1.0, we must use Eq. 3.33 or the tabular procedure using Eqs. 3.31 and 2.41 and Table 3.4. Equation 3.29 with $C_D = 0.44$ can be used for motion with $1000 < \text{Re} < 2 \times 10^5$, although this is beyond the range of usual aerosol particle motion.

3.8   STIRRED SETTLING

The equations for V_{TS} permit us to evaluate how aerosol concentration will change with time in a room or a container. Unfortunately, the real case is likely to be very complicated, so we consider here the two idealized situations shown in Fig. 3.5; tranquil and stirred settling of monodisperse aerosols. For monodisperse aerosols, the real case lies between these two idealized situations, usually closer to the stirred settling case than to the tranquil settling case.

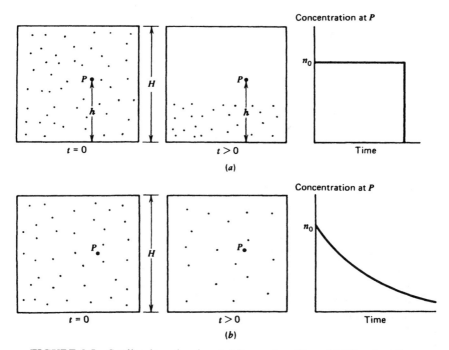

FIGURE 3.5   Settling in a chamber. (a) Tranquil settling. (b) Stirred settling.

In tranquil settling, there is no motion of the gas, and each particle's motion is solely its gravitational settling in the stagnant air mass. Consider a cylindrical chamber of height H containing at time $t = 0$ a number concentration n_0 of uniformly distributed monodisperse particles. If we neglect diffusion, all the particles will be settling with the same constant velocity V_{TS}. After a time t, there will be an upper region with no particles and a lower region containing an aerosol with a flat upper boundary having the original number concentration. (See Fig. 3.5a.) Some fraction of the particles will have deposited on the floor of the chamber. At any point within the chamber the concentration will remain constant, until it abruptly drops to zero as the aerosol boundary moves past the point. The time required for the aerosol boundary to reach a point P located a distance h above the chamber floor is $(H - h)/V_{TS}$. The concentration becomes zero everywhere in the chamber after a time equal to H/V_{TS} has elapsed.

At the other extreme, stirred settling (Fig. 3.5b), the aerosol is stirred vigorously, and it is assumed that the concentration is always uniform throughout the chamber. Diffusion, resuspension, and deposition on the walls are assumed to be negligible. The particle settling velocity is superimposed on the vertical components of the convective velocity. Because the up and down components of the convective velocity are equal and cancel out over a time period that is long compared with their period, every particle will have an average *net* velocity equal to V_{TS}. The concentration decreases with time, but the rate of removal also decreases with time, because it is proportional to the number of particles left suspended in the chamber, which, although the particles are uniformly mixed, is decreasing. The fraction removed, dn/n, during an interval of time dt that is brief compared with the time for convective motion is

$$\frac{dn}{n} = \frac{V_{TS}dt}{H}$$

Integrating for an initial condition of $n = n_0$ at $t = 0$ gives

$$\frac{n(t)}{n_0} = \exp\left(\frac{-V_{TS}t}{H}\right) \tag{3.36}$$

The concentration of particles undergoing stirred settling decays exponentially with time and, as such, never reaches zero. In stirred settling, the concentration reaches $1/e$ of the original concentration in the time required for complete removal in the tranquil settling situation (H/V_{TS}).

For polydisperse aerosols, the situation is more complicated. In both types of settling, the average size of the particles remaining in suspension decreases with time. For tranquil settling, there is no distinct upper boundary. For stirred settling, Eq. 3.36 can be applied to each particle size in the size distributions. Equations characterizing particle concentrations as a function of time can be obtained by integrating Eq. 3.36 over the size distribution.

EXAMPLE

The concentration of wood dust in a room with a 3-m ceiling height is 10 mg/m^3. What will be the concentration after one hour? Assume ideal monodisperse stirred settling of particles with aerodynamic diameters of 4 μm.

$$V_{TS} = \frac{\rho_0 d^2 g}{18\eta} = \frac{1000(4 \times 10^{-6})^2 9.81}{18(1.81 \times 10^{-5})} = 4.8 \times 10^{-4} \ \text{m/s}$$

$$\left[V_{TS} = \frac{1(4 \times 10^{-4})^2 981}{18(1.81 \times 10^{-4})} = 4.8 \times 10^{-2} \ \text{cm/s}\right]$$

$C_m \propto n$ for monodisperse aerosols, and $t = 60 \times 60 = 3600$ s. Hence,

$$C_{3600} = C_0 \exp\left(\frac{-V_{TS}t}{H}\right)$$

$$C_{3600} = 10 \exp\left(\frac{-4.8 \times 10^{-4} \times 3600}{3}\right) = 10 \exp(-0.576) = 5.6 \ \text{mg/m}^3$$

$$\left[C_{3600} = 10 \exp\left(\frac{-4.8 \times 10^{-2} \times 3600}{300}\right) = 5.6 \ \text{mg/m}^3\right]$$

3.9   INSTRUMENTS THAT RELY ON SETTLING VELOCITY

Several instruments rely on particle settling to measure the aerodynamic diameter of a particle or to fractionate particles into two or more size categories. The simplest is the *sedimentation cell*, which is similar to that adapted by Millikan for his oil-drop experiments. (See Section 15.3.) Aerosol particles are introduced into a sealed volume of less than 1 cm^3 and illuminated through windows by an intense beam of light. The particles are viewed by a horizontal microscope in a direction perpendicular to the light beam. The particles, which appear as tiny specks of light, are timed individually as they settle between calibrated lines marked on the eyepiece of the microscope. The true settling distance is known from calibration and is usually less than 1.0 mm. The aerodynamic diameter is calculated directly from the measured settling velocity.

The Brownian motion of particles less than 0.3 μm in diameter causes significant variation in the measured settling velocity, and such particles may wander out of the field of view. Particles larger than 5 μm settle too fast for accurate measurement. Care must be taken to minimize thermal convection caused by the heat of the light beam. Care must also be taken to prevent an operator bias toward the selection of larger and brighter particles.

For larger particles, 5–50 μm, the aerodynamic diameter can be measured directly using a settling tube (Wall et al, 1985). Typically, this is a vertical glass tube about 7 mm in diameter and 0.3–0.8 m long, illuminated along its axis by a low-power laser. The tube is marked off in 10-cm or other convenient intervals. Particles are seen as points of light, and their settling time over a known distance is measured with a stopwatch. The choice of tube diameter is a compromise between reducing convection currents inside the tube and reducing tube wall effects. This method provides an absolute measurement of the aerodynamic diameter and requires no calibration.

A *vertical elutriator* is a device used to remove particles larger than a certain aerodynamic diameter from an aerosol stream. The aerosol flows upwards at low velocity in a vertical duct. Particles having a V_{TS} greater than the duct velocity cannot be carried out of the duct and are thereby removed from the aerosol stream. The situation can be thought of as a contest between the particle and the airstream, with the faster one controlling the outcome ("winning"). These devices work satisfactorily for rough separation of large particles, but the distribution of gas velocities in the duct makes it difficult to obtain a precise cutoff. The large particles, having a net downward velocity, may "filter" out smaller particles moving upwards. A vertical elutriator used for cotton dust sampling is described in Section 11.5.

The *horizontal elutriator* can be used either as a separator to fractionate an aerosol stream or as an aerosol spectrometer to measure the distribution of particle size. For separation, an aerosol is passed at low velocity through a horizontal duct having a rectangular cross section. Particle settling is perpendicular to the gas streamlines. Particles reaching the floor of the duct are removed from the aerosol stream. This situation also may be viewed as a contest that results in an attenuation in particle concentration that is greater for the larger particles than for the smaller ones. As shown in Fig. 3.6, particles having a $V_{TS} > H\overline{V}_x/L$, where $\overline{V}_x = \overline{U}$, will be

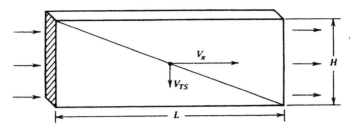

FIGURE 3.6 Horizontal elutriator, side view showing limiting trajectory for complete capture.

completely removed if the flow is laminar. Smaller particles will have a fraction $V_{TS}L/\bar{V}_xH$ of their number removed. Horizontal plates evenly spaced in the duct will reduce H and increase the number collected for a given particle size. The horizontal elutriator is one of the methods used for respirable mass sampling. (See Section 11.5.)

A similar approach is utilized in an elutriation spectrometer to measure the distribution of the particle sizes. Two flow streams are used: a clean airstream and an aerosol stream. A horizontal laminar flow of clean air at an average velocity $\bar{V}_x$, called the winnowing stream or carrier gas stream, is established in the duct (Fig. 3.7). At the beginning of the separation section, the aerosol is introduced as a thin stream or film along the upper surface of the duct. Particles of different sizes settle at different rates and deposit at different locations along the duct floor. There is a unique location along the floor where each particle size will deposit. The distance from the inlet to that location for a particle with a settling velocity V_{TS} is

$$L_d = \frac{H\bar{V}_x}{V_{TS}} \tag{3.37}$$

where H is the height of the duct. The bottom of the duct is lined with glass slides or foil that can be removed in sections and analyzed for the number or mass of particles deposited. Each section corresponds to a range of V_{TS} that defines a range

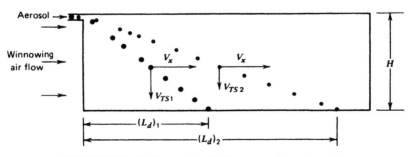

FIGURE 3.7 Horizontal elutriation aerosol spectrometer.

of aerodynamic diameters. The distribution of number (or mass) as a function of aerodynamic diameter is determined from the fraction of the total number (or mass) in each size range. A filter can be placed at the end of the duct to capture particles smaller than the limiting size. Because $V_{TS} \propto d_a^2$, the duct must be excessively long to collect small particles. One approach is to use a duct whose width increases along its length to reduce $\overline{V}_x$ and permit smaller particles to be separated in an instrument of convenient length. Because gravitational settling of aerosol particles is a slow process, the carrier gas flow rate must be low, and consequently, these instruments are sensitive to convection currents.

The principle of the *aerosol centrifuge* is similar to that of the horizontal elutriator except that the force of gravity is replaced by centrifugal force. The elutriation duct is wrapped around a cylinder with the collection surface at the periphery and is rotated. The duct is often arranged in a spiral to permit the aerosol and carrier gas air to be introduced along the rotational axis and exit at the periphery. The outer side of the channel is lined with a deposition foil, which is removed, segmented, and analyzed for particle mass to determine the mass distribution as a function of aerodynamic diameter. Because centrifugal force can easily be made much greater than the force of gravity, these devices can operate at a higher sampling flow rate and separate smaller particle sizes than the horizontal elutriator can. They are practical for particle sizes of 0.1–15 μm aerodynamic diameter. The radial component of flow velocity causes a secondary flow (eddies), which distorts the flow stream and limits the resolution for small particle sizes. Some aerosol centrifuges are high-resolution instruments that can separate particles differing in aerodynamic diameter by only a few percent. Characteristics of aerosol centrifuges are summarized by Marple et al. (1993).

3.10 APPENDIX: DERIVATION OF STOKES'S LAW

The Navier–Stokes equation for incompressible flow (flow for which $\nabla \cdot \mathbf{V} = 0$) can be written, neglecting gravity and buoyancy forces, as

$$\rho\left(\frac{\partial U}{\partial t} + u\frac{\partial U}{\partial x} + v\frac{\partial U}{\partial y} + w\frac{\partial W}{\partial z}\right) = -\frac{\partial p}{\partial x} + \eta\left(\frac{\partial^2 U}{\partial x^2} + \frac{\partial^2 U}{\partial y^2} + \frac{\partial^2 U}{\partial z^2}\right) \quad (3.38)$$

for the x direction, and similar equations can be written for the y and z directions. In operator notation, this becomes

$$\rho\frac{D\mathbf{V}}{dt} = -\nabla p + \eta\nabla^2(\mathbf{V}) \quad (3.39)$$

where $\mathbf{V}$ is the three-dimensional velocity vector of the fluid at a fixed point (x, y, z) or (r, θ, ϕ) in space and $D\mathbf{V}/dt$ *is* the substantial (total) derivative of $\mathbf{V}$ (i.e., $D\mathbf{V}/dt = \partial\mathbf{V}/\partial t + \mathbf{V} \cdot \nabla\mathbf{V}$). Equation (3.39) is called the *Eulerian description of fluid motion.* For steady flow, the first term in the substantial derivative is zero. To use the Navier–Stokes equation in this derivation, we assume that the particle is held fixed and the fluid is moved past it at a velocity $V = V_\infty$, where V_∞ is the velocity of the fluid well away from the particle. The fluid velocity is assumed to be in the z direction in spherical coordinates. (See Fig. 3.8.) The $V \cdot \nabla V$ term is the acceleration (inertial) term and represents the resistance to flow caused by the inertia of the gas that has to be accelerated around the particles. Because Stokes's derivation is restricted to low Reynolds numbers, the inertial terms are small compared with the viscous terms and can be neglected. Thus, we get the simplified (linear) form of the Navier–Stokes equation for steady, incompressible flow at low Reynolds number:

$$\nabla p = \eta \nabla^2 V \tag{3.40}$$

The analytical solution of this equation, giving the flow field around a sphere of radius a in spherical coordinates (see Fig. 3.8), has been found. [See, for example, Bird et al. (1960) or Landau and Lifshitz (1959).] The velocity distribution is

$$V_r = V_\infty \left[1 - \frac{3}{2}\left(\frac{a}{r}\right) + \frac{1}{2}\left(\frac{a}{r}\right)^3 \right] \cos\theta \tag{3.41}$$

$$V_\theta = -V_\infty \left[1 - \frac{3}{4}\left(\frac{a}{r}\right) - \frac{1}{4}\left(\frac{a}{4}\right)^3 \right] \sin\theta \tag{3.42}$$

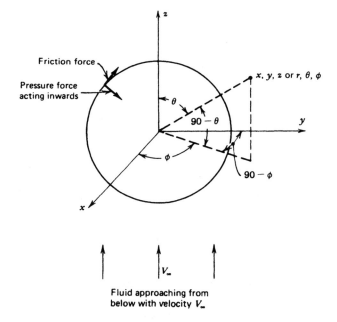

FIGURE 3.8 Forces acting on a stationary sphere in a flowing fluid.

Because of symmetry, V is the same for any value of ϕ; that is, V does not depend on ϕ. The velocity distribution satisfies the boundary conditions, namely, $V_r = V_\theta = 0$ at the surface of the sphere ($r = a$), and V_r and V_θ go to V_∞ for $r \gg a$. This velocity distribution, Eqs. 3.41 and 3.42, causes normal and tangential forces p and τ, respectively. At any point in the fluid near the sphere, these forces are given by

$$p = p_0 - \frac{3\eta V_\infty}{2a}\left(\frac{a}{r}\right)^2 \cos\theta \tag{3.43}$$

$$\tau = \frac{3\eta V_\infty}{2a}\left(\frac{a}{r}\right)^4 \sin\theta \tag{3.44}$$

where p_0 is the ambient pressure. The normal force p goes to p_0 for $r \gg a$, as expected.

To calculate the net force exerted by the fluid on the sphere, we integrate the pressure (normal) force and friction (tangential) force over the surface of the sphere ($r = a$).

The z component of the pressure force is $-p \cos\theta$, which is multiplied by an incremental area (see Fig. 3.9) on the surface of the sphere, $(a \sin\theta\, d\phi)(a\, d\theta) = a^2 \sin\theta\, d\theta\, d\phi$, and integrated over the surface to give the total pressure force. Substituting $r = a$ into Eq. 3.43 and setting up the integration gives

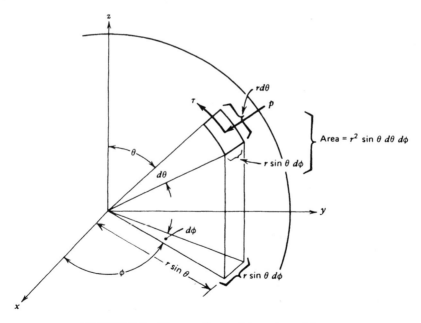

FIGURE 3.9   Diagram for the integration of forces.

$$F_n = \int_S -p \cos \theta \, dS = \int_0^{2\pi} \int_0^\pi \left(-p_0 + \frac{3\eta V_\infty}{2a} \cos^2 \theta \right) a^2 \sin \theta \, d\theta \, d\phi \tag{3.45}$$

Completing the integration gives the form component of the Stokes drag:

$$F_n = 2\pi\eta V_\infty a \tag{3.46}$$

The tangential shear stress τ acts tangentially at the surface of the sphere. The component in the z direction is $\tau \sin \theta$. Integration of this force over the surface of the sphere gives the net frictional component of the Stokes drag,

$$F_\tau = \int_S \tau \sin \theta \, ds = \int_0^{2\pi} \int_0^\pi \left(\frac{3}{2} \frac{\eta V_\infty}{a} \sin^3 \theta \right) a^2 \, d\theta \, d\phi \tag{3.47}$$

$$F_\tau = 4\pi\eta V_\infty a \tag{3.48}$$

The form and friction components are combined to give the total force acting on the particle due to its motion relative to the fluid, or Stokes's law:

$$F_D = 3\pi\eta V_\infty d \tag{3.49}$$

PROBLEMS

3.1   What is the drag force on a 50-μm-diameter particle traveling at 1 m/s in still air? Use the graph of C_D versus Re, Fig. 3.1. What is the drag force when the particle is traveling at 0.01 m/s?
ANSWER: 1.2×10^{-8} N, 8.2×10^{-11} N [1.2×10^{-3} dyn, 8.2×10^{-6} dyn].

3.2   Calculate the drag force on a 12-μm-diameter particle moving at 0.1 m/s using Newton's drag equation and Fig. 3.1. Check your result by calculating the drag force using Stokes's equation.
ANSWER: 2.0×10^{-10} N [2.0×10^{-5} dyn].

3.3   An old occupational hygiene rule of thumb is that a 10-μm silica (quartz) particle settles at a rate of 1 ft/min (0.5 cm/s). What is the true settling velocity for such a particle? The density of silica is 2600 kg/m³ [2.6 g/cm³].
ANSWER: 0.58 cm/s.

3.4   Derive an expression for the aerodynamic diameter of a cylindrically shaped particle of diameter D and length L in terms of D, L, χ, ρ_p.

3.5   For a 0.15-μm-diameter spherical particle ($\rho_p = 2500$ kg/m³), determine the Cunningham correction factor and the terminal settling velocity at standard conditions.
ANSWER: 2.11, 3.6×10^{-6} m/s [3.6×10^{-4} cm/s].

3.6 Calculate the terminal settling velocity of a 0.2-µm steel sphere ($\rho_p = 7600$ kg/m^3) at atmospheric pressure and a temperature of (a) 20°C and (b) 200°C. ANSWER: 1.7×10^{-5} m/s, 1.5×10^{-5} m/s [1.7×10^{-3} cm/s, 1.5×10^{-3} cm/s].

3.7 In 1883 the volcano Krakatoa exploded, injecting dust 32 km up into the atmosphere. Fallout from this explosion continued for 15 months. If one assumes that the settling velocity of the particles was constant and neglects slip correction, what was the minimum particle size present? Assume that the particles are rock spheres with a density of 2700 kg/m^3. ANSWER: 3.2 µm.

3.8 What is the settling velocity of a 0.5-µm silica sphere at 20°C and an ambient pressure of 150 kPa? The density of silica is 2600 kg/m^3. ANSWER: 2.4×10^{-5} m/s [0.0024 cm/s].

3.9 Air is dried by bubbling it through concentrated sulfuric acid ($\rho_p = 1.84$ g/cm^3). A l-m tube 0.1 m in diameter contains 1.5 L of acid. When the bubbles burst at the liquid surface, they form droplets. What is the largest droplet that can be carried out of this system? The airflow rate is 10 L/min. Neglect slip correction. ANSWER: 19.5 µm.

3.10 Calculate the equivalent volume diameter, the aerodynamic diameter, the terminal settling velocity, and the drag force at terminal velocity for (a) a 2-µm-diameter standard-density sphere and (b) a doublet (cluster of two) composed of 2-µm-diameter standard-density spheres. Neglect slip correction. Compare your results in a table.
ANSWER:

	d_e (µm)	d_a (µm)	V_{TS} (m/s [cm/s])	F_D (N [dyn])
(a)	2	2	1.2×10^{-4} [0.012]	4.1×10^{-14} [4.1×10^{-9}]
(b)	2.52	2.38	1.7×10^{-4} [0.017]	8.2×10^{-14} [8.2×10^{-9}]

3.11 Derive the expression

$$d_a = n^{1/3} \left(\frac{\rho}{\chi} \right)^{1/2} d$$

where d_a is the aerodynamic diameter of a *cluster* of n spheres, each having a diameter d and density ρ, and χ is the shape factor for the cluster.

3.12 What is the drag force on a doublet composed of 2-µm-diameter spheres traveling at its terminal settling velocity? The particle density is 3000 kg/m^3.

Neglect slip correction. (*Hint*: There is an easy way to do this; review the derivation of Eq. 3.13.)

ANSWER: 2.5×10^{-13} N [2.5×10^{-8} dyn].

3.13 What is the settling velocity of an asbestos fiber in the shape of a cylinder 1 μm in diameter and 10 μm long at standard conditions? Assume random orientation and a density of 2500 kg/m³, and neglect slip correction.

ANSWER: 3.2×10^{-4} m/s [0.032 cm/s].

3.14 Air with a uniform velocity U is flowing through a settling chamber with dimensions L, H, and W, as shown in Fig. 3.10. Assume that there are no internal partitions, there is no reentrainment, the flow is laminar, the aerosol is uniformly distributed in the incoming air, and slip correction is negligible. Derive an expression that gives the diameter of the smallest particle that will be collected with 100% efficiency. This diameter will be a function of U, L, H, η, g, and ρ_p.

3.15 (a) Derive an expression for the collection efficiency (as a function of particle size) for a multilayer settling chamber (Fig. 3.10) with N equispaced sections. Assume uniform flow, a uniform distribution of aerosol particles at the inlet, and negligible slip correction. (b) Show that the collection efficiency for a given particle size is proportional to the total area of the collection surfaces.

3.16 Air is flowing with a uniform laminar flow in a horizontal duct with a square cross section. The duct is 1 ft × 1 ft × 100 ft long, and the air flows at a rate of 1000 ft³/min. If an aerosol of 10-μm iron-oxide spheres (ρ_p = 5.2 g/cm³) is uniformly distributed across the inlet, what fraction of entering particles will leave the 100-foot section? Assume that a particle sticks if it hits the bottom of the duct.

ANSWER: 0.69.

3.17 It is desired to recover particles larger than 200 μm emitted by a cement kiln, using a horizontal-flow settling chamber having a residence time of 24 s.

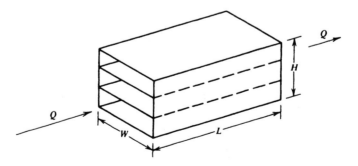

FIGURE 3.10 Diagram of settling chamber.

What is the maximum height of the chamber? Assume $\rho_p = 3100$ kg/m³ [3.1 g/cm³].
ANSWER: 39 m.

3.18 What is the terminal settling velocity of a 100-μm-diameter lead sphere ($\rho_p = 11,300$ kg/m³) in still air?
ANSWER: 1.8 m/s [180 cm/s].

3.19 What particle size ($\rho_p = 4000$ kg/m³) is too large to be carried out of a vertical exhaust duct 14 inches in diameter that handles 200 ft³/min?
ANSWER: 111 μm.

3.20 What is the terminal settling velocity of a 40-μm-diameter uranium sphere ($\rho_p = 19,000$ kg/m³)?
ANSWER: 0.74 m/s [74 cm/s].

3.21 A spherical steel particle ($\rho_p = 7800$ kg/m³) is observed to settle in still air at a rate of 2.0 m/s. What is its diameter?
ANSWER: 134 μm.

3.22 You measure airborne particulate concentrations of 20 and 6 mg/m³ one and two hours, respectively, after an explosion in a well-stirred reactor vessel 14 ft high. Assuming simple stirred settling of a monodisperse aerosol, what was the concentration immediately after the explosion? What is the particle size (aerodynamic diameter)?
ANSWER: 65 mg/m³, 6.9 μm.

3.23 What change in concentration occurs during a 5-min interval for 10-μm particles undergoing stirred settling in a chamber 1.0 m high? Assume that the particle density is 3000 kg/m³ [3 g/cm³].
ANSWER: 93%.

REFERENCES

Allen, M. D., and Raabe, O. G., "Re-Evaluation of Millikan's Oil Drop Data for the Motion of Small Particles in Air," *J. Aerosol Sci.*, **6**, 537–547 (1982).

Allen, M. D., and Raabe, O. G., "Slip Correction Measurements of Spherical Solid Aerosol Particles in an Improved Millikan Apparatus," *Aerosol Sci. Tech.*, **4**, 269–286 (1985).

Bird, B. R., Stewart, W. E., and Lightfoot, E. N., *Transport Phenomena*, Wiley, New York, 1960.

Cheng, Y-S., Yeh, H-C, and Allen, M. D., "Dynamic Shape Factor of Plate-Like Particles," *Aerosol Sci. Tech.*, **8**, 109–123 (1988).

Clift, R., Grace, J. R., and Weber, M. E., *Bubbles, Drops, and Particles*, Academic Press, New York, 1978 (p. 112).

Cunningham, E., "On the Velocity of Steady Fall of Spherical Particles through Fluid Medium," *Proc. R. Soc.*, **A-83**, 357–365 (1910).

Dahneke, B. A., "Slip Connection Factors for Spherical Bodies—III: The Form of the General Law," *J. Aerosol Sci.*, **4**, 163–170 (1973).

Dahneke, B. A., "Viscous Resistance of Straight-Chain Aggregates of Uniform Spheres," *Aerosol Sci. Tech.*, **1**, 179–185 (1982).

Davies, C. N., "Particle Fluid Interaction," *J. Aerosol Sci.*, **10**, 477–513 (1979).

Davies, C. N., "Definitive Equation for the Fluid Resistance of Spheres," *Proc. Phys. Soc.*, **57**, 322 (1945).

Fuchs, N. A., *The Mechanics of Aerosols*, Pergamon, Oxford, U.K., 1964.

Johnson, D. L., Leith, D., and Reist, P. C., "Drag on Non-Spherical, Orthotropic Aerosol Particles," *J. Aerosol Sci.*, **18**, 87–97 (1987).

Landau, L., and Lifshitz, E., *Fluid Mechanics*, Pergamon, London, 1959.

Marple, V.A., Rubow, K.L., and Olsen, B.A., "Inertial, Gravitational, Centrifugal, and Thermal Collection Techniques," in Willeke, K., and Baron, P.A. (Eds.), *Aerosol Measurement*, Van Nostrand Reinhold, New York, 1993 (p. 228).

Mercer, T. T., *Aerosol Technology in Hazard Evaluation*, Academic Press, New York, 1973.

Wall, S., John, W., and Rodgers, D., "Laser Settling Velocimeter: Aerodynamic Size Measurement of Large Particles," *Aerosol Sci. Tech.*, **4**, 81–87 (1985).

4 Particle Size Statistics

The size of the particles of a monodisperse aerosol is completely defined by a single parameter, the particle diameter. Most aerosols, however, are polydisperse and may have particle sizes that range over two or more orders of magnitude. Because of this wide size range and the fact that the physical properties of aerosols are strongly dependent on particle size, it is necessary to characterize these size distributions by statistical means. For the purposes of this chapter, we neglect the effect of particle shape and consider only spherical particles.

We begin by characterizing size distributions in a general way without specifying the type or shape of the distribution used. Later we apply these concepts to the most common aerosol size distribution, the lognormal distribution.

4.1 PROPERTIES OF SIZE DISTRIBUTIONS

In this section, we characterize size distributions and their properties by using examples based on a specific set of particle size data. The result of a careful size analysis might be a list of 1000 particle sizes. In some situations, keeping the data in this form may be desirable—for example, if the list is stored in a computer. In most situations, however, we would like to have a picture of how the particles are distributed among the various sizes and to be able to calculate several different kinds of statistics that describe the properties of the aerosol. For that purpose, a list of 1000 numbers is an awkward format, so it is necessary to resort to descriptive statistics to summarize the information.

The first step in such a summarization is to divide the entire size range into a series of successive particle size intervals and determine the number of particles (the count) in each interval. The intervals must be contiguous and cover the entire size range, so that no particles are left out. If there are 10 size intervals, our list of 1000 numbers is reduced to 20 numbers, the lower (or upper) size limits and the count for each interval, as shown in Table 4.1. The upper size limit of each interval coincides with the lower limit of the next-higher interval. If a particle size falls exactly on the interval limit, it is grouped in the higher interval. These grouped data are much easier to deal with and give the first glimpse of the shape of the size distribution.

One graphical representation of grouped data is the histogram, shown in Fig. 4.1, where the width of each rectangle represents the size interval and the height represents the number of particles in the interval. Unfortunately, the figure gives a dis-

TABLE 4.1   Example of Grouped Data

Size Range[a] (μm)	Count	Fraction/μm	Percent	Cumulative Percent
0–4	104	0.026	10.4	10.4
4–6	160	0.080	16.0	26.4
6–8	161	0.0805	16.1	42.5
8–9	75	0.075	7.5	50.0
9–10	67	0.067	6.7	56.7
10–14	186	0.0465	18.6	75.3
14–16	61	0.0305	6.1	81.4
16–20	79	0.0197	7.9	89.3
20–35	90	0.0060	9.0	98.3
35–50	17	0.0011	1.7	100.0
>50	0	0.0	0.0	100.0
Total	1000		100.0	

[a]Intervals are equal to or greater than the lower limit and less than the upper limit.

torted picture of the size distribution because the height of any interval is dependent on the width of that interval. Thus, doubling an interval's width results in roughly twice as many particles falling into that interval and the interval growing to twice its height. To prevent this distortion, the histogram is normalized for interval width by dividing the number of particles in each interval by the width of that interval. As shown in Fig. 4.2, the height of each rectangle now equals the

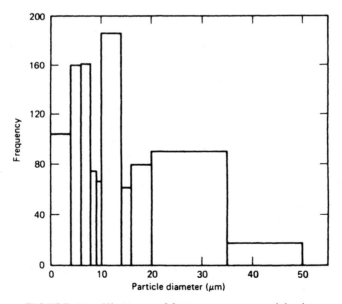

FIGURE 4.1   Histogram of frequency versus particle size.

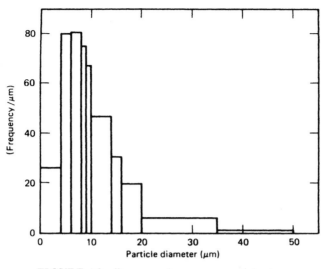

FIGURE 4.2 Frequency/μm versus particle size.

number of particles per unit of size interval (number/μm), and the heights of intervals with different widths are comparable. Furthermore, the area of each rectangle is proportional to the number, or frequency, of particles in that size range. This relationship can be seen from the units of the graph: The height h_i' in number/μm times the width Δd_i in μm gives an area equal to the number of particles in the interval. The total area of all the rectangles is the total number of particles in the sample N.

$$N = \sum_i (h_i' \Delta d_i) \tag{4.1}$$

The histogram is usually standardized for sample size by dividing the heights of the rectangles h_i (in frequency/μm) by the total number of particles observed in the sample, giving heights as fraction/μm. The area of each rectangle in the units of the graph, $h_i \Delta d_i$, is equal to the fraction f_i of particles in that size range, and the total area is equal to 1.0. This change allows direct comparison of histograms obtained from samples of different size.

$$f_i = \frac{n_i}{N} = (h_i \Delta d_i) \tag{4.2}$$

$$\sum f_i = \sum_i (h_i \Delta d_i) = 1.0 \tag{4.3}$$

As shown in Fig. 4.3, the shape of the distribution is the same as that shown in Fig. 4.2, but the ordinate is now the fraction of the total number of particles per unit of size interval.

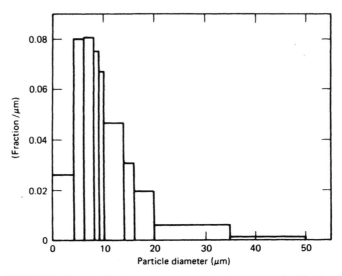

FIGURE 4.3   Fraction/μm versus particle size, count distribution.

Finally, by using many rectangles and drawing a smooth curve through their tops, we obtain the particle size distribution curve that is the graphical representation of the frequency function, or probability density function. Figure 4.4 is an accurate picture of how the particles are distributed among the various sizes; it has the same characteristics as Fig. 4.3, but may be amenable to mathematical interpretation.

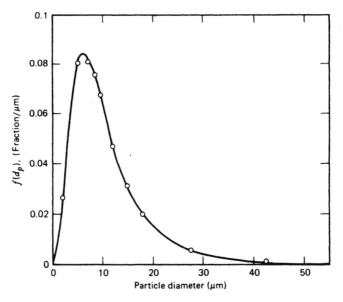

FIGURE 4.4   Frequency distribution curve.

Because the frequency function is such an important way of characterizing distributed quantities, we review its properties here. Although we describe the frequency function in terms of the particle size distribution, we can extend the analysis by analogy to any other distributed quantity, such as particle mass or velocity. The fraction of the total number of particles df having diameters between d_p and $d_p + dd_p$ is

$$df = f(d_p)dd_p \qquad (4.4)$$

where $f(d_p)$ is the frequency function and dd_p is a differential interval of particle size. This function is the mathematical representation of the curve shown in Fig. 4.4. The area under the curve is always

$$\int_0^\infty f(d_p)dd_p = 1.0 \qquad (4.5)$$

which is equivalent to Eq. 4.3. In thinking about frequency functions, always keep in mind which interval is under consideration. The interval may be from zero to infinity, as in Eq. 4.5, or it may be between given sizes a and b, or it may even be the tiny interval dd_p.

The area under the frequency function curve between two sizes a and b equals the fraction of particles whose diameters fall within this interval. Mathematically, this is expressed as

$$f_{ab} = \int_a^b f(d_p)dd_p \qquad (4.6)$$

which is equivalent to Eq. 4.2. Note that the fraction of particles having diameters *exactly* equal to diameter b is zero, because the interval width is zero:

$$f_{bb} = \int_b^b f(d_p)dd_p = 0 \qquad (4.7)$$

Size distribution information can also be presented as a cumulative distribution function, $F(d_p)$, which is defined by

$$F(a) = \int_0^a f(d_p)dd_p \qquad (4.8)$$

$F(a)$ gives the fraction of the particles having diameters less than a. Note that the cumulative distribution function has the size interval built into its definition. Figure 4.5 shows the cumulative distribution function for the data given in Table 4.1. The fraction less than a particular size is obtained from the location on the vertical axis associated with the point on the curve directly above that particular size.

An alternative cumulative distribution function can be constructed by considering the number of particles whose size is greater than, instead of less than, a particular size. A plot of this function contains the same information as is in Fig. 4.5, but has the S-shaped curve reversed. In this book we consider only the more common variety, the "less than indicated size" cumulative distribution.

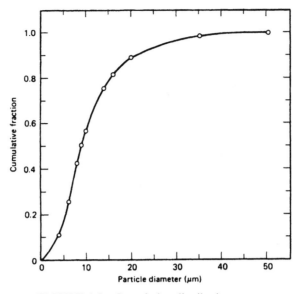

FIGURE 4.5   Cumulative distribution curve.

The cumulative distribution enables one to determine readily quantitative information about the particle size distribution. The fraction less than a given size can be read directly from the graph. The fraction of particles having diameters between sizes a and b (Eq. 4.6) can be determined directly by subtracting the cumulative fraction for size a from that for size b.

$$f_{ab} = F(b) - F(a) \qquad (4.9)$$

The slope of the cumulative distribution function at any point is equal to the value of the frequency function at that size.

$$f(d_p) = \frac{dF(d_p)}{dd_p} \qquad (4.10)$$

As a means of further summarizing the grouped data, a mathematical distribution function is assumed and parameters are calculated that define this function for a particular data set. Most distribution functions require two parameters: one that identifies the location, or center, of the distribution and one that characterizes the width, or spread, of the distribution. These parameters will be discussed more fully in connection with specific distribution functions in subsequent sections.

The most commonly used quantities for defining the location of a distribution are the mean, mode, median, and geometric mean. These and others to be discussed are included in the general term "average." The mean, or arithmetic average, $\overline{d}_p$ is simply the sum of all the particle sizes divided by the number of particles. The mean for listed data, for grouped data, and for a frequency function is given by

$$\bar{d}_p = \frac{\Sigma d}{N} = \frac{\Sigma n_i d_i}{\Sigma n_i} = \int_0^\infty d_p f(d_p) dd_p \qquad (4.11)$$

where n_i is the number of particles in group i, having a midpoint of size d_i, and where $N = \Sigma n_i$, i.e., the total number of particles, and the summations are over all intervals. The midpoint size d_i can be the geometric midpoint—the square root of the product of the upper and lower limits of the interval—or the arithmetic midpoint—the mean of the limits. The former is often preferred, but causes a distortion when the first interval has a lower limit of zero. The arithmetic midpoint is used here for simplicity.

The *median* is defined as the diameter for which one-half the total number of particles are smaller and one-half are larger. The median is also the diameter that divides the frequency distribution curve into equal areas, and the diameter corresponding to a cumulative fraction of 0.5. The *mode is* the most frequent size, or the diameter associated with the highest point on the frequency function curve. The mode can be determined by setting the derivative of the frequency function equal to zero and solving for d. For symmetrical distributions such as the normal distribution, the mean, median, and mode will have the same value, which is the diameter of the axis of symmetry. For an asymmetrical or skewed distribution, these quantities will have different values. The median is commonly used with skewed distributions, because extreme values in the tail have less effect on the median than on the mean. Most aerosol size distributions are skewed, with a long tail to the right, as shown in Fig. 4.4. For such a distribution,

$$\text{mode} < \text{median} < \text{mean} \qquad (4.12)$$

The *geometric mean*, d_g, is defined as the Nth root of the product of N values,

$$d_g = (d_1 d_2 d_3 \cdots d_N)^{1/N} \qquad (4.13)$$

For grouped data with I intervals,

$$d_g = (d_1^{n_1} d_2^{n_2} d_3^{n_3} \cdots d_I^{n_I})^{1/N} \qquad (4.14)$$

where $n_1, n_2, n_3, \ldots, n_I$ are the number of particles in intervals 1 through I and d_1, $d_2, d_3, \ldots, d_I$ are the midpoints, or some other characteristic diameter for intervals 1 through I.

The geometric mean can also be expressed in terms of $\ln(d)$ by converting Eq. 4.14 to natural logarithms:

$$\ln d_g = \frac{\Sigma n_i \ln d_i}{N} \qquad (4.15)$$

$$d_g = \exp\left(\frac{\Sigma n_i \ln d_i}{N}\right) \qquad (4.16)$$

For a monodisperse aerosol, $\bar{d}_p = d_g$; otherwise, $d_g < \bar{d}_p$. The geometric mean will be discussed further in Section 4.4; it is used widely for characterizing aerosols having lognormal size distributions.

EXAMPLE

Using Fig. 4.4 or 4.5 or Table 4.1, determine the mean, median, mode, and fraction between 5 and 15 μm for the given particle size distribution.

Mean. Use Eq. 4.11 and columns 1 and 2 of Table 4.1. Use midpoint values from column 1 to obtain

$$\bar{d}_p = \frac{\Sigma n_i d_i}{\Sigma n_i} = \left(\frac{1}{1000}\right)[(2 \times 104) + (5 \times 160) + (7 \times 161) + \cdots + (42.5 \times 17)]$$

$$\bar{d}_p = \frac{1}{1000}(11176) = 11.2 \mu m$$

Median. Use Fig. 4.5. The particle size corresponding to the cumulative fraction of 0.5 is approximately 9 μm. For this data set, one can also use Table 4.1. In the table, the upper limit of the interval with a cumulative percent of 0.5 is 9 μm.

Mode. Use Fig. 4.4. The size associated with the highest point on the curve is approximately 6 μm.

Fraction between 5 and 15 μm. Use Fig. 4.5 and Eq. 4.9. From the graph, the fraction less than 15 μm is 0.79 and that less than 5 μm is 0.18. From the equation,

$$f_{5-15\mu m} = 0.79 - 0.18 = 0.61$$

4.2 MOMENT AVERAGES

The concepts described in Section 4.1 apply equally well to particle size distributions and to other quantities commonly characterized by number statistics—for example, dollars, populations, or the dimensions of machined parts. What makes particle size statistics more complicated is that we frequently measure or need to know some quantity that is proportional to particle size raised to a power (moment), such as surface area, which is proportional to d^2, or mass, which is proportional to d^3. There is no counterpart of this in number statistics, for what is the meaning of $\$^2$ or (people)^3? The need to use moment averages for particle statistics arises because aerosol size is frequently measured indirectly. For example, if you have a basket of apples of different sizes, you could determine the average size by measuring each apple with calipers, summing the results, and dividing by the total number of apples. This procedure is direct measurement. If, however, each apple were weighed on a balance and the weights summed and divided by their number, the average mass would be obtained.

$$\overline{m} = \frac{\Sigma m_i}{N} \qquad (4.17)$$

Note that $\overline{m}$ could, of course, be obtained more easily by weighing the whole basket and counting the apples. In either case, this is the first step of indirect size measurement. Assuming a spherical apple, the average mass is related to the size (diameter) of the apple with average mass $d_{\overline{m}}$ by

$$\overline{m} = \frac{\pi}{6} \rho_p d_{\overline{m}}^3 \qquad (4.18)$$

Combining Eqs. 4.17 and 4.18 gives the diameter of average mass for a basket of apples, or, a sample of particles,

$$d_{\overline{m}} = \left(\frac{6}{\rho_p \pi N} \Sigma m \right)^{1/3} = \left(\frac{6}{\rho_p \pi N} \frac{\pi \rho_p \Sigma d^3}{6} \right)^{1/3} = \left(\frac{\Sigma d^3}{N} \right)^{1/3} \qquad (4.19)$$

The diameter of average mass (the third-moment average) has a different value from the mean diameter (the first-moment average). Nonetheless, it is a very useful quantity, because it provides the link between the number and the aggregate mass of aerosol particles in a sample and between the number concentration C_N and the mass concentration C_m. The total mass of the particles in a sample, M, is

$$M = N\overline{m} = N \frac{\rho \pi}{6} d_{\overline{m}}^3$$

and

$$C_m = C_N \overline{m} = C_N \frac{\rho \pi}{6} d_{\overline{m}}^3 \qquad (4.20)$$

The preceding discussion deals with only one of a whole family of moment averages. The diameter of a particle with average surface area or average settling velocity is defined similarly. The general form for the diameter of an average property proportional to d^p is

$$d_{\overline{p}} = \left(\frac{\Sigma d^p}{N} \right)^{1/p} \qquad (4.21)$$

or, for grouped data,

$$d_{\overline{p}} = \left(\frac{\Sigma n_i d_i^p}{N} \right)^{1/p} \qquad (4.22)$$

These moment averages represent means calculated for different moments (powers of d), converted back to units of diameter. They are moments about zero and should not be confused with moments about the mean, which define variance and skewness. For the same distribution, the higher the moment, the larger the moment average.

$$\bar{d} < d_{\bar{s}} < d_{\bar{m}} \tag{4.23}$$

The second-moment average $d_{\bar{s}}$ is identical to the rms average (square root of the mean of the squares) discussed in Chapter 2. For the lognormal size distribution—the most common aerosol size distribution—there are explicit equations for converting between any of the average diameters discussed in this chapter. These moment averages are based on the count distribution, the fraction of the total particle count in each size category. In the next section we explore other distributions, such as the distribution of mass.

4.3   MOMENT DISTRIBUTIONS

The count, or number, distribution discussed in the preceding sections is only one of a whole family of distributions. Another member is the distribution of mass (the third-moment distribution), or, more correctly, the mass distribution as a function of particle size. While the number distribution gives the fraction of the total number of particles in any size range, the distribution of mass gives the fraction of the total mass of the particles contributed by particles in any size range. A graph of the mass distribution shows how the mass is distributed among the various particle sizes. The distribution of mass and the distribution of count for the same sample of particles have different means, medians, geometric means, graphical representations, and probability density functions. The median of the distribution of mass is called the *mass median diameter* (MMD), to distinguish it from the count median diameter (CMD) discussed in Section 4.2. The latter is the particle size for which half the total *number* of particles are larger and half are smaller. The MMD is defined as the diameter for which half the *mass* is contributed by particles larger than the MMD and half by particles smaller than the MMD. It is the diameter that divides the graphical representation of the distribution of *mass* into two segments of equal area.

A cumulative distribution function similar to Fig. 4.5 can be constructed for the distribution of mass. The vertical axis is changed to represent the fraction of total mass attributable to particles less than the indicated size. Figures 4.6 and 4.7 show the distribution of mass and the cumulative distribution of mass for the particle size data given in Table 4.1 and shown in Figs. 4.1–4.5.

As a simple illustration of different moment distributions, consider an aerosol composed of equal numbers of two particle sizes, one in which 50 particles are 1 μm in diameter and 50 are 10 μm in diameter. While half the number are in each size category, most of the mass is represented by the 10-μm particles. Each 10-μm

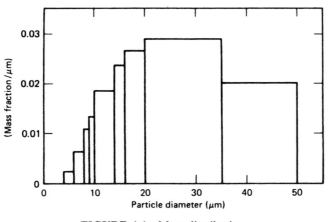

FIGURE 4.6   Mass distribution.

particle has 1000 times the mass of a 1-μm particle. The 10-μm particles contribute 50,000 units of mass to the total mass of the particles, whereas the 1-μm particles contribute only 50 units of mass; thus, the mass is distributed such that 99.9% is in the 10-μm particles and 0.1% in the 1-μm particles. A similar analysis for surface area attributes 99% of the surface area to the 10-μm particles and 1% to the 1-μm particles. The distinction between count and mass distributions can be put in

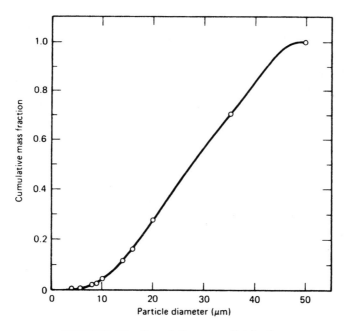

FIGURE 4.7   Cumulative mass distribution.

even simpler terms: A bowl containing nine grapes and an apple is 90% grapes by count, but 90% apples by mass.

The most confusing aspect of moment distributions (also called weighted distributions) is the mean value, the mass mean diameter (for the mass distribution), and the distinction between the mass mean diameter and the diameter of average mass, described in the previous section. The arithmetic mean of the number distribution, or the count mean diameter, (Eq. 4.11), can be written for grouped data with I intervals as

$$\bar{d} = \frac{\Sigma n_i d_i}{N} = \left(\frac{n_1}{N} d_1 + \frac{n_2}{N} d_2 + \cdots + \frac{n_I}{N} d_I \right) \qquad (4.24)$$

where N is the total number of particles. In the averaging process, the characteristic diameter for each size group d_i is given a weight equal to n_i/N, or the fraction of the total number of particles in that group.

For the distribution of mass, which describes the fraction of the total mass in the various size ranges, the mass mean diameter can be written in a completely analogous fashion. Let m_i be the mass of all the particles in group i (midpoint diameter d_i) and let M be the total mass for all groups.

$$\text{mass mean diameter} = d_{\text{mm}} = \left(\frac{m_1}{M} d_1 + \frac{m_2}{M} d_2 + \cdots + \frac{m_I}{M} d_I \right) \qquad (4.25)$$

Equation 4.25 can be rewritten for spherical particles of uniform density.

$$d_{\text{mm}} = \frac{\Sigma m_1 d_i}{M} = \frac{(\pi \rho_p/6)\Sigma n_i d_i^3 d_i}{(\pi \rho_p/6)\Sigma n_i d_i^3} = \frac{\Sigma n_i d_i^4}{\Sigma n_i d_i^3} \qquad (4.26)$$

The mass mean diameter, although completely analogous to the mean of the count distribution, has quite a different appearance. Nonetheless, it has units of length (d^1), as do the count mean diameter and the diameter of average mass. As with all mean diameters, the mass mean diameter has only a mathematical definition with no associated physical meaning. In the calculation of the diameter of average mass, a coarse and a fine particle are given equal representation in the averaging process, but the quantity averaged is the mass. In calculating the mass mean diameter, the quantity that is averaged is the diameter, but it is *weighted* according to its mass contribution to the total mass of the particles.

The surface area mean diameter d_{sm} can be written in a form analogous to Eq. 4.26 with group surface area s_i and total surface area S.

$$d_{\text{sm}} = \frac{\Sigma s_1 d_i}{S} = \frac{\Sigma n_i d_i^3}{\Sigma n_i d_i^2} \qquad (4.27)$$

This average diameter is also called the *Sauter diameter* or the *mean volume–surface diameter*.

The numerator and denominator of the rightmost side of Eq. 4.27 are moment sums and are closely related to moment averages. The numerator can be written in terms of the diameter of average mass. From Eq. 4.22 for grouped data,

$$d_{\bar{m}} = \left(\frac{\Sigma n_i d_i^3}{N}\right)^{1/3} \tag{4.28}$$

$$\Sigma n_i d_i^3 = N(d_{\bar{m}})^3 \tag{4.29}$$

Similarly, for the diameter of average surface,

$$\Sigma n_i d_i^2 = N(d_{\bar{s}})^2 \tag{4.30}$$

Substituting into Eq. 4.27 gives

$$d_{sm} = \frac{N(d_{\bar{m}})^3}{N(d_{\bar{s}})^2} = \frac{(d_{\bar{m}})^3}{(d_{\bar{s}})^2} \tag{4.31}$$

Thus, each weighted mean diameter can be written as the ratio of two moment averages.

Weighted mean diameters can also be expressed in terms of aggregate quantities. If M and S are the total mass and surface area, respectively, of an aerosol sample, then

$$M = \left(\frac{\rho_p \pi}{6}\right) \Sigma n_i d_i^3 \tag{4.32}$$

$$S = \pi \Sigma n_i d_i^2 \tag{4.33}$$

and Eq. 4.27 becomes

$$d_{sm} = \left(\frac{6}{\rho_p}\right) \frac{M}{S} \tag{4.34}$$

Equation 4.34 illustrates a characteristic of weighted mean diameters: An average particle size (d_{sm}) can be determined if the particle density is known by simply weighing a sample of particles and measuring the surface area of the sample. There are gas adsorption and radioactive coating methods to do this.

In addition to mass, surface, and count distributions, there is the less commonly used distribution of length. This distribution expresses the fraction of the total length, if all particles are laid side by side, in any size range. In fact, there can be a moment distribution for any property that is proportional to a constant power of diameter. The general form for the distribution function giving the fraction of a quantity proportional to d_p^x in an interval between d_p and $d_p + dd_p$ is

$$df_x = f_x(d_p)d_p = \frac{d_p^x f(d_p)}{\int_0^\infty d_p^x f(d_p)dd_p} dd_p \tag{4.35}$$

There is a set of moment averages for each moment distribution. The general form for the p moment average of the q moment distribution is

$$(d_{qm})_{\bar{p}} = \left(\frac{\Sigma n_i d_i^q d_i^p}{\Sigma n_i d_i^q}\right)^{1/p} \tag{4.36}$$

When $q = 0$, the weighting is by number; when $p = 1$ and $q = 0$, we have Eq. 4.11, the arithmetic mean; when $p = 1$ and $q = 3$, we have the mass mean diameter, Eq. 4.26, and when $p = 3$ and $q = 0$ we have the diameter of average mass, Eqs. 4.19 and 4.22. The names used here are common, but hardly universal. Some authors use mean mass diameter for the diameter of average mass, but this is easily confused with the mass mean diameter. The latter can best be remembered as the mass (distribution) mean diameter or mass (weighted) mean diameter. One often encounters the terms *CMAD* and *MMAD*, which refer to the count and mass medians of the distributions of count or mass with respect to aerodynamic diameter. Graphs for these distributions are constructed by plotting frequency or fraction versus aerodynamic diameter. The term *AMD* refers to the activity median diameter and is the median of the distribution of radioactivity or toxicological or biological activity with respect to size.

EXAMPLES

1   As an example of moment averaging, suppose that you have a list of heights for 20 geometrically similar people and you wish to select the person with the average surface area for a heat stress experiment. Surface area is difficult to measure, but height is easy to measure. Assume that surface area is proportional to height squared, h^2, and is related to height by some constant (which we do not need to know). Then the person with the average surface area will have a height given by

$$h_{\overline{SA}} = \left(\frac{\Sigma h^2}{N}\right)^{1/2}$$

which is the square root of the arithmetic average of height squared, or the second-moment average.

Similarly, if we wanted the person with average mass, we could determine that person's height by taking the cube root of the arithmetic average of the heights cubed:

$$h = \left(\frac{\Sigma h^3}{N}\right)^{1/3}$$

If we desired the total weight to determine whether all 20 people could ride to-gether on an elevator, we need only weigh that person whose height is given by the previous equation (or interpolate between the two nearest people) and multiply his or her weight by 20. The person with the median height will also have the median weight, because the order will not change by cubing heights.

2   An example of a moment distribution is that obtained from an analysis of the grain size of sand by sieving. The sand is passed through a stack of sieves ar-ranged in order of decreasing mesh size. The mass of sand retained on each sieve is determined. From the characteristics of the screens, we can assign a diam-eter to each mass of sand retained. With these data, we can determine the dis-tribution of mass and calculate the mass mean diameter of the sand particles as follows:

Calculation of Mass Mean Diameter

Sieve Opening (mm)	Midsize of Collected Sand, d_i (mm)	Mass Collected, m_i (g)	Percent of Total Mass (%)	$m_i d_i$ (g·mm)
1.0	1.5	8	8.2	12.0
0.50	0.75	34	35.1	25.5
0.25	0.375	40	41.2	15.0
0.125	0.188	12	12.4	2.26
Final Pan	0.062	3	3.1	0.19
	Total	97	100.0	54.9

$$d_{mm} = \frac{\Sigma m_i d_i}{M} = \frac{54.9}{97} = 0.57 \text{ mm}$$

3   A simple numerical example:

n	d	nd	nd^2	nd^3	nd^4
3	1.0	3	3	3	3
5	2.0	10	20	40	80
2	6.0	12	72	432	2592
10	9.0	25	95	475	2675

Arithmetic average, or count mean diameter = $\Sigma nd/N = 2.50$

Median = 2.0

Mode = 2.0

Diameter of average surface = $(\Sigma nd^2/N)^{1/2} = (9.5)^{1/2} = 3.08$

Diameter of average volume = $(\Sigma nd^3/N)^{1/3} = (47.5)^{1/3} = 3.62$

Length mean diameter = $\Sigma nd^2/\Sigma nd = 95/25 = 3.80$

Surface mean diameter = $\Sigma nd^3/\Sigma nd^2 = 475/95 = 5.00$

Volume mean diameter = $\Sigma nd^4/\Sigma nd^3 = 2675/475 = 5.63$

Note that the surface mean diameter can also be calculated by Eq. 4.31:

$$d_{sm} = (d_{\overline{m}})^3/(d_{\overline{s}})^2 = 3.62^3/3.08^2 = 5.00$$

4.4   THE LOGNORMAL DISTRIBUTION

The preceding sections have focused on the properties of distributions in general without considering any particular type of distribution. In this section, we describe the characteristics and applications of the lognormal distribution for aerosol particle size analysis. As discussed below, the normal distribution, although widely used elsewhere, is not suitable for most aerosol particle size distributions.

The characteristics of other distributions that have been applied to aerosol particle size, such as the Rosin–Rammler, Nukiyama–Tanasawa, power law, exponential, and Khrgian–Mazin distributions are given in the appendix to this chapter. These distributions apply to special situations and find limited application in aerosol science. They (and the lognormal distribution) have been selected empirically to fit the wide range and skewed shape of most aerosol size distributions.

The normal distribution function is rarely used to describe aerosol particle size distributions because most aerosols exhibit a skewed (long tail at large sizes) distribution function. The normal distribution is, of course, symmetrical. It can be applied to monodisperse test aerosols, to certain pollens and spores, and to specially prepared polystyrene latex spheres. The number frequency function is given by

$$df = \frac{1}{\sigma\sqrt{2\pi}} \exp\left(-\frac{(d_p - \overline{d}_p)^2}{2\sigma^2}\right) dd_p \tag{4.37}$$

where $\overline{d}_p$ is the arithmetic mean diameter and σ is the standard deviation, defined for grouped data as

$$\sigma = \left(\frac{\Sigma n_i(d_i - \overline{d}_p)^2}{N - 1}\right)^{1/2} \tag{4.38}$$

A second problem in applying the normal distribution to any quantity that varies over a wide range, such as aerosol particle size, is that such a wide normal distribution requires a certain fraction of the particles to have negative size, a physical impossibility. This problem combined with the frequently observed skewed shape of the distribution, led to the use of the logarithmic transformation of size data to obtain the lognormal distribution. With the logarithm of particle size (actually, the logarithm of (d_p/d_0), where d_0 is 1 μm) plotted, instead of size, along the horizontal axis, the frequency distribution shows the symmetrical form of the normal distribution. This has been done in Fig. 4.8, using a logarithmic scale and the

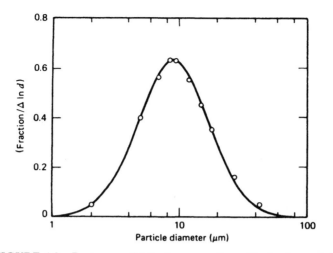

FIGURE 4.8   Frequency distribution curve (logarithmic size scale).

size data given in Table 4.1. Note that the logarithmic scale has no negative values, thus overcoming the problem of negative particle size.

The lognormal distribution has been applied to such diverse distributions as incomes, populations (of both people and bacteria), prices of stocks listed on the New York Stock Exchange, concentrations of environmental contaminants, and particle sizes of aerosols. There is no fundamental theoretical reason why particle size data should approximate the lognormal distribution, but it has been found to apply to most single-source aerosols. (See Appendix 2 at the end of the Chapter.) Mixtures of lognormal distributions will not be lognormal.

The lognormal distribution is most useful in situations where the distributed quantity can have only positive values and covers a wide range of values—that is, where the *ratio* of the largest to the smallest value is greater than about 10. When this range is narrow, the lognormal distribution approximates the normal distribution. The lognormal distribution is used extensively for aerosol size distributions because it fits the observed size distributions reasonably well and its mathematical form is convenient for dealing with the moment distributions and moment averages described in the preceding sections.

Because the logarithm of d_p is normally distributed, the lognormal distribution frequency function can be formed by replacing $\bar{d}_p$ and σ in Eq. 4.37 with their logarithmic counterparts. Thus, $\bar{d}_p$ is replaced by the arithmetic mean of $\ln d$, defined by Eq. 4.15 as the geometric mean diameter d_g,

$$\ln d_g = \frac{\Sigma n_i \ln d_i}{N} \tag{4.39}$$

Either the natural logarithm or the logarithm to the base 10 can be used, but the natural logarithm is more common and is used here.

The standard deviation σ is replaced by the standard deviation of the logarithms, called the *geometric standard deviation* σ_g, or GSD.

$$\ln \sigma_g = \left(\frac{\Sigma n_i (\ln d_i - \ln d_g)^2}{N - 1} \right)^{1/2} \tag{4.40}$$

The geometric standard deviation is a dimensionless quality with a value equal to or greater than 1.0.

For the count distribution, the geometric mean diameter d_g is customarily replaced by the count median diameter, or CMD. The geometric mean is the arithmetic mean of the distribution of $\ln d_p$, which is a symmetrical normal distribution, see Fig. 4.8, and hence, its mean and median are equal. The median of the distribution of $\ln d_p$ is also the median of the distribution of d_p, as the order of values does not change in converting to logarithms. Thus, for a lognormal count distribution, $d_g = \text{CMD}$. The frequency function can be expressed as

$$df = \frac{1}{\sqrt{2\pi} \ln \sigma_g} \exp\left(-\frac{(\ln d_p - \ln \text{CMD})^2}{2(\ln \sigma_g)^2} \right) d \ln d_p \tag{4.41}$$

which gives the fraction of the particles having diameters whose logarithms lie between $\ln d_p$ and $\ln d_p + d \ln d_p$. It is more convenient, however, to express the frequency function in terms of the particle size d_p rather than $\ln d_p$. Rewriting Eq. 4.41 in terms of d_p, using the fact that $d \ln d_p = dd_p/d_p$, gives

$$df = \frac{1}{\sqrt{2\pi} \, d_p \ln \sigma_g} \exp\left(-\frac{(\ln d_p - \ln \text{CMD})^2}{2(\ln \sigma_g)^2} \right) dd_p \tag{4.42}$$

The cumulative distribution for the lognormal distribution is obtained as described in Section 4.1. The cumulative plot shown in Fig. 4.9 is the same as Fig 4.5, but diameter is plotted on a logarithmic scale. Note that the median size is the same for both figures.

For the normal distribution, 95% of the particles fall within a size range defined by $\bar{d} \pm 2\sigma$. For the lognormal count distribution, the distribution is normal with respect to $\ln d$, so that 95% of the particles fall within a size range defined by

$$\exp(\ln \text{CMD} \pm 2 \ln \sigma_g) \tag{4.43}$$

This range is asymmetrical and goes from CMD/σ_g^2 to $\text{CMD} \times \sigma_g^2$. For $\sigma_g = 2.0$, 95% of the particles have sizes between one-fourth and four times the count median diameter.

All moment distributions of any lognormal distribution will be lognormal and have the same geometric standard deviation. This means that they will have the same shape when they are plotted on a logarithmic scale. The lognormal distribution is the only common distribution that has this property. Figure 4.10 shows the distri-

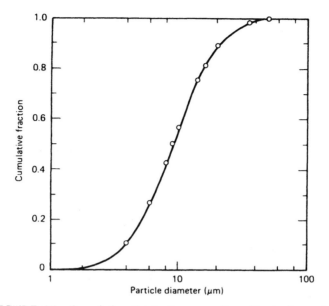

FIGURE 4.9 Cumulative distribution curve (logarithmic size scale).

bution of count and mass plotted on the same logarithmic diameter scale. For the mass distribution, we replace the geometric mean diameter with the mass median diameter MMD, analogous to what we did for the count distribution. The mass distribution has the same shape as the count distribution but is displaced along the size axis by a constant amount equal to MMD/CMD. The ratio MMD/CMD can be calculated knowing only the GSD. This calculation is covered in Section 4.6.

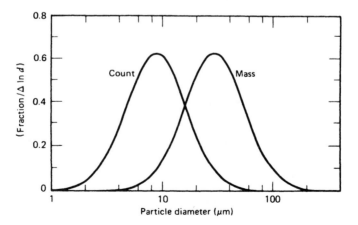

FIGURE 4.10 Count and mass distributions (logarithmic size scale).

4.5   LOG-PROBABILITY GRAPHS

Much of the practical application of the lognormal distribution to particle size analysis is facilitated by using log-probability graphs. In the most common form of these
graphs, the axes of Fig. 4.9 are changed as shown in Fig. 4.11, and the cumulative
fraction (or percent) scale is converted to a probability scale. The probability scale
compresses the percent scale near the median (50% point) and expands the scale
near the ends such that a cumulative plot of a lognormal distribution will yield a
straight line, as shown in Fig. 4.12. When the particle diameter scale is logarithmic, the graph is called a *log-probability plot*. When the size scale is linear, the
cumulative graph will yield a straight line for a normal distribution and is called a
probability plot. Both types of graph paper are commercially available. In either case
the median size can be read directly from the graph, as with any cumulative plot.
A log-probability plot can be constructed on arithmetic graph paper by plotting the
logarithm of diameter versus the probit of the cumulative percentages. The latter,
obtained from a table of probits, gives the linear displacement from the midpoint
(50%) in units of standard deviations. The slope of the straight line is related to the
geometric standard deviation. A line with a steep slope is associated with a wide
distribution, and a line with a shallow slope represents a narrow distribution. A horizontal line in Fig. 4.12 characterizes a monodisperse aerosol—that is, one in which
all particles have the same size.

For any *normal* distribution, one standard deviation represents the *difference*
between the size associated with a cumulative count of 84.1% and the median size
(a cumulative count of 50%) or between the 50% cumulative size and the 15.9%
cumulative size.

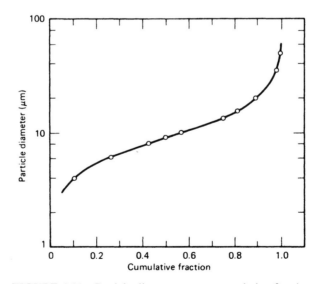

FIGURE 4.11   Particle diameter versus cumulative fraction.

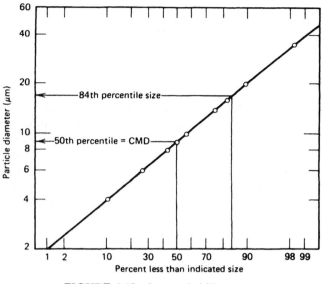

FIGURE 4.12   Log-probability graph.

$$\sigma = d_{84\%} - d_{50\%} \tag{4.44}$$

For a lognormal distribution, which is normal with respect to ln d, these differences become ratios

$$\ln \sigma_g = \ln d_{84\%} - \ln d_{50\%}$$
$$= \ln(d_{84\%}/d_{50\%}) \tag{4.45}$$

$$\mathrm{GSD} = \sigma_g = \frac{d_{84\%}}{d_{50\%}} = \frac{d_{50\%}}{d_{16\%}} = \left(\frac{d_{84\%}}{d_{16\%}}\right)^{1/2} \tag{4.46}$$

The geometric standard deviation, being the ratio of two sizes, has no units and must always be greater than or equal to 1.0. The CMD (50% cumulative size) and the GSD can be determined directly from a log-probability plot and completely define a lognormal distribution.

The first step in the graphical procedure for determining a lognormal size distribution is to determine the number of particles in each of a consecutive series of size intervals. We then express these numbers as a percent of the total and (starting with the smallest size interval), sequentially add the percent for each interval, to obtain the cumulative frequency in percent, as shown in the last column of Table 4.1. Next we plot the datum for each interval on log-probability paper as the percentage less than the upper size limit versus the upper size limit for that interval. If a straight line fits the data, the distribution can be represented by a lognormal distribution and the CMD and GSD can be determined from the 50th percentile size and Eq. 4.46. While there will usually be some scatter about the straight line, the trend of the data must be linear. If the distribution is not lognormal—that is, if it does not display a linear trend on a log-probability plot—the CMD can still be determined, but Eq. 4.46 holds only for a lognormal distribution.

Frequently, in aerosol sampling there is an aerodynamic size above which particles are aerodynamically unable to enter the sampling apparatus. This is called the *aerodynamic cutoff size* and means that the cumulative line on the log-probability plot will curve near its upper end, so that it never exceeds the cutoff size.

A similar limit may exist for the lower end of the size distribution if sizing is done by optical microscopy. Because of the optical limitations explained in Section 20.2, particles less than about 0.3 μm in diameter are not included in the size distribution. This 0.3-μm limit, sometimes called the *optical cutoff*, curves the lower end of the size distribution line so that it never goes below 0.3 μm. An example of data having both cutoffs is shown in Fig. 4.13. These cutoffs are artifacts of the sampling and measurement system. If the cutoffs affect only a small fraction of the distribution, it is acceptable to ignore them when fitting a straight line to the data on a log-probability plot.

Figure 4.13 also shows the effect of a mixture of two lognormal distributions having the same GSD but different median sizes. This is an example of a *bimodal distribution*, because its frequency function has two peaks. It is often possible to

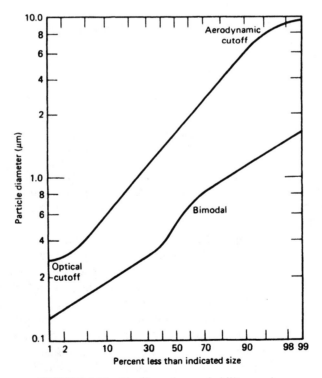

FIGURE 4.13   Nonlinear log-probability graphs.

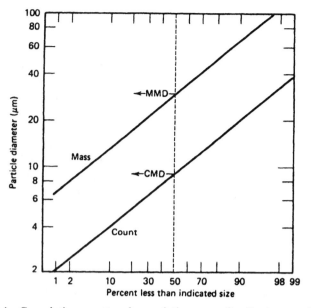

FIGURE 4.14 Cumulative count and cumulative mass distributions on log-probability graph.

represent a bimodal distribution by two overlapping lognormal distributions. (See Knutson and Lioy (1995).)

One great advantage of the lognormal distribution is that, for a given distribution, the geometric standard deviation remains constant for all moment distributions. If the GSD is the same, the ratio $d_{84\%}/d_{50\%}$ will be the same, for each distribution, and the cumulative distribution lines for other moment distributions will be parallel to the cumulative count distribution line on a log-probability graph. The separation between the lines (and between median diameters) is a function only of GSD. (See Section 4.6.) Figure 4.14 shows both count and mass distributions for the data given in Table 4.1, plotted as cumulative distributions on log-probability graph paper. The horizontal axis refers to cumulative percent of mass for the mass distribution line and cumulative percent of count for the count distribution line.

4.6 THE HATCH–CHOATE CONVERSION EQUATIONS

The real power and utility (and perhaps the popularity) of the lognormal distribution comes from the fact that any type of average diameter discussed in this chapter can easily be calculated for any known lognormal distribution—that is, a distribution for which one average diameter and the GSD are known. This is useful because it is frequently necessary to measure one characteristic of the size distribution, such as the number distribution, when what is really needed is another characteristic, such as the mass distribution or the diameter of average mass.

If a detailed size distribution is available, it is possible to calculate the other moment distributions, as well as their means, medians, and moment averages, using the equations given in Sections 4.1–4.3. This procedure is tedious and may be inaccurate if there is insufficient detail. If, and only if, the distribution is lognormal, these quantities can be calculated directly using the lognormal conversion equations originally derived by Hatch and Choate (1929) and called the *Hatch–Choate equations*.

We consider here three types of equations that convert the CMD to (1) another median diameter, (2) a weighted mean diameter, and (3) a diameter of average property. Although these equations are written for CMDs, they can be combined to convert any type of average diameter to any other type of average diameter. The examples that follow are for the most common case, conversions between number and mass distribution. Analogous equations can be written for any moment distribution, such as the distribution of surface area or of settling velocity.

All the conversion equations that follow have the same form, namely,

$$d_A = CMD \exp(b \ln^2 \sigma_g) \tag{4.47}$$

where b is a constant that depends only on the type of conversion—that is, on what type of average diameter d_A is. For a given type of conversion, the ratio of the two diameters depends only on the value of σ_g. A derivation of these lognormal conversion equations is given in Appendix 3 to this chapter.

To convert the CMD to the median diameter of the qth moment distribution (qMD), we use the general conversion equation

$$q\text{MD} = CMD \exp(q \ln^2 \sigma_g) \tag{4.48}$$

where qMD is the median of the qth moment distribution. For this conversion, b in Eq. 4.47 is set equal to q. For example, the MMD is obtained from Eq. 4.47 by setting $b = q = 3$:

$$\text{MMD} = CMD \exp(3 \ln^2 \sigma_g) \tag{4.49}$$

To convert the CMD to the *mean* diameter of the qth moment distribution, d_{qm}, we use the conversion equation

$$d_{qm} = CMD \exp\left[\left(q + \frac{1}{2}\right) \ln^2 \sigma_g\right] \tag{4.50}$$

For example, the mass mean diameter is given by

$$d_{mm} = CMD \exp(3.5 \ln^2 \sigma_g) \tag{4.51}$$

To convert the CMD to the diameter of average, d^p, where $p = 1$ for length, 2 for surface, and 3 for mass or volume, we use the equation

$$d_{\bar{p}} = CMD \exp\left(\frac{p}{2} \ln^2 \sigma_g\right) \tag{4.52}$$

Thus, the diameter of the particle with average mass (that size particle whose mass multiplied by the total number of particles gives the total mass) is

$$d_{\bar{m}} = \text{CMD} \exp(1.5 \ln^2 \sigma_g) \qquad (4.53)$$

The mode, $\hat{d}$, is given by

$$\hat{d} = \text{CMD} \exp(-1 \ln^2 \sigma_g) \qquad (4.54)$$

The general conversion equation for any average diameter described by Eq. 4.36—that is, the pth moment average of the qth moment distribution—is given by

$$(d_{qm})_{\bar{p}} = \text{CMD} \exp\left[\left(q + \frac{p}{2}\right) \ln^2 \sigma_g\right] \qquad (4.55)$$

For use with the logarithm to the base 10, the $(q + p/2)$ term is multiplied by 2.303, and $\log_{10}$ and 10^x are used. When $q = 0$, Eq. 4.55 becomes the conversion equation for the diameter of average d^p (Eq. 4.52). When $p = 0$, Eq. 4.55 becomes the conversion equation for the median of the qth moment distribution (qMD). When $p = 1$, we get the conversion equation for the mean of the qth moment distribution, d_{qm}. Table 4.2 gives the value of b to be used in Eq. 4.47 for the most common conversions. Figure 4.15 shows the relative positions of several average diameters on an arithmetic plot of the count distribution data given in Table 4.1. The larger the coefficient in the conversion equation, the further that converted average diameter is from the CMD. Also, the larger the GSD, the greater the ratio between the MMD and the CMD, as shown in Fig. 4.16.

Table 4.3 summarizes information about the various types of averages we have discussed and gives the coefficient b for the lognormal conversion equation, Eq. 4.47. There is an obvious pattern to the values of b, and the table can be extended by analogy in any direction. However, as will be discussed in the next section, care must be taken to ensure that the data are accurate and properly represent a lognormal distribution in the region of the converted average diameter.

TABLE 4.2 Coefficients for Equation 4.47 for the Most Common Conversions

To Convert from CMD to[a]	Use, for the value of b in Eq. 4.47,
Mode, $\hat{d}$	-1
Count mean diameter, $\bar{d}$	0.5
Diameter of average mass, $d_{\bar{m}}$	1.5
Mass median diameter, MMD	3
Mass mean diameter, d_{mm}	3.5

[a]Diameters listed replace d_A in Eq. 4.47.

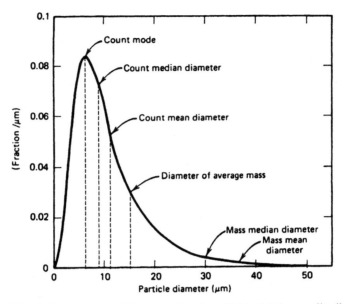

FIGURE 4.15 Arithmetic plot of the size data from Table 4.1 (count distribution with CMD = 9.0 μm and GSD = 1.89), showing the locations of various average diameters.

EXAMPLE

Calculate the MMD and the diameter of average mass, $d_{\overline{m}}$, for a lognormal size distribution with CMD = 3.5 μm and GSD (σ_g) = 2.2.

For the MMD, use Eq. 4.47 with $b = 3$:

$$MMD = CMD \ exp(b \ \ln^2 \sigma_g) = 3.5 \ exp(3 \ (\ln 2.2)^2)$$

$$MMD = 3.5 \ exp(3 \times 0.622) = 3.5 \ e^{1.865} = 3.5 \times 6.46 = 23 \ \mu m$$

For $d_{\overline{m}}$, $b = 1.5$, and

$$d_{\overline{m}} = CMD \ exp(1.5 \ \ln^2(2.2)) = 3.5 \ e^{0.932} = 8.9 \ \mu m$$

4.7 STATISTICAL ACCURACY

Great care must be exercised in calculating the mass mean or mass median diameter based on count data. As shown in Fig. 4.15, the distribution of mass is centered (MMD) well out on the tail of the number distribution. Thus, it is important to ensure that this region of the number distribution is accurately represented. When there are errors or scatter in the data, calculated values for different moment distri-

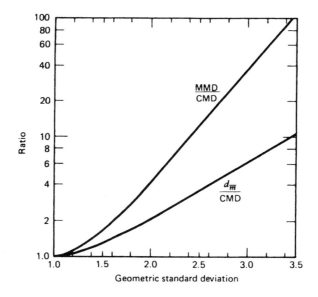

FIGURE 4.16 Ratios of mass median diameter and diameter of average mass to count median diameter versus geometric standard deviation.

butions should be considered estimates and used cautiously. For example, a 10% error in the 84% size changes the GSD from 2.0 to 2.2 and the calculated MMD by 53%. The converse is also true: When deriving count sizes from mass data, care must be taken to ensure that the lower end of the mass distribution is statistically accurate.

Most of the particles in a sample will be in a relatively few size intervals, and intervals in the tail of a distribution will contain few particles. To ensure statistical reliability of the measurements at the tail, a common criterion is to observe at least 10 particles in every size interval of importance to the distribution (either mass or count) curve. An efficient procedure for doing this for microscopic sizing is described in Section 20.1.

Confidence intervals for the CMD and GSD can be constructed to provide a measure of the range in which the true CMD and GSD values are likely to lie. Thus, the 95% confidence interval for the CMD expresses the likelihood (19 chances out of 20) that the true CMD lies within the range defined by the confidence interval of a random sample.

For a lognormal distribution the distribution of ln d is normal, and a confidence interval for ln CMD (and CMD) can be constructed using the standard deviation of ln CMD, $\sigma_{\ln \text{CMD}}$,

$$\sigma_{\ln \text{CMD}} = \left[\frac{\ln^2 \sigma_g}{N} + \sigma_{EE}^2 \right]^{1/2} \tag{4.56}$$

TABLE 4.3 Names, Defining Equations, and Coefficients[a] for the Lognormal Conversion Equations for Common Types of Average Diameters

	Types of Average[b]			
Distribution (d^q)	$\ln [p=0]$: Median, Geometric	$p=1$: Diameter	$p=2$: Area, V_{TS}	$p=3$: Volume, Mass
Count (d^0)	Count median diameter, geometric mean: $CMD = \exp\left[\dfrac{\Sigma n \ln d}{N}\right]$ $b=0$	Count mean diameter: $\bar{d} = \dfrac{\Sigma nd}{N}$ $b=0.5$	Diameter of average surface: $d_{\bar{s}} = \left[\dfrac{\Sigma nd^2}{N}\right]^{1/2}$ $b=1$	Diameter of average volume, diameter of average mass: $d_{\bar{m}} = \left[\dfrac{\Sigma nd^3}{N}\right]^{1/3}$ $b=1.5$
Length (d^1)	Length median diameter: $LMD = \exp\left[\dfrac{\Sigma nd \ln d}{\Sigma nd}\right]$ $b=1$	Length mean diameter: $d_{lm} = \dfrac{\Sigma nd^2}{\Sigma nd}$ $b=1.5$		
Area (d^2)	Surface median diameter: $SMD = \exp\left[\dfrac{\Sigma nd^2 \ln d}{\Sigma nd^2}\right]$ $b=2$	Surface mean diameter, Sauter diameter, Mean volume–surface diameter: $d_{sm} = \dfrac{\Sigma nd^3}{\Sigma nd^2}$ $b=2.5$		
Volume (d^3) or mass (d^3)	Volume median diameter, mass median diameter: $MMD = \exp\left[\dfrac{\Sigma nd^3 \ln d}{\Sigma nd^3}\right]$ $b=3$	Volume mean diameter, mass mean diameter: $d_{mm} = \dfrac{\Sigma nd^4}{\Sigma nd^3}$ $b=3.5$		

[a] b values for Eq. 4.47 and $b = q + p/2$ for Eq. 4.55.
[b] p = moment of diameter.

The first term on the right-hand side of Eq. 4.56, the random-sampling error, reflects the variation in ln CMD from sample to sample, assuming that each sample is taken from the same stable aerosol, without any experimental error. The term σ_{EE} reflects the variation in ln CMD that results from experimental error due to measurement. In most particle sizing situations N is large, so random sampling errors are small compared with experimental errors. The confidence interval (CI) for ln CMD is given by

$$95\% \text{ CI for ln CMD} \cong \text{ln CMD} \pm 2\sigma_{\text{ln CMD}} \qquad (4.57)$$

Confidence limits calculated by Eq. 4.57 are logarithms and can be converted to diameters to get the 95% confidence interval for CMD.

In most situations, there is no simple way to calculate the effect of experimental error on ln CMD or ln GSD. Nevertheless, confidence intervals for CMD and GSD can still be determined by making repeated size distribution measurements. The mean and standard deviation of the distribution of CMDs and GSDs are then calculated in the usual way, and confidence intervals for the CMD and GSD are calculated by

$$95\% \text{ CI for CMD} = \overline{\text{CMD}} \pm t\sigma_{\text{CMD}} \qquad (4.58)$$

$$95\% \text{ CI for GSD} = \overline{\text{GSD}} \pm t\sigma_{\text{GSD}} \qquad (4.59)$$

where t is the t-test value, obtained from statistical tables ($P = 0.05$ and degrees of freedom = $N' - 1$, where N' is the number of replications). On the log-probability plot, the confidence interval of CMD defines the range of vertical positions of the size distribution line, and the confidence interval of GSD defines the range of slopes. Statistical tests such as the t-test and F-test can be used to compare the parameters of one aerosol size distribution with those of another. These tests are discussed by Cooper (1993), Land (1988), Herdan (1960), and Aitchison and Brown (1957).

The cumulative log-probability plot assumes that the data cover the size range from zero to infinity. If there are significant numbers of particles lost by truncation at either end of the size range, the cumulative plot should not be used, but a frequency histogram can still be constructed without bias.

Because of the expansion of the scale at the ends of log-probability graph paper, an error in cumulative frequency is about four times wider on the graph at 5% and 95% than it is at 50%. The rule of thumb for fitting a straight line on a log-probability plot by eye is to give the most weight to the central region (20–80%) and little weight to those points at less than 5% and greater than 95%. The use of least squares regression to find the best fitting straight line on an arithmetic plot of ln(d) versus probit of the cumulative frequency is not recommended, because it overemphasizes the tails of the distribution. There are iterative computer procedures for finding the best fitting line that give proper weight to all points in the distribution. [See Raabe (1978) and Kottler (1951).]

Because of the convenience of the lognormal distribution, there is a tendency to try to fit a straight line to points that clearly do not follow a straight line. This is a risky procedure, especially if other average diameters are to be calculated using the lognormal conversion equations. When the data do not follow the lognormal distribution, it is often better not to assume any distribution function and to use percentile diameters from the cumulative plot to describe the distribution. Statistical tests such as the Kolmogorov-Smirnov (number of intervals > 10) or the chi-square goodness of fit (for count data) can be used to determine whether the distribution departs significantly from a lognormal distribution. [See Cooper (1993) and Waters et al. (1991).]

Calculation of a particle size distribution from measured data, especially when the measurement method does not sharply delineate size ranges, is called *data inversion* or *deconvolution*. The subject is complicated and beyond the scope of this text. [See Knutson and Lioy (1995) and Cooper (1993) for a summary of different approaches.]

4.8 APPENDIX 1: DISTRIBUTIONS APPLIED TO PARTICLE SIZE

The most commonly used distribution for characterizing aerosol particle size is the lognormal distribution described in Sections 4.4–4.7. Several less common distributions, however, have been found useful for specific types of aerosols. In the equations that follow, a and b are empirical constants having different values for each distribution.

1 The Rosin–Rammler distribution [Rosin and Rammler (1933)], originally devised for sizing crushed coal, is applicable to coarsely dispersed dusts and sprays. It is particularly useful for size distributions that are more skewed than the lognormal distribution and for sieve analysis. The weight fraction between d_p and $d_p + dd_p$ is given by

$$df_m = abd_p^{b-1} \exp(-ad_p^b)dd_p \tag{4.60}$$

where a depends on the fineness of the particles and b depends only on the their material. This distribution can be used when cutoff points for the smallest and largest diameters are well defined. Expressed in differential form, as in Eq. 4.60, the Rosin–Rammler distribution has the same form as the Weibull distribution.

2 The Nukiyama–Tanasawa distribution [Nukiyama and Tanasawa (1939)] is used for sprays having extremely broad size ranges. The number of particles with diameters between d_p and $d_p + dd_p$ is given by

$$df = ad_p^2 \exp(-b/d_p^3)dd_p \tag{4.61}$$

where the empirical constants a and b are functions of a nozzle constant c.

3 The power function distribution has been applied to the size distribution of atmospheric aerosols. (See Chapter 14.) The number of particles with sizes between d_p and $d_p + dd_p$ is given by

$$df = ad_p^{-b}dd_p \tag{4.62}$$

4 The exponential distribution has been applied to powdered materials. The number of particles with sizes between d_p and $d_p + dd_p$ is given by

$$df = a \exp(-bd_p)dd_p \tag{4.63}$$

The total number of particles is a/b.

5 The size distributions of cloud droplets are described by the Khrgian–Mazin distribution (Pruppacher and Klett, 1997). The number of droplets per unit volume with sizes between d_p and $d_p + dd_p$ is given by

$$df = ad_p^2 \exp(-bd_p)dd_p \tag{4.64}$$

Total number concentration is $2a/b^3$, and the mean diameter is $3/b$.

4.9 APPENDIX 2: THEORETICAL BASIS FOR AEROSOL PARTICLE SIZE DISTRIBUTIONS

Section 4.4 states that there is no fundamental theoretical basis for aerosols having a lognormal size distribution. While this is true in general, several authors have noted that when the particle formation process follows the "law of proportionate effect," a lognormal size distribution results. This law requires changes in size to occur in steps, with the particle size for each step being a random multiplicative factor of the size of the previous step. Breakage (crushing) of solid particles is an example of such a process. Brown and Wohletz (1995) give physically based derivations of the Weibull and Rosin–Rammler distributions based on fractal cracking, a breakage process that produces a branched tree of cracks that looks the same on any scale. (See Chapter 20.) Brown and Wohletz note that certain of the resulting distributions are very similar to the lognormal distribution.

4.10 APPENDIX 3: DERIVATION OF THE HATCH–CHOATE EQUATIONS

The diameter of average d^p, where $p = 2$ for surface area and 3 for mass, is given by Eq. 4.22 as

$$d_{\bar{p}} = (\overline{d^p})^{1/p} = \left(\int_0^\infty d^p f(d)dd \right)^{1/p} \tag{4.65}$$

To evaluate $d_{\bar{p}}$ for a lognormal distribution, we must first express the quantity $d^p f(d)$ in terms of the CMD, σ_g, and p. Expressing $f(d)$ in terms of Eq. 4.42 and making the substitution $d^p = \exp(p \ln d)$ gives

$$d^p f(d) = \frac{e^{p \ln d}}{\sqrt{2\pi} \, d \ln \sigma_g} \exp\left(\frac{-(\ln d - \ln \text{CMD})^2}{2 \ln^2 \sigma_g} \right) \tag{4.66}$$

Combining and expanding the exponent yields

$$d^p f(d) = \frac{1}{\sqrt{2\pi}\, d \ln \sigma_g}$$

$$\times \exp\left(\frac{+2(\ln^2 \sigma_g)p \ln d - \ln^2 d + 2(\ln d)\ln \text{CMD} - \ln^2 \text{CMD}}{2 \ln^2 \sigma_g} \right) \quad (4.67)$$

We can complete the square in the exponential term of Eq. 4.67 by multiplying the entire equation by

$$1 = \left[\exp\left(p \ln \text{CMD} + \frac{p^2}{2} \ln^2 \sigma_g \right) \right]\left[\exp\left(-p \ln \text{CMD} - \frac{p^2}{2} \ln^2 \sigma_g \right) \right] \quad (4.68)$$

The second factor of Eq. 4.68 is combined with the exponential term in Eq. 4.67 to give a new exponential term,

$$\frac{-1}{2 \ln^2 \sigma_g} [\ln^2 d - 2(\ln d)(\ln \text{CMD} + p \ln^2 \sigma_g)$$

$$+ \ln^2 \text{CMD} + 2p(\ln \text{CMD}) \ln^2 \sigma_g + (p \ln^2 \sigma_g)^2] \quad (4.69)$$

which is equal to

$$\frac{-[\ln d - (\ln \text{CMD} + p \ln^2 \sigma_g)]^2}{2 \ln^2 \sigma_g} \quad (4.70)$$

Thus,

$$d^p f(d) = \exp\left(p \ln \text{CMD} + \frac{p^2}{2} \ln^2 \sigma_g \right)$$

$$\times \left[\frac{1}{\sqrt{2\pi}\, d \ln \sigma_g} \exp\left(-\frac{[\ln x - (\ln \text{CMD} + p \ln^2 \sigma_g)]^2}{2 \ln^2 \sigma_g} \right) \right] \quad (4.71)$$

We can now substitute Eq. 4.71 into Eq. 4.65 and integrate the resulting expression. The first exponential in Eq. 4.71 is a constant term. The bracketed term is identical in form to the frequency function of a lognormal distribution. Equation 4.65 becomes

$$d_{\bar{p}} = \left(\overline{d^p} \right)^{1/p} = \exp\left[p \ln \text{CMD} + \frac{p^2}{2} \ln^2 \sigma_g \right]^{1/p} \quad (4.72)$$

$$\ln d_{\bar{p}} = \frac{1}{p} \ln \overline{d^p} = \frac{1}{p}\left[p \ln \text{CMD} + \frac{p^2}{2} \ln^2 \sigma_g\right] \tag{4.73}$$

$$\ln d_{\bar{p}} = \ln \text{CMD} + \left(\frac{p}{2}\right)\ln^2 \sigma_g \tag{4.74}$$

and $d_{\bar{p}}$, the diameter of average d^P, is

$$d_{\bar{p}} = \text{CMD} \exp\left[\frac{p}{2}\ln^2 \sigma_g\right] \tag{4.75}$$

which is identical to Eq. 4.52.

To obtain the conversion equation for the median (qMD) of the qth moment distribution, we start with the distribution function, Eq. 4.35. Substituting Eq. 4.71, with p replaced by q, for the numerator and using the procedure just outlined in Eqs. 4.66–4.72 to evaluate the integral of Eq. 4.71 for the denominator, we get

$$df_q = \frac{1}{\sqrt{2\pi}\, d \ln \sigma_g} \exp\left(-\frac{[\ln d - (\ln \text{CMD} + q \ln^2 \sigma_g)]^2}{2 \ln^2 \sigma_g}\right) \tag{4.76}$$

This equation has the form of a lognormal distribution, Eq. 4.42, with a median diameter given by

$$\ln q\text{MD} = \ln \text{CMD} + q \ln^2 \sigma_g$$
$$q\text{MD} = \text{CMD} \exp(q \ln^2 \sigma_g) \tag{4.77}$$

which is identical to Eq. 4.48.

We can use Eq. 4.71 to get d_{qm}, the mean of the qth moment distribution. This mean is the ratio of two moments of diameter and has the general form

$$d_{qm} = \frac{\int_0^\infty x^{q+1} f(x)\, dx}{\int_0^\infty x^q f(x)\, dx} \tag{4.78}$$

Both the numerator and denominator of Eq. 4.78 are equivalent to the integrals of Eq. 4.71. Evaluating Eq. 4.78 using the procedure followed in Eqs. 4.66–4.72 gives

$$d_{qm} = \frac{\exp\left[(q + 1) \ln \text{CMD} + \frac{(q+1)^2}{2} \ln^2 \sigma_g\right]}{\exp\left[q \ln \text{CMD} + \frac{q^2}{2} \ln^2 \sigma_g\right]} \tag{4.79}$$

$$\ln d_{qm} = \ln \text{CMD} + \frac{[(q+1)^2 - q^2] \ln^2 \sigma_g}{2} \tag{4.80}$$

$$d_{qm} = CMD \exp[(q + \frac{1}{2}) \ln^2 \sigma_g] \tag{4.81}$$

which is identical to Eq. 4.50.

PROBLEMS

4.1 For the following particle size distribution data, calculate the arithmetic mean, geometric mean, count median, and diameter of the particle with average surface area.

Particle size (μm)	Number
1	3
3	5
5	2
8	1

ANSWER: 3.27 μm, 2.67 μm, 3.0 μm, 3.84 μm.

4.2 A frequency function has a constant value between 1.5 and 6.5 μm and is zero elsewhere. Determine the arithmetic mean, number median, diameter of average mass, and surface mean diameter. [*Hint*: Divide the distribution into five equal intervals with midpoints at 2, 3, 4, 5, and 6 μm.]
ANSWER: 4.0 μm, 4.0 μm, 4.45 μm, 4.89 μm.

4.3 An aerosol has a lognormal particle size distribution with a mass median diameter of 10.0 μm and a geometric standard deviation of 2.5. What is the count median diameter? Assume ρ_p = 3000 kg/m³ [3.0 g/cm³].
ANSWER: 0.81 μm.

4.4 You are given the following data, obtained by sequential sieving of a sample of granite dust:

Sieve Opening (μm)	Mass Captured on Sieve (g)		
200	4.0	0.05	5
100	21.6	0.27	95
50	38.4	0.48	68
40	8.0	10%	20
Final Pan ∠40	8.0	10%.	10
Total	80.0		

Determine the mass median diameter and geometric standard deviation of this distribution using log-probability graph paper. Use the appropriate conversion equation to determine the count median diameter.
ANSWER: 80 μm, 1.75, 31 μm.

4.5 An aerosol with a lognormal size distribution has a count median diameter of 2.0 μm and a geometric standard deviation of 2.2. If the mass concentration is 1.0 mg/m³, what is the number concentration? Assume spherical particles with ρ_p = 2500 kg/m³ [2.5 g/cm³].
ANSWER: 5.8×10^6/m³ [5.8/cm³].

4.6 Consider an aerosol with a lognormal size distribution (GSD = 1.8). The size distribution is in a range where slip correction can be neglected and Stokes's law holds. If the diameter of the particle of average settling velocity is determined to be 6.0 μm, what is the count median diameter?
ANSWER: 4.2 μm.

4.7 An aerosol with a lognormal size distribution has a count median diameter of 0.3 μm and a GSD of 1.5. If the number concentration is 10^6 particles/cm³, what is the mass concentration? The particles are spheres with a density of 4500 kg/m³ [4.5 g/cm³].
ANSWER: 130 mg/m³.

4.8 An aerosol is known to have a lognormal size distribution with a geometric standard deviation σ_g. The average time required for the particles to settle a known distance h is t. Derive an expression for the diameter of average settling time. Neglect slip correction. Also, give the equation for converting this diameter to the count median diameter. Assume spherical particles.
ANSWER: $d_{\bar{t}} = (\Sigma d^{-2}/N)^{-0.5}$, $b = +1$.

4.9 An aerosol is mixed with radon gas, resulting in a surface coating of radioactive radon decay products on the particles. The aerosol is then divided into eight aerodynamic size groups, and the radioactivity of each size group is measured. How can this information be used to calculate the count median diameter if the distribution is lognormal? All particles have the same density and are geometrically similar. Assume that log-probability graph paper is available.

4.10 For particles less than 0.05 μm, light extinction efficiency is proportional to d^4. If an aerosol is lognormally distributed with a CMD of 0.01 μm and a GSD of 1.8, what is the diameter of average extinction efficiency?
ANSWER: 0.020 μm.

4.11 Measurements of specific surface indicate that 16% of the total specific surface is contributed by particles less than 0.3 μm and 84% by particles less than 1.5 μm. Assuming that the particle size distribution is lognormal, calculate (without the aid of log-probability graph paper) the specific surface median diameter, the CMD, and the GSD. The specific surface is the surface area per gram of particles.
ANSWER: 0.67 μm, 1.28 μm, 2.24.

4.12 Specific surface is defined as surface area per gram of particles. Set up the conversion equation that will give the diameter of the particle with average specific surface in terms of the CMD and σ_g.
ANSWER: $b = -\frac{1}{2}$.

REFERENCES

Aitchison, J., and Brown, J. A. C., *The Lognormal Distribution Function*, Cambridge University Press, Cambridge, U.K., 1957.

Brown, W. K., and Wohletz, K. H., "Derivation of the Weibull distributions based on physical principles and its connection to the Rosin–Rammler and lognormal distribution," *J. Appl. Phys.*, **52**, 493–502 (1995).

Cooper, D. W., "Methods of Size Distribution Data Analysis and Presentation, in Willeke, K., and Baron, P. A., (Eds), *Aerosol Measurement*, Van Nostrand, Reinhold, New York, 1993.

Crow, E. L., and Shimizu, K. (Eds.), *Lognormal Distributions: Theory and Applications*, Marcel Dekker, New York, 1988.

Hatch, T., and Choate, S. P., "Statistical Description of the Size Properties of Non-Uniform Particulate Substances," *J. Franklin Inst.*, **207**, 369 (1929).

Herdan, G., *Small Particle Statistics*, 2d ed., Academic Press, New York, 1960.

Knutson, E. O., and Lioy, P. J., Measurement and Presentation of Aerosol Size Distributions, in Cohen, B. S., and Hering, S. V., (Eds.), *Air Sampling Instruments*, 8th ed., ACGIH, Cincinnati, 1995.

Kottler, F. J., *J. Franklin Inst.*, **251**, 449, 515 (1951).

Land, C. E., "Hypothesis Tests and Interval Estimates," in Crow, E. L., and K. Shimizu, K. (Eds.), *Lognormal Distributions*, Marcel Dekker, New York, 1988.

Mercer, T. T., *Aerosol Technology in Hazard Evaluation*, Academic Press, New York, 1973.

Nukiyama, S., and Tanasawa, Y., *Trans. Soc. Mech. Eng. (Japan)*, **5**, 63 (1939).

Pruppacher, H. R., and Klett, J. D., *Microphysics of Clouds and Precipitation*, 2d ed., Kluwer, Dordrecht, the Netherlands, 1997.

Raabe, O. G., "Particle Size Analysis Utilizing Grouped Data and the Log-Normal Distribution," *J. Aerosol Sci.*, **2**, 289–303 (1971).

Raabe, O. G., *Env. Sci. Tech.*, **12**, 1162 (1978).

Rosin, P., and Rammler, E., *J. Inst. Fuel*, **7**, 29 (1933).

Silverman, L., Billings, C., and First, M., *Particle Size Analysis in Industrial Hygiene*, Academic Press, New York, 1971.

Waters, M. A., Selvin, S., and Rappaport, S. M., "A Measure of Goodness-of-Fit for the Lognormal Model Applied to Occupational Exposures," *Am. Ind. Hyg. Assoc. J.*, **52**, 493–502 (1991).

5 Straight-Line Acceleration and Curvilinear Particle Motion

In Chapter 3 we analyzed the simplest kind of particle motion: steady, straight-line motion. We consider here more complicated kinds of particle motion, namely, two types of particle acceleration: straight-line acceleration under the influence of constant and varying forces; and motion along a curved path, or curvilinear motion. These types of motion are important for describing the aerosol collection mechanisms that operate in a fibrous filter, a lung, or a cascade impactor. To analyze such motion we introduce three new tools: the relaxation time, stopping distance, and Stokes number.

5.1 RELAXATION TIME

The terminal velocity of a particle moving in the Stokes region is directly proportional to the net external force F acting on the particle. Here, "external force" means a force acting remotely on a particle, such as gravity, centrifugal or electrostatic forces. The drag force is not considered an external force. The constant of proportionality is the mechanical mobility B defined by Eq. 3.16:

$$V_{TF} = BF \qquad (5.1)$$

When the external force is the force of gravity, Eq. 5.1 becomes

$$V_{TS} = BF_G = Bmg \qquad (5.2)$$

The product of particle mass and mobility, mB, occurs frequently in aerosol mechanics and is a useful quantity for the analysis of complex particle motion. This quantity is called the *relaxation time* of the particle and is given the symbol τ. It has units of time. In terms of particle diameter, the relaxation time is

$$\tau = mB = \rho_p \frac{\pi}{6} d^3 \left(\frac{C_c}{3\pi\eta d} \right) = \frac{\rho_p d^2 C_c}{18\eta} = \frac{\rho_0 d_a^2 C_c}{18\eta} \qquad (5.3)$$

The term "relaxation time" is used because, as will be shown, it characterizes the time required for a particle to adjust or "relax" its velocity to a new condition of forces. Relaxation time is analogous to a characteristic acceleration time for a car, such as the "0 to 60" (mph) time. It depends only on the mass and mobility of the

111

TABLE 5.1   Relaxation Time for Standard Density Particles at Standard Conditions

Particle Diameter (μm)	Relaxation Time (s)
0.01	7.0×10^{-9}
0.1	9.0×10^{-8}
1.0	3.5×10^{-6}
10.0	3.1×10^{-4}
100	3.1×10^{-2}

particle and is not affected by the nature or magnitude of the external forces acting on the particle. Although it is useful to think of relaxation time as a particle property, it includes viscosity and slip correction and is thus affected by the temperature and pressure of the surrounding gas. The use of relaxation time as defined by Eq. 5.3 is restricted to particle motion in the Stokes region, where Re < 1. Table 5.1 compares relaxation times for a range of particle sizes. Relaxation time increases rapidly with particle size because it is proportional to the square of diameter, Eq. 5.3.

Relaxation time can be used to simplify the calculation of a particle's terminal settling velocity. Substituting for τ in Eq. 3.21 gives

$$V_{TS} = \tau g \qquad (5.4)$$

The terminal velocity of a particle is simply the product of τ and the acceleration caused by an external force. For any constant external force F acting on a particle of mass m, the terminal velocity is

$$V_{TF} = \tau \frac{F}{m} \qquad (5.5)$$

Another way of considering τ is that it is equal to the terminal velocity of a particle subjected to unit acceleration.

5.2   STRAIGHT-LINE PARTICLE ACCELERATION

The equations for a particle's terminal settling velocity that are derived in Chapter 3 ignore the acceleration of the particle and consider only the equilibrium condition, when the forces acting on the particle are balanced and the velocity of the particle is constant. We now consider the acceleration of a particle that is released with zero initial velocity in still air. The particle quickly reaches its terminal settling velocity, and we wish to know how long that takes and the nature of the acceleration process.

Newton's second law of motion must hold at every instant during the acceleration process.

$$\Sigma F = \frac{d[mV(t)]}{dt} \tag{5.6}$$

where $V(t)$ *is* the instantaneous particle velocity at time t. For all situations except an evaporating (or growing) droplet, the particle mass is constant, and Eq. 5.6 becomes

$$\Sigma F = m\frac{dV(t)}{dt} = ma(t), \tag{5.7}$$

where $a(t)$ is the instantaneous acceleration. The forces acting on a settling particle are the constant force of gravity and the drag force, which depends on the particle velocity at any instant. The analysis of aerosol particle acceleration is facilitated by the fact that particle acceleration is considered "noninertial"; that is, it does not involve additional acceleration of the air. Therefore, the drag force at any time is given by Stokes's law, Eq. 3.18, using the instantaneous velocity $V(t)$ of the accelerating particle. If we assume that the direction of the force of gravity is positive and neglect slip correction for the moment, Eq. 5.7 becomes

$$F_G - F_D = mg - 3\pi\eta V(t)d = m\frac{dV(t)}{dt} \tag{5.8}$$

Multiplying both sides of Eq. 5.8 by the particle mobility B gives

$$mBg - 3\pi\eta V(t)Bd = mB\frac{dV(t)}{dt} \tag{5.9}$$

Making the substitutions $\tau = mB$ and $B = (3\pi\eta d)^{-1}$ yields

$$\tau g - V(t) = \tau\frac{dV(t)}{dt} \tag{5.10}$$

Replacing τg with V_{TS} and rearranging, we obtain

$$\int_0^t \frac{dt}{\tau} = \int_0^{V(t)} \frac{dV(t)}{V_{TS} - V(t)} \tag{5.11}$$

Integrating both sides of Eq. 5.11 between equivalent limits gives

$$\frac{t}{\tau} = -\ln[V_{TS} - V(t)] + \ln V_{TS} \tag{5.12}$$

Multiplying through by −1 and exponentiating both sides, we get

$$e^{-t/\tau} = \frac{V_{TS} - V(t)}{V_{TS}} \tag{5.13}$$

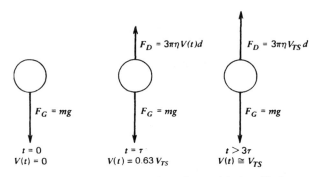

FIGURE 5.1   Acceleration of a particle in still air.

Solving for $V(t)$ yields the desired equation:

$$V(t) = V_{TS}(1 - e^{-t/\tau}) \tag{5.14}$$

Equation 5.14 gives the velocity $V(t)$ of a particle at any time t after it is released (at $t = 0$) in still air in a gravitational field. A check of this equation reveals that it satisfies the boundary conditions of the original problem: $V(t) = 0$ for $t = 0$ and $V(t) = V_{TS}$ for $t \gg \tau$. As shown in Figs. 5.1 and 5.2, the particle reaches 63% of its terminal settling velocity after an elapsed time of τ and 95% of its terminal settling velocity after $t = 3\tau$. From a mathematical point of view, $V(t)$ never reaches V_{TS}. However, from a practical point of view, ($\pm 5\%$ accuracy) the particle velocity reaches V_{TS} when $t = 3\tau$ and is constant after that. As shown in Table 5.2, the time taken to reach terminal velocity is very brief—less than 1 ms for particles with aerodynamic diameters less than 10 μm.

Equation 5.14 is a special case of the general equation of motion that results from a slightly more complicated derivation of $V(t)$ for constant external forces acting on a particle along a given axis. If V_0 is the initial velocity of the particle at time $t = 0$ and V_f is the particle's equilibrium final velocity, or terminal velocity due to

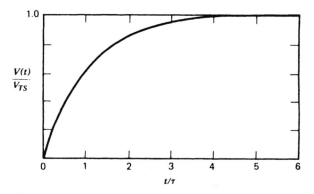

FIGURE 5.2   Velocity versus time for an accelerating particle.

TABLE 5.2   Time Required for Particles of Standard Density to Reach Their Terminal Velocity at Standard Conditions

Particle Diameter (μm)	Time to Reach Terminal Velocity[a] (ms)
0.01	0.00002
0.1	0.00027
1.0	0.011
10.0	0.94
100	92

[a]Defined as 3τ.

the balancing of the forces acting on the particle, then the general equation of motion is

$$V(t) = V_f - (V_f - V_0)e^{-t/\tau} \qquad (5.15)$$

For a particle released in still air and accelerating to its terminal settling velocity, V_0 is zero, V_f is V_{TS}, and Eq. 5.15 is identical to Eq. 5.14.

The application of Eq. 5.15 to situations in which the initial velocity and all forces act along the same axis requires only that the initial and final particle velocities be known. In most cases involving constant forces, these velocities are known or can be determined by a simple analysis of the physical situation. Equation 5.15 characterizes the adjustment of the particle's velocity as it changes from the initial to the final condition. The particle velocity at any time is equal to the final velocity minus a term that is proportional to the magnitude of the adjustment and the time-dependent term. The latter depends only on the ratio of elapsed time to particle relaxation time. Thus, the initial difference in velocity decreases by $1/e$ after an elapsed time of τ.

Several common situations can be evaluated by Eq. 5.15. For a particle released into a horizontal stream of air moving with a velocity U, the initial velocity of the particle is zero and the final velocity is U if gravity is neglected. Substituting these quantities into Eq. 5.15 results in an expression identical to Eq. 5.14, but with V_{TS} replaced by U. A similar expression is obtained if the air velocity is vertical and settling is included. In this case, the final velocity is $V_{TS} + U$ for positive velocities in the downward direction.

For the case of a particle injected with a horizontal velocity V_0 into an airstream moving along the same axis with a velocity U, the initial velocity is V_0 and the final velocity is U if gravity is neglected. The particle will either speed up or slow down to the airstream velocity if U and V_0 are in the same direction, or slow down, reverse direction, and accelerate to the airstream velocity if U and V_0 are in opposite directions.

In the cases just described, the time required to reach the terminal condition is, for practical purposes, equal to 3τ and is independent of the magnitude of the velocities or their differences. As shown in Table 5.2, even a 100-μm particle will reach its terminal velocity in less than 0.1 s, and a 10-μm particle will do so in less than 1 ms. In most situations of practical interest particle motion occurs over much longer times, and *the assumption that a particle comes instantly to its terminal velocity introduces a negligible error.* This assumption permits a relatively easy analysis of a wide variety of aerosol mechanics problems.

EXAMPLE

How long will it take a 30-μm glass sphere (ρ_p = 2500 kg/m^3) to reach a velocity equal to 50% of its terminal settling velocity if it is released from rest in still air?

$$V_0 = 0 \quad V_f = V_{TS}$$

$$\tau = \frac{\rho_p d_p^2}{18\eta} = \frac{2500 \times (30 \times 10^{-6})^2}{18 \times 1.81 \times 10^{-5}} = 0.0069 \text{ s}$$

$$= \left[\frac{2.5 \times (30 \times 10^{-4})^2}{1.8 \times 1.81 \times 10^{-4}} = 0.0069 \text{ s} \right]$$

Substituting into Eq. 5.15 yields

$$V(t) = 0.5V_{TS} = V_{TS} - (V_{TS} - 0)e^{-t/\tau}$$

$$0.5 = 1 - e^{-t/\tau}$$

$$\ln 0.5 = -\frac{t}{\tau} = \frac{t}{0.0069}$$

$$t = 0.0048 \text{ s} = 4.8 \text{ ms}$$

In the preceding analysis, we considered only constant external forces. Particle motion in response to a *changing* external force acting along an axis can be analyzed readily if the rate at which the particle adjusts to changes in the force is more rapid than the rate of change of the force. If a force changes by $1/e$ in a time t_c that is much larger than the particle relaxation time τ, then the particle can be assumed to adjust instantly, and its velocity at any time is given by

$$V(t) = \tau \frac{F(t)}{m} \quad \text{for } \tau \ll t_c \tag{5.16}$$

Examples of situations in which this concept may be applied are an aerosol flowing in a duct of changing dimensions and a particle in a centrifugal field in which the centrifugal force increases with the radial position of the particle.

5.3 STOPPING DISTANCE

The analysis of particle acceleration due to constant external forces can be carried one step further by replacing $V(t)$ in Eq. 5.15 with dx/dt and integrating to get the displacement along the x-axis as a function of time, $x(t)$:

$$\int_0^{x(t)} dx = \int_0^t V_f \, dt - \int_0^t (V_f - V_0)e^{-t/\tau} \, dt \tag{5.17}$$

$$x(t) = V_f t - (V_f - V_0)\tau(1 - e^{-t/\tau}) \tag{5.18}$$

Equation 5.18 is an expression that gives the position $x(t)$ of an accelerating particle at any time t. We can use this equation to determine the instantaneous position of a particle in any of the situations analyzed in Section 5.2. An important case not previously analyzed is the determination of the maximum distance a particle with an initial velocity V_0 will travel in still air in the absence of external forces. For such a process, $V_f = 0$ and $t \gg \tau$, so Eq. 5.18 becomes

$$S = V_0 \tau \tag{5.19}$$

where S is the total distance traveled—the *stopping distance*, or inertial range. On a displacement scale, the stopping distance represents a measure of a particle's effective initial momentum, which is diminished to zero by air friction over a distance equal to the stopping distance. In Section 5.1 τ was defined as the product mB, so the stopping distance can be defined in terms of the product of a particle's mobility and its initial momentum:

$$S = BmV_0 \tag{5.20}$$

In a mathematical sense it will take an infinite time for a particle to travel its entire stopping distance, but in a practical sense the particle travels 95% of that distance very rapidly, in a time equal to 3τ, as shown in Fig. 5.3. The stopping distance represents the ultimate distance a particle will travel in still air if an external force acting on the particle were suddenly turned off, as might be the case for an electrical force. A more important use of the stopping distance is for a particle moving with an airstream that is abruptly turned 90°. In this case, the stopping distance represents the distance the particle continues to travel in its original direction and thus can be thought of as a measure of the "persistence" of the particle. Equation 5.19 provides another meaning for the relaxation time τ; it is the stopping distance for unit initial velocity. The stopping distance finds application in characterizing curvilinear motion and determining how far particles can be thrown during various mechanical operations such as sawing or grinding.

The preceding discussion of stopping distance assumes that the entire motion of a particle takes place within the Stokes region. Stopping distance is most important for large particles with high velocities that frequently have motion, at least initially, outside the Stokes region. This situation is quite difficult to analyze be-

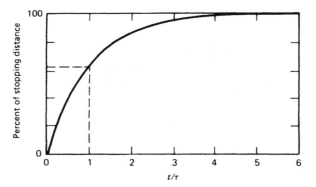

FIGURE 5.3 Particle displacement versus time for a particle with an initial velocity V_0 in still air.

cause the proportionality between the drag force and the velocity changes with velocity. An empirical expression given by Mercer (1973) is accurate within 3% for calculating stopping distances for particles having an initial Reynolds number Re_0 < 1500. For these particles,

$$S = \frac{\rho_p d}{\rho_g}\left[Re_0^{1/3} - \sqrt{6}\ \arctan\!\left(\frac{Re_0^{1/3}}{\sqrt{6}} \right) \right] \qquad (5.21)$$

where $Re_0 = \rho_g V_0 d/\eta$, and arctan is expressed in radians. Table 5.3 gives stopping distances for standard–density particles with diameters of 0.01–100 μm having an arbitrary initial velocity of 10 m/s [1000 cm/s]. Even at such a high initial velocity, stopping distances are small and particles slow rapidly. This rapidly stopping serves to underscore the idea that one cannot throw an aerosol particle—even a big one—very far. In many situations, such as with aerosol spray cans or natural con-

TABLE 5.3   Stopping Distance, Initial Reynolds Number, and Time to Travel 95 Percent of the Stopping Distance for Standard Density Spheres with an Initial Velocity of 10 meters per Second

Particle Diameter (μm)	Re_0	Stopping Distance,[a] $V_0 = 10$ m/s (mm)	Time to Travel 95% of Stopping Distance[a] (s)
0.01	0.0066	7.0×10^{-5}	2.0×10^{-8}
0.1	0.066	9.0×10^{-4}	2.7×10^{-7}
1.0	0.66	0.035	1.1×10^{-5}
10.0	6.6	2.3^{b}	$8.5 \times 10^{-4\,b}$
100	66	127^{b}	0.065^{b}

[a]Calculated using Eq. 5.19.
[b]Calculated using Eq. 5.21.

vection, a particle's motion is dominated by the motion of the gas, and neither Eq. 5.19 nor Eq. 5.21 can be used to determine its motion.

EXAMPLE

What is the stopping distance of a 1-μm aluminum-oxide sphere thrown from a grinding wheel at 9 m/s into still air. The particle density is 4000 kg/m^3 [4.0 g/cm^3]. Neglect slip correction.

$Re_0 = 66,000 \times 10^{-6} \times 9 = 0.6$, so we use Eq. 5.19:

$$S = \tau V_0 = \frac{\rho d^2}{18\eta} V_0 = \frac{4000 \times (10^{-6})^2 \times 9}{18 \times 1.81 \times 10^{-5}} = 0.00011 \text{ m} = 0.11 \text{ mm}$$

$$= \left[\frac{4.0 \times (10^{-4})^2 \times 900}{18 \times 1.81 \times 10^{-4}} = 0.011 \text{ cm} \right]$$

What is the stopping distance for a 16-μm aluminum-oxide sphere under the same conditions?

$Re_0 = 66,000 \times 16 \times 10^{-6} \times 9 = 9.6$, so we use Eq. 5.21:

$$S = \left(\frac{4000 \times 16 \times 10^{-6}}{1.20} \right) \left(9.6^{1/3} - \sqrt{6} \arctan(9.6^{1/3}/\sqrt{6}) \right)$$

$$= 0.053(2.13 - 2.45 \arctan(2.13/2.45))$$

$$= 0.053(2.13 - 1.75) = 0.020 \text{ m} = 20 \text{ mm}$$

5.4 CURVILINEAR MOTION AND STOKES NUMBER

A particle is said to have *curvilinear motion* when it follows a curved path. This type of motion can result from two quite different situations:

1 In still air or in uniformly moving air, curvilinear motion will result when a particle responds to forces along two (or more) axes if it is accelerating along one axis or if one or more of the forces vary with time or position. The motion along each axis is characterized by the equations of motion described in the previous section.

2 When flowing air converges, changes directions, or passes around an obstacle, it will have curved streamlines and cause suspended particles to have curvilinear motion.

The first situation can be analyzed by separating the forces into x and y components and analyzing each component separately, using the techniques described in Sections 5.1–5.3. One can do this because, within the Stokes region, the forces in the x and y (and z) directions are independent and separable; that is, the particle motion in the x direction does not affect the resisting force in the y direction. If the particle Reynolds number exceeds unity along any axis, the motion of the particle along that axis affects the drag force along the other axes, and a much more complicated analysis is required.

An example of the first kind of curvilinear motion is the trajectory of a particle injected horizontally with a velocity V_0 into still air. The equations of motion for the x and y directions are,

$$x(t) = V_0\tau(1 - e^{-t/\tau})$$
$$y(t) = V_{TS}t - V_{TS}\tau(1 - e^{-t/\tau})$$

$$(5.22)$$

By solving for x and y at $t = \tau$, 2τ, 3τ, and so forth, the trajectory of the particle can be determined, as shown in Fig. 5.4.

The situation is more complicated for flow around an obstacle. Very small particles with negligible inertia will follow the gas streamlines perfectly. Large and heavy particles will tend to continue in a straight line, regardless of what the gas flow does. The particles of interest are those whose motion lies between these extremes. To analyze this type of motion, one first defines the flow field—the pattern of streamlines for the gas flow around the obstacle. This can be a difficult fluid mechanics problem. Once the flow field is defined—that is, once the velocity and direction of the flow are known for every point near the obstacle—one determines the actual particle trajectory as it proceeds through the flow field. The analysis can be done analytically only for very simple geometries, such as flow around a spherical or cylindrical obstacle. With more complicated shapes, it can be done numerically by following a series of incremental steps through the flow field. The forces on the particles are assumed constant during each step and are reevaluated for each step. Usually, this analysis is done for a large number of starting positions of the particle relative to the obstacle.

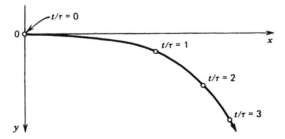

FIGURE 5.4 Example of particle trajectory for settling particle with an initial horizontal velocity.

Curvilinear motion is characterized by a dimensionless number called the *Stokes number* (Stk). It is the ratio of the stopping distance of a particle to a characteristic dimension of the obstacle. For example, for flow perpendicular to a cylinder of diameter d_c, the Stokes number is

$$\text{Stk} = \frac{S}{d_c} = \frac{\tau U_0}{d_c}, \quad \text{for Re}_0 < 1.0 \tag{5.23}$$

where U_0 is the undisturbed air velocity well away from the cylinder and $\text{Re}_0 = \rho_g d_p U_0 / \eta$. When $\text{Re}_0 > 1.0$, Eq. 5.21 can be used to calculate S for subsequent use in Eq. 5.23 (Israel and Rosner, 1983). The Stokes number is also the ratio of the particle relaxation time to the transit time past an obstacle, or the ratio of the time τ it takes a particle to adjust to the time d_c / U_0 available for adjustment. When Stk $\gg$ 1, particles continue moving in a straight line when the gas turns; when Stk $\ll$ 1, particles follow the gas streamlines perfectly. Because the characteristic dimension d_c in Eq. 5.23 can be defined somewhat differently for different applications, the definition of Stokes number may be application specific.

For geometrically similar particle motion to occur around different-sized cylinders, two conditions must be met: (1) the flow Reynolds numbers for the two situations must be equal, and (2) the Stokes numbers must be equal. Equality of the Reynolds numbers ensures that the gas flows are similar, and equality of the Stokes numbers ensures that the particle motion in the flow fields are also similar. The Stokes number is the ratio of a particle's "persistence" to the size of the obstacle. As the Stokes number approaches zero, particles track the streamlines perfectly. As the Stokes number increases, the particles resist changing their directions as the gas streamlines change directions. The Stokes number is used to characterize *inertial impaction*—the inertial transfer of particles onto surfaces, described in the next section.

5.5 INERTIAL IMPACTION

Impaction is a special case of curvilinear motion that finds extensive application in the collection and measurement of aerosol particles. Because of its importance, impaction has been analyzed, both theoretically and experimentally, more thoroughly than any other aerosol separation process. In the first half of this century, impaction was a common method for collecting dust for the evaluation of occupational environments. Since the 1960s, cascade impactors—instruments based on impaction—have been used extensively for the measurement of particle size distributions by mass.

All inertial impactors operate on the same principle. As shown in Fig. 5.5, an aerosol is passed through a nozzle and the output stream (jet) directed against a flat plate. The flat plate, called an *impaction plate*, deflects the flow to form an abrupt 90° bend in the streamlines. Particles whose inertia exceeds a certain value are unable to follow the streamlines and collide (impact) on the flat plate. For the mo-

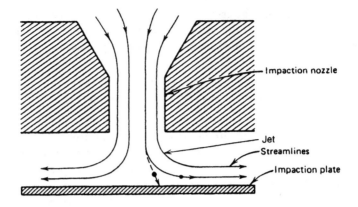

FIGURE 5.5   Cross-sectional view of an impactor.

ment, we assume that particles stick to the surface if they hit it. Smaller particles can follow the streamlines and avoid hitting the impaction plate. They remain airborne and flow out of the impactor. Thus, an impactor separates aerosol particles into two size ranges; particles larger than a certain aerodynamic size are removed from the airstream, and those smaller than that size remain airborne and pass through the impactor.

Impactor theory seeks to explain the shape of the curve of collection efficiency E_I versus particle size. (see Fig. 5.6.) The parameter that governs collection efficiency is the Stokes number, or impaction parameter, which is defined for an impactor as the ratio of the particle stopping distance at the average nozzle exit velocity U to the jet radius, $D_j/2$:

$$\text{Stk} = \frac{\tau U}{D_j/2} = \frac{\rho_p d_p^2 U C_c}{9\eta D_j} \tag{5.24}$$

This Stokes number is slightly different from Eq. 5.23, being defined in terms of the nozzle radius $D_j/2$ rather than a characteristic dimension of an obstacle. For an impactor with a rectangular nozzle, the jet half-width should be used in place of the jet radius in Eq. 5.24.

Although $E_I = f(\text{Stk})$, there is no simple relationship between impactor collection efficiency and Stokes number. Theoretical determination of the characteristic efficiency curve for an impactor requires numerical analysis using a computer. First, the pattern of streamlines in the vicinity of the jet is determined by solving the Navier–Stokes equations for a particular impactor geometry. Then, for a given particle size, particle trajectories are determined for each entering streamline. The efficiency for that particle size is determined by the fraction of the trajectories that intercept the impaction plate. This process is repeated for many particle sizes in order to generate the characteristic impactor efficiency curve such as the one shown in Fig. 5.6. Collection efficiency curves for impactors are often plotted in a gen-

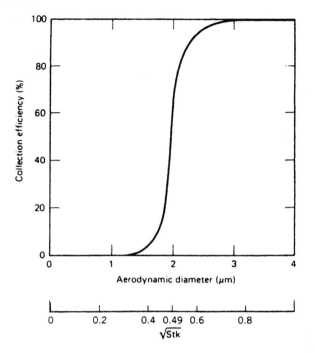

FIGURE 5.6 Typical impactor efficiency curve.

eral form as efficiency versus the square root of the Stokes number $[(Stk)^{0.5}]$, which is directly proportional to particle size. Experimental calibration requires efficiency measurements made with a series of monodisperse aerosols. However, this need only be done once for each impactor design, because all geometrically similar impactors, that meet the recommended design criteria given later in this section, will have the same collection efficiency when operated at the same Stokes number.

The *simplified and approximate* analysis that follows serves to illustrate the process of impaction and the importance of the relevant parameters. It is necessary to make the simplifying assumption that the flow velocity is uniform in the jet and that the streamlines are arcs of a circle with their centers at A, as shown in Fig. 5.7, a cross-sectional view of a rectangular jet impactor. Because of symmetry, only one-half of the impactor needs to be considered. A particle exiting the nozzle along a streamline experiences a centrifugal force causing it to move toward the impaction plate. If these departures are slight, as will be the case for the limiting condition that separates impaction from no impaction, the particle will depart from its original streamline with a constant radial velocity V_r while traversing the curved part of the streamline. This velocity is given by

$$V_r = \tau a_r = \frac{\tau U^2}{r} \tag{5.25}$$

where r is the radius of curvature of the streamline. We assume that the gas veloc-

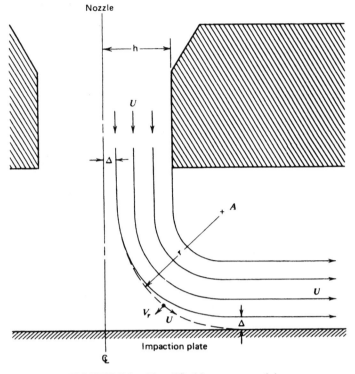

FIGURE 5.7   Simplified impactor model.

ity U remains constant in the curved-streamline region. The total radial displacement Δ of the particle from its original streamline is the product of its radial velocity and the time required to traverse the curved portion (quarter circle) of the streamline.

$$\Delta = V_r t = \frac{\tau U^2}{r}\left(\frac{2\pi r}{4U}\right) = \frac{\pi}{2}\tau U \tag{5.26}$$

At the end of the quarter circle, each particle will have moved a distance Δ from its original streamline in the direction of the impaction surface. As shown in Fig. 5.7, any particle passing through the nozzle within a distance Δ of the jet centerline will end up impacting on the plate. Particles located more than a distance Δ from the centerline will be shifted from their original streamlines, but not enough to reach the impaction surface. Because Fig. 5.7 is a transverse section of a rectangular nozzle, the impaction efficiency E_I, the fraction of entering particles collected—is equal to the ratio of the lengths Δ and h.

$$E_I = \frac{\Delta}{h} = \frac{\pi\tau U}{2h} = \frac{\pi}{2}(\text{Stk}) \tag{5.27}$$

This simplified analysis is not accurate enough to characterize impactor efficiency fully, but it does show that the Stokes number defined by Eq. 5.24 is the relevant parameter for characterizing impaction.

It might seem that the characteristic dimension for the Stokes number for an impactor should be the distance between the jet and the impaction plate, but this is not the case. The streamlines at the nozzle exit are not strongly affected by the spacing between the nozzle and the plate, because the jet of aerosol expands only slightly until it reaches within about one jet diameter of the impaction plate. Hence, the characteristic dimension for an impactor is the jet radius, or half width, rather than the spacing between the nozzle and the plate.

For most impactors, a complete curve of collection efficiency versus particle size is not necessary. Impactors that have a "sharp cutoff" curve approach the ideal (from the standpoint of particle size classification) step-function efficiency curve, in which all particles greater than a certain aerodynamic size are collected and all particles less than that size pass through. The size in question is called the *cutoff size, cutoff diameter, cut point, cutsize, cutoff,* or, d_{50}. As a practical matter, most well-designed impactors can be assumed to be ideal and their efficiency curves characterized by a single number Stk_{50}, the Stokes number that gives 50% collection efficiency. Stk_{50} is the location of the ideal cutoff curve that best fits the actual cutoff curve. As shown in Fig. 5.8, this is equivalent to assuming that the mass of particles larger than the cutoff size that get through (the upper shaded area) equals the mass of particles below the cutoff size that are collected (the lower shaded area). Table 5.4 gives Stk_{50} for two types of impactor jets meeting the criteria for Reynolds number and geometry outlined later in this section. All round jet impactors meeting these design criteria will have the same Stk_{50} value, regardless of the nozzle di-

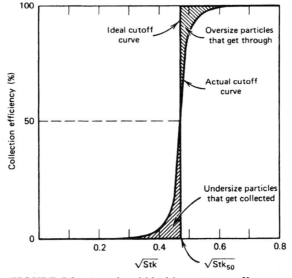

FIGURE 5.8 Actual and ideal impactor cutoff curves.

TABLE 5.4   Stokes Number for 50-Percent Collection Efficiency for Impactors[a]

Type of Impactor	Stk_{50}	$(Stk_{50})^{0.5}$
Circular jet	0.24	0.49
Rectangular jet	0.59	0.77

[a]Impactors meeting recommended design criteria.

ameter or velocity. The simplified model of the rectangular nozzle impactor just analyzed gives a Stk_{50} of 0.32, significantly smaller than the correct value of 0.59.

Equation 5.24 can be rearranged to give the particle diameter having 50% collection efficiency, d_{50}, in terms of the values of Stk_{50} given in Table 5.4.

$$d_{50}\sqrt{C_c} = \left[\frac{9\eta D_j (Stk_{50})}{\rho_p U}\right]^{1/2} \qquad (5.28)$$

Equation (5.28) can be also expressed in terms of the jet flow rate Q rather than the jet velocity. For a round jet impactor,

$$d_{50}\sqrt{C_c} = \left[\frac{9\pi\eta D_j^3 (Stk_{50})}{4\rho_p Q}\right]^{1/2} \quad \text{substituted} \atop @ Q = VA \qquad (5.29)$$

and for a rectangular jet impactor with jet width W and length L,

$$d_{50}\sqrt{C_c} = \left[\frac{9\eta W^2 L (Stk_{50})}{\rho_p Q}\right]^{1/2} \qquad (5.30)$$

Because C_c is a function of d_{50}, Eqs. 5.28–5.30 cannot be conveniently solved for particle diameter. Furthermore, the slip correction must be evaluated for the conditions downstream of the nozzle, a region where the pressure will be below atmospheric. The downstream pressure p_d is equal to the inlet pressure p_u minus the dynamic pressure:

$$p_d = p_u - \frac{\rho_g U^2}{2} \qquad (5.31)$$

For conventional impactors, d_{50} can be estimated from $d_{50}\sqrt{C_c}$ using the following empirical equations.

$$d_{50} = d_{50}\sqrt{C_c} - 0.078 \quad \text{for } d_{50} \text{ in } \mu m \qquad (5.32)$$

This equation is accurate within 2% for $d_{50} > 0.2$ μm and pressure from 91 to 101 kPa [0.9 to 1.0 atm]. For d_{50} in meters, the last term is 7.8×10^{-8} [for d_{50} in centimeters, 7.8×10^{-6}]. Impactor calibration (the cutoff diameter) is usually given in terms of an aerodynamic d_{50} (ρ_p = 1000 kg/m³ [1.0 g/cm³]), and the results of impactor measurements are expressed in terms of aerodynamic diameter. If an impactor is operated at a flow rate that differs from its calibration flow rate, its cutoff size can be adjusted using Eq. 5.29 or Eq. 5.30 together with Eq. 5.32.

The recommended design criteria that produce the desirable sharp cutoff can be stated as follows. The Reynolds number of the gas flow in the nozzle throat should be between 500 and 3000. For a given cutoff size, the Reynolds number can be controlled by using multiple nozzles in parallel. The ratio of the separation distance (the distance between the nozzle and the impaction plate) to the jet diameter or width should be 1 to 5 for circular nozzles and 1.5 to 5 for rectangular nozzles, with lower values preferred.

An examination of Eq. 5.28 reveals that a submicrometer cutoff diameter requires a small-diameter nozzle operating at a high velocity. Practical limitations on these parameters for conventional impactors limit the smallest cutoff sizes to 0.2–0.3 μm. Micro-orifice impactors extend this limit to 0.06 μm by using large numbers of chemically etched nozzles as small as 50 μm in diameter. Another approach, which can have d_{50} values as small as 0.05 μm, is the low-pressure impactor. These impactors operate at very low absolute pressures, 3 to 40 kPa [0.03 to 0.4 atm]. This has the effect of greatly increasing the slip correction factor, which reduces the cutoff size according to Eq. 5.28. Volatilization of droplets at these low pressures can change the particle size. A large vacuum pump may be required because of the low exit pressure. Both types of impactors have higher jet velocities than conventional impactors, so particle bounce can be a problem.

EXAMPLE

What is the cutoff diameter d_{50} of a round jet impactor operating at 1.0 L/min if the jet diameter is 1 mm. Assume that the impactor meets recommended design criteria.

$$Q = 1.0 \text{ L/min} = 0.001/60 = 1.67 \times 10^{-5} \text{ m}^3/\text{s} = [16.7 \text{ cm}^3/\text{s}]$$

Substituting into Eq. 5.29 yields

$$d_{50}\sqrt{C_c} = \left(\frac{9\pi 1.81 \times 10^{-5}(10^{-3})^3(0.24)}{4 \times 1000 \times 1.67 \times 10^{-5}} \right)^{1/2} = 1.36 \times 10^{-6} \text{ m} = 1.36 \text{ μm}$$

$$= \left[\left(\frac{9\pi 1.81 \times 10^{-4}(0.1)^3(0.24)}{4 \times 1 \times 16.7} \right)^{1/2} = 1.36 \times 10^{-4} \text{ cm} = 1.36 \text{ μm} \right]$$

and by Eq. 5.32,

$$d_{50} = d_{50}\sqrt{C_c} - 0.078 = 1.36 - 0.078 = 1.28 \text{ μm}$$

The operating pressure downstream of conventional impactors with submicrometer cutoff sizes may be about 0.9 atm. This pressure may cause an error if droplet aerosols of volatile materials are being sampled. Solid particles, particularly large ones, may bounce when they hit the impaction plate and be carried through with the smaller particles. This problem is discussed in the next section.

5.6 CASCADE IMPACTORS

By adding a downstream filter to the impactor shown in Fig. 5.5, one can collect all the particles that escape impaction. Sampling an aerosol with such an impactor can provide information about its particle size distribution. The mass of the particles collected on the impactor plate and of those collected on the filter are determined by weighing them before and after sampling. The impactor separates the sampled particulate mass into two particle size ranges: that contributed by particles larger than the cutoff size (collected on the impaction plate) and that contributed by particles smaller than the cutoff size (collected on the filter). For example, assume that an impactor with a 5-μm cutoff size collects 30% of the aerosol mass on the impactor plate and 70% on the filter. Then 30% of the aerosol mass is from particles greater than 5 μm in aerodynamic diameter and 70% is from particles less than 5 μm. Thus, this measurement provides one point on the cumulative distribution curve, namely, 70% of the particulate mass is associated with particles less than 5μm. By operating the impactor at several flow rates, each corresponding to a different cutoff diameter, several points on the cumulative mass distribution curve can be obtained. There are practical limitations on the range of flow rates that can be used, and the aerosol size distribution must remain constant for all the samples. The latter problem can be overcome by simultaneously operating several impactors with different cutoff sizes.

The use of several impactors in parallel is, however, not common, because of the complexity of controlling multiple flow rates. The more common approach is to operate several impactors in series, arranged in order of decreasing cutoff size with the largest cutoff size first. This configuration is called a *cascade impactor*. Each separate impactor is called an *impactor stage*, as shown in Fig. 5.9. The cutoff size is reduced at each stage by decreasing the nozzle size. Reducing D_j increases U, and both serve to reduce d_{50} according to Eq. 5.28. Since the same volume of gas flows through each stage, only one flow needs to be controlled. Each stage is fitted with a removable impaction plate for gravimetric (or chemical) determination of the collected particles. The last stage in a cascade impactor is usually followed by a filter that captures all particles less than the cutoff size of that stage.

In its operation, each stage is assumed to capture all particles reaching it that are larger than its cutoff size. Because the aerosol flows in sequence through successive stages, the particles captured on the impaction plate of a given stage represent all particles smaller than the cutoff size of the previous stage and larger than the cutoff size of the given stage. Keep in mind that an impactor stage consists of a nozzle section and the impaction plate its jets impinge on, as shown in Fig. 5.9.

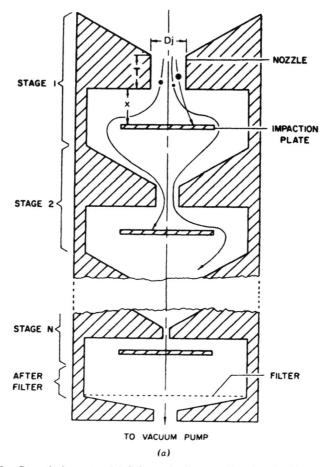

TO VACUUM PUMP

(a)

FIGURE 5.9 Cascade impactor (a) Schematic diagram. Reprinted with permission from *Aerosol Measurement,* by Dale Lundgren et al. Copyright 1979 by the Board of Regents of the State of Florida. (b) Eight stage Anderson ambient cascade impactor with nozzle plate and impaction plate shown at left.

The sequential separation divides the entire distribution of particles into a series of contiguous groups according to their aerodynamic diameters. From the gravimetric measurements of each stage, the fraction of the total mass in each aerodynamic size range can be determined. The mass distribution is determined from these data by procedures described in Chapter 4. Table 5.5 shows the data reduction required to obtain a cumulative mass distribution plot for cascade impactor data. To avoid confusion, it is best to define the *range* of particle sizes collected in each stage (column 7) and to plot *cumulative mass percent versus the upper size limit for each range.* The mass median diameter is obtained directly from the cumulative plot as described in Chapter 4.

FIGURE 5.9 Continued.

Data reduction is based on the assumption that each stage has the ideal cutoff characteristics. This assumption imposes an additional requirement that the cutoff sizes for successive stages be far enough apart that there is negligible overlap in their collection curves. If they overlap, a much more complicated analysis is required, using the exact shape of the collection efficiency curve. Also it is assumed that the nonideal cutoff effects cancel; that is, the mass of oversized particles getting through each stage equals the mass of undersized ones that are collected. This may not always be the case, and if it is not, it will cause a distortion of the size distribution.

Hering (1995) and Marple et al. (1993) have tabulated information on more than 40 commercially available cascade impactors. Table 5.6 gives similar information on the characteristics of selected cascade impactors as examples of the most common types. The cutoff size ranges given are for typical flow rates and can be changed by operating the impactor at a different flow rate according to Eq. 5.28.

The preceding discussion has assumed that the particles stick if they strike the surface of the impaction plate. For liquid particles, this is nearly always correct. Hard, solid particles, however, may bounce when they strike the impaction plate, or they may adhere and subsequently blow off. The effect is the same: Some of the mass of a certain particle size ends up on a lower stage and is thus attributed to a smaller particle size range. Furthermore, once a particle bounces, it is likely to continue to bounce in subsequent stages, because the impaction velocity is greater in

TABLE 5.5 Example of Cascade Impactor Data Reduction

Stage Number	Initial Mass (mg)	Final Mass (mg)	Net Mass (mg)	Mass Fraction (%)	d_{50} (μm)	Size Range of Collected Particles (μm)	Cumulative Mass Fraction[a] (%)
1	850.5	850.6	0.1	0.6	9.0	>9.0	100.0
2	842.3	844.1	1.8	11.0	4.0	4.0–9.0	99.4
3	855.8	861.0	5.2	31.7	2.2	2.2–4.0	88.4
4	847.4	853.6	6.2	37.8	1.2	1.2–2.2	56.7
5	852.6	855.1	2.5	15.2	0.70	0.70–1.2	18.9
Downstream filter	78.7	79.3	0.6	3.7	0	0–0.70	3.7
			16.4	100.0			

[a]Cumulative mass fraction is plotted against the upper limit of each size range, to construct a cumulative mass distribution curve.

TABLE 5.6 Characteristics of Selected Commercial Cascade Impactors

Type	Make and Model	Flow Rate (L/min)	Number of Stages	Jets/ Stage	Range[a] of d_{50} (μm)
Ambient	Mercer (02–130)[b]	1	7	1	0.3–4.5
Ambient	Multijet CI (02–200)[b]	10	7	1–12	0.5–8
Ambient	One ACFM Ambient[c]	28	8	400	0.4–10
Ambient (HiVol)	Series 230[c]	1420	5	9 slots	0.5–7.2
Personal	Marple Model 298[c]	2	8	4 slots	0.5–20
Source Test	In-Stack Mark 3[c]	21	8	20 slots	0.3–12
Low Pressure	Low Pressure[c]	3	12	1	0.08–35
Micro-Orifice	MOUDI[d]	30	9	≤2000	0.06–16.7
Viable	Viable[c]	28	6	400	0.65–7

[a]Range of d_{50} is for the flow rate given in column 3.
[b]In-Tox Products, Albuquerque, NM.
[c]Graseby Andersen, Inc., Smyrna, GA.
[d]MSP Corporation, Minneapolis, MN.

these stages. Whether or not a particle bounces depends on its material, its velocity, and the type of impaction surface. This subject is covered in Section 6.4. Coating the impaction plate with a thin film of oil or grease reduces bounce, as shown in Fig. 5.10. Antibounce coatings that have been used successfully include silicone stopcock or high-vacuum grease, Apiezon L grease, petroleum jelly, and silicone oil. To obtain a thin film, these materials are usually dissolved in a solvent such as toluene or cyclohexane, painted, spread, or sprayed onto the impaction surface, and allowed to dry. Baking may be required to achieve weight stability. Even with an antibounce coating, bounce can occur when the impaction surface becomes sufficiently loaded with particles. This is because incoming particles have an increased probability of impacting on a deposited particle rather than the coating. Impaction surfaces made up of a porous metal or a coarse membrane filter saturated with oil serve as a reservoir and allow the oil to "wick" through the deposited particles to maintain an effective antibounce coating.

The use of fibrous filters as a collection surface is unsatisfactory, because some of the flow goes through the filter and particles may still bounce off the fibers. Plastic film or small-pore membrane filters do not reduce particle rebound, although these surfaces may be convenient for analytical purposes.

Particles can be deposited in the passageways between stages of a cascade impactor. Such deposition is called *interstage loss* and represents another operational problem with cascade impactors. For conventional cascade impactors, interstage

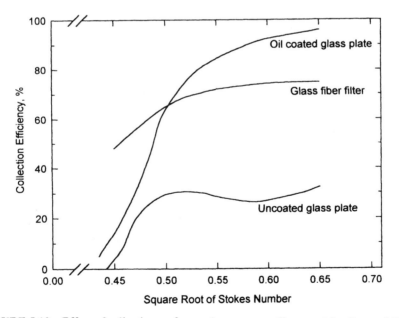

FIGURE 5.10   Effect of collection surface on impactor cutoff curve. After Rao and Whitby (1978).

losses are primarily a problem of large particles being lost in the first two stages. Particles are lost primarily by inertial removal at bends in the flow path. Since these interstage losses depend on particle size and are not included in the collected mass, they distort the size distribution toward smaller sizes. Interstage losses can be reduced by designing the impactor to minimize sharp bends in the interstage flow path of the first few stages or by operating the impactor at a lower flow rate.

As particles deposit on the impaction plate, they form a spot which grows into a conical mound that changes the flow geometry and the cutoff size. This also reduce the effectiveness of grease coatings, and sets a practical upper limit on the mass that can be collected on any stage. The lower limit is usually set by the analytical technique used, such as the sensitivity of the balance used. Some impactors are equipped with a precollector to collect large amounts of the large particles, so that accurate measurements can be made of the size distribution in the smaller particle size range. Precollectors do not need to have a sharp cutoff, because as long as the precollector collects particles larger than the cutoff size of the first stage, the mass collected by the precollector can be combined with that collected on the first stage to give the total mass larger than the first-stage cutoff size. Some impactors have a collection surface that slowly rotates during operation, to prevent excessive buildup of deposited particles.

5.7   VIRTUAL IMPACTORS

Another approach to overcoming the problems of bounce and overloading is the virtual impactor. In this type of impactor, the impaction plate is replaced by a collection probe, as shown in Fig. 5.11. Particles with inertia sufficient to cause them to impact on an impaction plate of a conventional impactor are thrown into the collection probe. These particles remain airborne and are carried by the minor air flow onto a filter. Smaller particles are carried radially away from the jet axis by the major air flow and avoid the collection probe. These particles are collected on a downstream filter, as occurs in a conventional impactor. Thus, in a virtual impactor, particles larger than the cutoff size and those smaller than the cutoff size are collected on separate filters. This technique simplifies various types of chemical analysis and reduces the possibility of interference from substrate coatings. The minor flow is usually 5–10% of the total flow and consequently includes 5–10% of the particles smaller than the stage cutoff size in addition to those larger than the cutoff size.

The calculation of Stokes number for a virtual impactor is the same as for a conventional impactor, using Eq. 5.24. The cutoff diameter can be calculated by Eqs. 5.28–5.32, but a different value for Stk_{50} is used. This value depends on the ratio of the minor to the total flow and is 0.48 for a 10% minor flow. The collection efficiency curve for a virtual impactor has nearly as sharp a cutoff as that for a conventional impactor, but levels off for small particles at the minor flow ratio, as shown in Fig. 5.12.

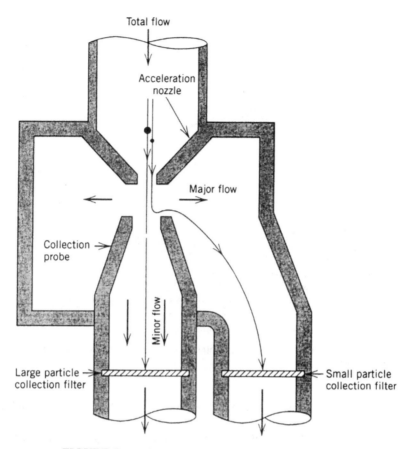

FIGURE 5.11   Schematic diagram of a virtual impactor.

Unlike conventional impactors, losses in virtual impactors are greatest for particle sizes near the cutoff size. (See Fig. 5.12.) Losses occur on the lip of the collection probe and are greater for liquid particles than for solid particles because of bounce. Losses increase as the minor flow ratio decreases. They can be reduced to a few percent. [See Loo and Cork (1988) and Chen and Yeh (1987).] The most important design requirements, expressed in terms of the nozzle diameter D_N, are (1) a collection probe diameter 30–50% greater than D_N, (2) a radius of $0.3 \times D_N$ on the inside of the collection probe inlet, (3) a nozzle protrusion of 2–$3 \times D_N$ from the base, and (4) a separation distance between the nozzle and collection probe of 1.0–$1.8 \times D_N$. Additional requirements are a minor flow rate of 5–15% of the total flow and a very careful alignment of the axis of the nozzle with that of the collection probe.

Because of the complexity of controlling two flows for each stage, virtual impactors usually have only one or two stages. The dichotomous sampler used for air pollution studies utilizes a virtual impactor to separate the "coarse" and "fine" frac-

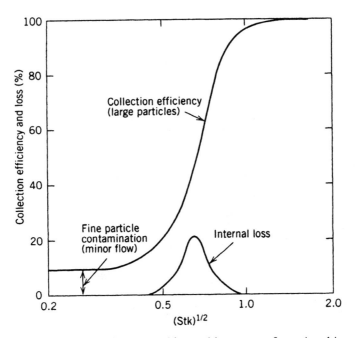

FIGURE 5.12   Collection efficiency and internal loss curves for a virtual impactor. After Chen and Yeh (1987).

tions of atmospheric aerosols. These units operate at 1 m³/h [16.7 L/min] and have a d_{50} of 2.5 μm.

The impactor part of a virtual impactor does not collect particles, but separates them into two streams according to their aerodynamic diameters. Thus, the virtual impactor concept can be used as a particle concentrator to increase the concentration of particles larger than the cutoff size. By setting d_{50} below the size of the particles of interest, all of these particles will end up in the minor flow, a much smaller volume than the original volume sampled. Thus, a concentrator with a minor-to-total-flow ratio of 10% will achieve an increase in concentration of the particles of interest in the minor flow by a factor of 10. If the concentrated stream is diluted with clean, dry air back to the original concentration, then a tenfold reduction of water vapor or other gases will be achieved.

5.8   TIME-OF-FLIGHT INSTRUMENTS

Time-of-flight instruments can provide real-time, high-resolution measurement of aerodynamic particle size and size distribution over a wide size range. As shown in Fig. 5.13, air is accelerated (> 10⁶ m/s² [10⁸ cm/s²]) in a converging nozzle to a high velocity (> 100 m/s [10,000 cm/s]) at the nozzle exit. Two narrowly focused laser beams are positioned in the jet about 100 μm apart. Particles, focused into the

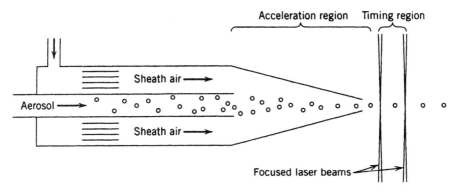

FIGURE 5.13   Schematic diagram of time-of-flight instrument.

center of the jet by the clean sheath air, are accelerated by the airflow in the nozzle. Small particles less than 0.3 μm can keep up with the accelerating air in the nozzle and exit with approximately the same velocity as the air. As a particle passes through the first laser beam, it creates a very brief (< 1 μs) pulse of scattered light that is detected by a photomultiplier tube. A similar pulse is generated when the particle passes through the second beam. The time interval between the two pulses is sensed electronically and is used to determine the average velocity of the particle as it passes through the timing zone—the space between the two laser beams. Large particles accelerate more slowly than the air and have not yet reached their final velocity (the air velocity) when they pass through the timing zone. The larger or heavier the particle, the more it lags behind the air and the lower its velocity in the timing zone. With suitable calibration, the particle's aerodynamic diameter can be determined from the magnitude of the lag. This is done electronically, and the particle size and size distribution are determined nearly in real time. Specifications for two commercially available time-of-flight instruments are given in Table 5.7.

There are important limitations to these instruments that arise primarily because particle motion is outside the Stokes region due to the high velocities. A particle

TABLE 5.7   Characteristics of Two Commercial Time-of-Flight Instruments[a]

Parameter	APS 33B[b]	Aerosizer[c]
Size range (μm)	0.5–30	0.5–200
Maximum number concentration (cm^{-3})	200	1100
Flow rate (L/min)	5.0	5.3
Spacing between laser beams (mm)	0.12	1.0
Nozzle exit velocity (m/s)	150	310

[a]Adapted from Cheng et al. (1993).
[b]®TSI, Inc., St. Paul, MN.
[c]®Amherst Process Instruments, Inc., Hadley, MA.

density effect occurs for particles whose density differs from that for which the instrument was calibrated, usually near 1000 kg/m³ [1.0 g/cm³]. Two particles with the same aerodynamic diameter and settling velocity in the Stokes region, as defined by Eq. 3.26, can have different lag times. The particle with the greater density will have less drag in the nozzle and consequently will lag the airflow by a greater degree than the one with the smaller density. Thus, the particle with greater density has a lower velocity in the timing region and is interpreted by the instrument as a larger particle. This effect occurs in a predictable way, so calibration curves can be generated for any particle density.

Nonspherical particles have the opposite effect. They have a greater drag in the nozzle than would be predicted based on their aerodynamic diameter, as defined for the Stokes region. Consequently, they go through the timing region faster and are interpreted as smaller particles. Liquid particles are flattened into an oblate spheroid by the nozzle flow. This increases their drag and velocity though the timing region, so that they are interpreted as smaller particles. There is no simple way to correct for the effect of nonspherical shape. Moderately nonspherical particles, with an axial ratio about 2:1, can have errors as large as a factor of two.

Another type of error occurs when a second particle arrives before the first has completed its time of flight. This situation is called *coincidence*, and it tends to confuse the electronics and may be interpreted as a signal from two smaller particles. Modern instruments have coincidence rejection circuitry to sense and reject such events. Coincidence sets a limit on the maximum number concentration that can be accurately measured. (See Table 5.7.)

Finally, there is an error if the instrument is operated at an ambient pressure different from its calibration pressure, usually near sea level. The lower ambient pressure at high altitude reduces the drag force in the nozzle, thereby reducing a particle's velocity through the timing region, and causes an overestimation of the particle's size. For the APS instrument characterized in Table 5.7, a 10% decrease in ambient pressure causes a 14% overestimate of particle size for particles larger than 1 μm. All of these errors are somewhat greater for the Aerosizer than the APS, because of the Aerosizer's higher jet velocity and acceleration.

PROBLEMS

5.1 What are the velocity, acceleration, and distance traveled of a 6-μm water droplet 100 μs after release in stationary air?

ANSWER: 6.5×10^{-4} m/s [0.065 cm/s], 4.1 m/s² [410 cm/s²], 3.7×10^{-8} m [3.7×10^{-6} cm].

5.2 A bursting bubble at the ocean surface ejects a 30-μm droplet vertically upwards at a velocity of 0.1 m/s [10 cm/s]. What is the maximum height this particle will reach in still air? [*Hint:* $V_f = -V_{TS}$; find t for $V(t) = 0$, find the displacement at t.]

ANSWER: 1.6×10^{-4} m [0.016 cm].

5.3 A vibrating orifice aerosol generator breaks up a fine stream of liquid flowing at 190 mm^3/min through a 23-μm-diameter orifice to form monodisperse aerosol droplets 21 μm in diameter. How far will one of these particles travel in still air?

ANSWER: 7.1 mm.

5.4 A spherical steel particle 20 μm in diameter is thrown from the rim of a 7-in.-diameter grinding wheel rotating at 3450 rpm. The density of steel is 7800 kg/m^3 [7.8 g/cm^3].
(a) What is the stopping distance of this particle if Stokes's law is assumed to hold?
(b) What is the actual stopping distance?

ANSWER: 0.31 m [31 cm], 0.15 m [15 cm].

5.5 A high-speed dentist's drill rotating at 50,000 rpm is 1 mm in diameter. How far can the drill throw a 40-μm piece of your tooth? Assume that the tooth particle is spherical and has a density of 2 g/cm^3.

ANSWER: 0.019 m [1.9 cm].

5.6 You wish to use a single-stage impactor to separate diesel smoke from coal dust in a sample of air from a coal mine so that the mass concentration of each can be determined. A typical mass size distribution from a diesel-equipped coal mine is bimodal, with one mode for the diesel smoke particles at 0.16 μm and another for coal dust at 7 μm. The saddle point between the modes is at an aerodynamic diameter of 0.8 μm. What nozzle diameter is required if four nozzles are used and the total sample flow rate is 2.5 L/min?

ANSWER: 0.64 mm.

5.7 You wish to use a cascade impactor to measure the particle size distribution in a stack at 500°F. How do you adjust the calibration cutoff sizes (d_{50}) for this operating condition? Assume that the original calibration was for 70°F and that you are using the correct (calibration) flow rate of 500°F gas. Neglect slip correction.

ANSWER: Increase d_{50} by 25%.

5.8 The fifth stage of the Andersen impactor has 400 holes, each 0.0135 inches in diameter. What is the theoretical d_{50} of this stage for standard-density spheres at an impactor flow rate of 2 ft^3/min?

ANSWER: 0.65 μm.

5.9 One stage of a cascade impactor designed to operate at 1.0 ft^3/min at standard conditions has a cutoff diameter of 5 μm. What will be the cutoff diameter for this stage if it is operated at 300°F, but flow is measured at 1.0 ft^3/min after the gas has cooled to standard conditions?

ANSWER: 4.8 μm.

5.10 What flow rate is required for a single-jet round-hole impactor to have an aerodynamic cutoff diameter of 0.7 μm? The jet diameter is 0.07 cm.

ANSWER: 0.063 m³/hr [1.04 L/min].

REFERENCES

Chen, B. T., and Yeh, H. C., "An Improved Virtual Impactor: Design and Performance," *J. Aerosol Sci.,*18, 203–214 (1987).

Cheng, V. S., Barr, E. B., Marshall, I. A., and Mitchell, J. P., "Calibration and Performance of an API Aerosizer," *J. Aerosol Sci.,* 24, 501–514 (1993).

Fuchs, N. A., *The Mechanics of Aerosols,* Pergamon, Oxford, U.K., 1964.

Hering, S.V., "Impactors, Cyclones, and Other Inertial and Gravitational Collectors," in Cohen B. S., and Hering S. V., (Eds.), *Air Sampling Instruments,* 8th ed., American Conference of Government Industrial Hygienists, Cincinnati, 1995.

Israel, R., and Rosner, D. E., "Use of a Generalized Stokes Number to Determine the Aerodynamic Capture Efficiency of Non-Stokesian Particles from a Compressible Gas Flow," *Aerosol Sci. Tech.,* 2, 45–51 (1983).

Lodge, J. P., and Chen, T. L., *Cascade Impactors: Sampling and Data Analysis,* American Industrial Hygiene Association, Akron, OH (1988).

Loo, B. W., and Cork, C. P., "Development of High Efficiency Virtual Impactors," *Aerosols Sci. Tech.,* 9, 167–176 (1988).

Marple, V.A., and Chien, C. M., "Virtual Impactor: A Theoretical Study," *Environ. Sci. Technol.,* 14, 976–984 (1980).

Marple, V. A., and Willeke, K., in Lundgren, D. A., et al. (Eds.), *Aerosol Measurement,* University Presses of Florida, Gainesville, FL, 1979.

Marple, V. A., Rubow, K. L., and Olsen, B. A., "Inertial, Gravitational, Centrifugal, and Thermal Collection Techniques," in Willeke, K., and Baron, B. A., (Eds.), *Aerosol Measurement,* Van Nostrand Reinhold, New York 1993.

Mercer, T. T., *Aerosol Technology in Hazard Evaluation,* Academic Press, New York, 1973.

Rader, D. J., and Marple, V. A., "Effect of Ultra-Stokesian Drag and Particle Interception on Impaction Characteristics," *Aerosol Sci. Tech.,* 4, 141–156 (1985).

Rao, A. K., and Whitby, K. T., "Non-ideal Collection Characteristics of Inertial Impactors— I. Single-Stage Impactors and Solid Particles," *J. Aerosol Sci.,* 9, 77–88 (1978).

6 Adhesion of Particles

Aerosol particles attach firmly to any surface they contact. This is one of the characteristics that distinguishes them from gas molecules and from millimeter-sized particles. Whenever aerosol particles contact one another, they adhere and form agglomerates. Filtration and other particle collection methods rely on the adhesion of particles to surfaces. The adhesive forces on micrometer-sized particles exceed other common forces by orders of magnitude.

Despite its importance, particle adhesion is poorly understood and its description is partly qualitative. Because it is such a complicated phenomenon, no complete theory accounts for all the factors that influence adhesion. Much of the experimental work on particle adhesion has been conducted with ideal surfaces under special conditions, such as a high vacuum, that have little relevance to real surfaces of practical interest. Section 6.1 describes the nature of adhesive forces and the variables affecting them.

6.1 ADHESIVE FORCES

The main adhesive forces are the van der Waals force, the electrostatic force, and the force arising from the surface tension of adsorbed liquid films. All of these forces are affected by the material, shape, surface roughness, and size of the particle; the material, roughness, and contamination of the surface; relative humidity; temperature; the duration of contact; and the initial contact velocity.

We consider first the theoretical description of adhesive forces. The most important forces are the London–van der Waals forces—the long-range attractive forces that exist between molecules. These forces are long range in comparison to chemical bonds, which are called short-range forces. Because of the shielding effects of adsorbed layers of water and organic molecules, chemical bonds are not important for the adhesion of aerosol particles under ambient conditions. The van der Waals forces arise because the random movement of electrons in any material creates momentary areas of concentrated charge called dipoles. At any instant these dipoles induce complementary dipoles in neighboring material, which in turn produce attractive forces, as shown in Fig. 6.1. Van der Waals forces decrease rapidly with separation distance between surfaces; consequently, their influence extends only a few molecular diameters away from a surface. At a submicroscopic level most surfaces are irregular, with peaks, called asperities, and valleys as shown in Fig. 6.2. At least initially, contact between a particle and a surface occurs only at a

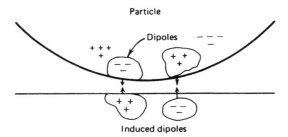

FIGURE 6.1   Van der Waals adhesive force.

few asperities. As shown in Fig. 6.2, most of the material is separated by an average distance x that depends on the scale of surface roughness. For smooth surfaces, this distance is usually assumed to be 0.0004 μm (0.4 nm), about the size of the molecules involved.

The net effect of the van der Waals forces is determined by integrating the forces between all pairs of molecules of a spherical particle near a flat surface. The resulting adhesive force between the particle and a plane surface is

$$F_{adh} = \frac{Ad}{12x^2} \qquad (6.1)$$

where A is the Hamaker constant, which depends on the materials involved and ranges from 6×10^{-20} to 150×10^{-20} J [6×10^{-13} to 150×10^{-13} ergs] for common materials. Equation 6.1 applies to ideal, hard materials with negligible flattening in the contact area. After initial particle contact, the van der Waals and electrostatic forces gradually deform the asperities to reduce the separation distance and increase the contact area until the attractive forces balance the forces resisting deformation. The deformation process may take as long as a few hours. The hardness of the materials involved controls the size of the ultimate area of contact and therefore, the strength of the adhesive force. Flattening can increase the adhesive force by up to

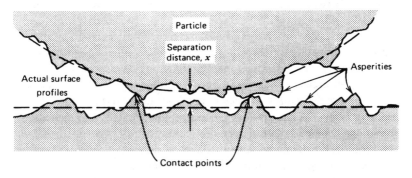

FIGURE 6.2   Submicroscopic surface contact geometry.

fifteen-fold in soft metals and more than a hundred-fold in plastics. [See Tsai et al. (1991).]

Most particles 0.1 μm or larger carry some small net charge q, which induces an equal and opposite charge in the surface. This gives an attractive electrostatic force of

$$F_E = \frac{K_E q^2}{x_q^2} \tag{6.2}$$

where K_E is a constant of proportionality (see Eq. 15.1) and x_q is the separation distance of opposite charges, which may be different from the separation distance of the surfaces.

Particles of insulating material at low humidities retain their charge and are held to surfaces by the attractive electrostatic force. The equilibrium charge (see Section 15.7) carried by particles larger than 0.1 μm is approximately proportional to $\sqrt{d}$, so the electrostatic adhesion force is also proportional to the first power of particle diameter.

Under normal conditions, most materials have adsorbed liquid molecules on their surface. An attractive force between a particle and a surface is created by the surface tension of the liquid drawn into the capillary space at the point of contact, as shown in Fig. 6.3. For relative humidities greater than 90% and ideal, smooth surfaces, this force is

$$F_s = 2\pi\gamma d \tag{6.3}$$

where γ is the surface tension of the liquid. For real surfaces at lower relative humidities, the force depends on the curvature of asperities at the points of contact, not on the particle diameter. This curvature varies greatly from particle to particle and gives rise to a distribution of adhesive forces for the same size particles.

The three adhesive forces just discussed are all proportional to the first power of particle diameter, and this is the form of most empirical expressions for adhesive force. Except for highly charged particles, van der Waals and surface tension forces are greater than electrostatic forces.

Experimental measurements of adhesive forces are made by determining the force required to separate a particle from a surface. These may be direct measurements using a fiber microbalance or centrifugal force, or they may be indirect measurements using vibration or air currents to remove the particles. For hard materi-

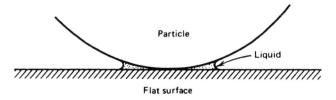

FIGURE 6.3 Adhesive force due to a liquid film.

als and clean surfaces a useful empirical expression for the adhesive force, based on direct measurement of glass and quartz particles ($> 20\mu m$) at 25°C (Corn, 1961), is

$$F_{adh} = 0.063 \ d[1 + 0.009(\%RH)] \tag{6.4}$$

where the force is in newtons, the particle diameter is in meters, and %RH is the relative humidity in percent. [For cgs units, F_{adh} is in dynes, d is in centimeters, and 63 replaces 0.063.]

6.2   DETACHMENT OF PARTICLES

The force required to detach particles from a surface can be measured by subjecting the particles to a centrifugal force normal to the surface in a centrifuge and determining the rotational speed required to detach the particles. These experiments yield a distribution of forces required to remove monodisperse particles from a surface. About 10 times as much force is required to remove 98% of the particles as that required for 50% removal. Particles can also be removed by air currents directed at the surface. (See Section 6.3.)

Adhesive forces are proportional to d, while removal forces are proportional to d^3 for gravitational, vibrational, and centrifugal forces and d^2 for air currents. These relationships suggest that as the particle size decreases, it becomes more difficult to remove particles from surfaces. This fact agrees with our intuition that large, visible particles, such as grains of sand, can be removed by shaking or air currents, but smaller ones, such as soot particles, cannot, although they may be removed by washing. The important point is that the adhesive force on a particle less than 10 μm is much greater than other forces on such a particle, as shown in Table 6.1. While individual particles less than 10 μm are not likely to be removed by common forces, a thick layer of such particles may be easily dislodged in large (0.1– 10 mm) chunks. The particles adhere tightly to each other, to form a large agglomerate that can be easily blown or shaken from the surface.

TABLE 6.1   Comparison of Adhesive, Gravitational, and Air Current Forces on Spherical Particles of Standard Density

Diameter (μm)	Force (N)		
	Adhesion[a]	Gravity	Air Current (at 10 m/s [1000 cm/s])
0.1	10^{-8}	5×10^{-18}	2×10^{-10}
1.0	10^{-7}	5×10^{-15}	2×10^{-9}
10	10^{-6}	5×10^{-12}	3×10^{-8}
100	10^{-5}	5×10^{-9}	6×10^{-7}

[a]Calculated by Eq. 6.4 for 50% RH.

6.3 RESUSPENSION

Closely related to adhesion and detachment of particles is the process of *re-suspension*, which can be defined as the detachment of a particle from a surface and its transport away from the surface. Resuspension may occur as a result of air jets, mechanical forces, the impaction of other particles, or electrostatic forces. *Reentrainment*, or *blow-off*, a more specific term, refers to resuspension by a jet of air. As with other aspects of adhesion, we understand the nature of the process, but are unable to predict reliably when resuspension will occur. The process is important for the buildup and removal of dust in ducts and on cooling coils and for fugitive dust emissions caused by vehicles on paved and unpaved roads.

Unlike static adhesion, described in the preceding sections, resuspension may involve rolling or sliding of the particle before it becomes airborne. Forces required to remove a particle under these circumstances are much less—approximately 1% of those required for static detachment in a centrifuge.

To analyze reentrainment by an air jet or by an airflow in a tube, we consider two cases: individual particles on a surface (a sparse monolayer) and an adhering dust layer wherein particles contact each other. While it is relatively easy to determine jet or duct velocity and whether a particle has been removed, it is difficult to determine the air velocity and resulting drag force on a particle adhering to a surface.

Reentrainment is a stochastic process in which a given condition of air velocity permits one to estimate only the fraction of particles of a given size that will be removed from a surface. Representative data from Corn and Stein (1965) for the bulk air velocity required to reentrain different sizes of glass beads are shown in Fig. 6.4. As expected, the larger the particle and the greater the air velocity, the greater the probability of reentrainment.

For the usual case of a turbulent airstream, there is a thin layer of laminar airflow at the surface called the *boundary layer* or *laminar sublayer*. Particles smaller than this layer are partially protected from reentrainment by being submerged in the boundary layer. Reentrainment of particles submerged in the boundary layer occurs as a result of occasional bursts of turbulent eddies penetrating through the boundary layer to detach particles. Thus, such reentrainment is time dependent. Figure 6.4 is for a one-minute exposure to the airstream; longer exposure times and pulsed exposures result in greater fractional reentrainment of particles of a given size.

The situation is different if a layer of particles is present on a surface. Two processes may occur. First, particles may be reentrained from the top surface of the layer as individual particles or small clusters. This is called *erosion*. Alternatively, a whole section of the layer can be reentrained in a process called *denudation*. Which process occurs depends on the relative strength of the adhesive forces between particles and between the particles and the surface. Erosion increases with duration of exposure to an air jet, while denudation is complete in less than a second. Studies by Zimon (1982) showed that denudation of layers of various mineral dusts ranging in particle size from 10 to 87 μm occurred at air velocities from 3 to 10 m/s [300 to 1000 cm/s].

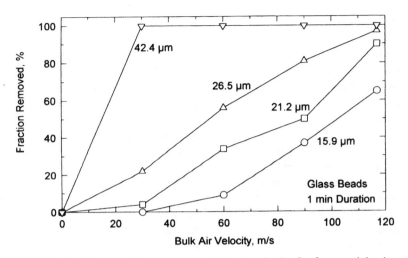

FIGURE 6.4 Particle reentrainment versus bulk air velocity for four particle sizes. Data from Corn and Stein (1965).

Monolayer particles subject to air currents can be resuspended by the impaction of other particles. Such resuspension occurs more readily than by air currents alone. The process is important for soil erosion, fugitive dust emission, and misclassification of particles in an impactor. John (1995) and coworkers measured the resuspension of 8.6-μm ammonium fluorescein particles impacted by 3.0-μm particles of ammonium fluorescein in a 40-m/s [4000-cm/s] air jet directed perpendicular to the surface. Resuspension showed wide variability, with a threshold of $\sqrt{\text{Stk}} = 0.81$ for Stk calculated for incoming particles at the jet velocity. This value corresponds to an impaction velocity of 9.3 m/s [930 cm/s]. The related process of powder dispersion is discussed in Section 21.3 as part of the process of aerosolizing powders.

6.4 PARTICLE BOUNCE

When a solid particle contacts a surface at low velocity—less than a few m/s—the particle loses its kinetic energy by deforming itself and the surface. The greater the velocity, the greater the deformation and the better the adhesion. At high velocities, part of the kinetic energy is dissipated in the deformation process (plastic deformation), and part is converted elastically to kinetic energy of rebound. If the rebound energy exceeds the adhesion energy—the energy required to overcome the adhesive forces—a particle will bounce away from the surface. Bouncing can occur for particle sizes that would adhere tightly in a static situation; it does not occur for droplets or easily deformed materials such as tar. Maximum rebound velocities occur at intermediate approach velocities.

The problem of particle bounce has been studied for solid particle collection in impactors and fibrous filters. The harder the particle material, the larger the particle, or the greater its velocity, the more likely bounce is to occur, although surface roughness plays a significant role. Coating surfaces with oil or grease increases the adhesion energy, the deformation, and the dissipative energy and greatly reduces the problem of bounce.

There are two approaches to defining the conditions at which bounce will occur. One approach defines the limiting adhesion or kinetic energy, and the other defines a critical velocity V_c for which bounce will occur if that velocity is exceeded. The latter has the form

$$V_c = \frac{\beta}{d_a} \tag{6.5}$$

where β is a constant that depends on the materials used and the geometry of the situation. For example, $\beta = 2 \times 10^{-6}$ m^2/s [0.02 cm^2/s] defines an upper limit of velocity for which bounce will *not* occur on uncoated metal impaction plates (Cheng and Yeh, 1979). Measurements of V_c by Wall et al. (1990) for ammonium fluorescein particles impacting four target materials gave β values that ranged from 7.4×10^{-6} to 2.9×10^{-5}, with a mean of 1.3×10^{-5} m^2/s [0.13 cm^2/s].

The kinetic energy required for bounce to occur when a particle collides with a surface, KE$_b$, is given by Dahneke (1971) as

$$\mathrm{KE}_b = \frac{d_p A(1 - e^2)}{2xe^2}, \tag{6.6}$$

where x is the separation distance, defined previously, A is the Hamaker constant, and e is the coefficient of restitution, which is equal to the ratio of the rebound velocity to the approach velocity. Representative values for e range from 0.73 to 0.81 (Wall et al., 1990). A and e depend only on the material of the particle and the surface. In practice, one relies on experimental determination of these constants. A regression by Ellenbecker et al. (1980) gives the probability of bounce P_b for irregular fly ash particles, with CMD = 0.14 µm, versus their approach kinetic energy in J as

$$P_b = 1 - 0.000224 \, (\mathrm{KE})^{-0.233} \tag{6.7}$$

For cgs units, KE is in ergs and 0.000224 is replaced by 0.00958. Equation (6.7) agrees reasonably well with measurements made by other investigators using sodium chloride and uranine particles up to 15 µm. The equation gives 0 and 50% probability of bounce at KEs of 2×10^{-16} J [2×10^{-9} erg] and 4×10^{-15} J [4×10^{-8} erg], respectively. These values correspond to approach velocities of 0.03 m/s [3 cm/s] and 0.12 m/s [12 cm/s] for 10-µm particles of standard density.

PROBLEMS

6.1 What air velocity is required to dislodge a 2-µm sphere from a 100-µm filter fiber at 90% relative humidity? Assume that the drag force is given by

Stokes's law and that the air velocity is unaffected by the presence of the fiber.

ANSWER: 670 m/s [6.7×10^4 cm/s].

6.2 What mainstream air velocity parallel to a surface is required to dislodge a 10-μm standard-density sphere from that surface? Assume a laminar boundary layer 0.1 mm thick having a linear velocity gradient. For simplicity, assume that the particle is dislodged when the drag force for the velocity at the centerline of the particle equals 0.01 times the adhesive force. The relative humidity is 50%. Assume that Stokes's law holds.

ANSWER: 110 m/s [11,000 cm/s].

6.3 What impactor jet velocity will permit sampling solid particles (ρ_p = 1000 kg/m^3 [1.0 g/cm^3]) up to 10 μm without bounce on an uncoated impaction plate? Use the critical velocity criterion (Cheng and Yeh, 1979) given in Section 6.4. Determine the number of jets and their diameter for an impactor with a d_{50} of 10 μm and a total flow rate of 1.0 L/min that will meet this critical velocity criterion.

ANSWER: 0.2 m/s [20 cm/s], 405, 0.51 mm.

6.4 What size particle can just be shaken from a surface with a peak acceleration of 1000 times the acceleration of gravity? Assume dry air and a particle of standard density.

ANSWER: 110 μm.

REFERENCES

Cheng, Y. S., and Yeh, H. C., "Particle Bounce in Cascade Impactors," *Env. Sci. Tech.*, **13**, 1392–1396 (1979).

Corn, M., "The Adhesion of Solid Particles to Surface, II," *J. Air Pol. Control Assoc.*, **11**, 566–584 (1961).

Corn, M., "Adhesion of Particles" in Davies, C. N. (Ed.), *Aerosol Science*, Academic Press, New York, 1966.

Corn, M., and Stein, F., "Re-entrainment of Particles from a Plane Surface," *Am. Ind. Hyg. Assoc. J.*, **26**, 325–336 (1965).

Dahneke, B., "The Capture of Aerosol Particles by Surfaces," *J. Colloid Interface Sci.*, **37**, 342–353 (1971).

Ellenbecker, M. J., Leith, D., and Price, J. M., "Impaction and Particle Bounce at High Stokes Numbers," *J. Air Pol. Control Assoc.*, **30**, 1224–1227 (1980).

John, W., "Particle–Surface Interactions: Charge Transfer, Energy Loss, Resuspension, and Deagglomeration," *Aerosol Sci. Tech.*, **23**, 2–24 (1995).

Krupp, H., "Particle Adhesion, Theory and Experiment," *Advan. Colloid Interface Sci.*, **1**, 113–239 (1967).

Tsai, C-J., Pui, D. Y. H., and Liu, B. Y. H., "Elastic Flattening and Particle Adhesion," *Aerosol Sci. Tech.*, **15**, 239–255 (1991).

Wall, S., John, W., Wang, H-C, and Goren, S., "Measurements of Kinetic Energy Loss for Particles Impacting Surfaces," *Aerosol Sci. Tech.*, **12**, 926–946 (1990).

Zimon, A. D., *Adhesion of Dust and Powder*, 2nd ed. (English translation), Consultants Bureau, New York, 1982.

7 Brownian Motion and Diffusion

In 1827, botanist Robert Brown first observed the continuous wiggling motion of pollen grains in water that we now call Brownian motion. About 50 years later, a similar motion was observed for smoke particles in air, and the connection between this motion and that predicted for gas molecules by the kinetic theory of gases was first made. In the early 1900s, Einstein derived the relationships characterizing Brownian motion, which were verified experimentally soon afterwards.

Excluding convection, thermal diffusion is the primary transport and deposition mechanism for particles less than 0.1 μm in diameter. Thermal diffusion is responsible for the collection of these particles in situations where the transport distances are small, such as in a filter or in the airways of a human lung. Where the physical scale is large, convective or eddy diffusion of parcels of aerosol greatly exceeds the transport of aerosol particles by thermal diffusion.

7.1 DIFFUSION COEFFICIENT

Brownian motion is the irregular wiggling motion of an aerosol particle in still air caused by random variations in the relentless bombardment of gas molecules against the particle. *Diffusion* of aerosol particles is the *net* transport of these particles in a concentration gradient. This transport is always from a region of higher concentration to a region of lower concentration. Both processes are characterized by the particle diffusion coefficient D. The larger the value of D, the more vigorous the Brownian motion and the more rapid the mass transfer in a concentration gradient. The diffusion coefficient is the constant of proportionality that relates the flux J of aerosol particles (the net number of particles traveling through unit cross section each second) to the concentration gradient dn/dx. This relationship is called *Fick's first law of diffusion*. In the absence of any external forces, Fick's law is

$$J = -D \frac{dn}{dx} \tag{7.1}$$

Equation 7.1 for aerosol particles is the same as Eq. 2.30 for gases, except that the concentration and flux are expressed in terms of the number of particles.

The diffusion coefficient of an aerosol particle can be expressed in terms of particle properties by the Stokes–Einstein derivation. In this derivation, the diffusion force on the particles, which causes their net motion down the concentration gradient, is equated to the force exerted by the gas resisting the particles' motion. The latter is given by Stokes drag, described in Section 3.2,

$$\text{diffusion force} = F_{\text{diff}} = \frac{3\pi\eta Vd}{C_c} \tag{7.2}$$

Einstein (1905) showed that (1) the observable Brownian motion of an aerosol particle is equivalent to that of a giant gas molecule; (2) the kinetic energy of an aerosol particle undergoing Brownian motion is the same as that of the gas molecules it is suspended in (KE = $(3/2)kT$); and (3) the diffusion force on a particle is the net osmotic pressure force on that particle.

Osmotic pressure is best understood by considering dissolved molecules in liquids. The partition shown in Fig. 7.1, a semipermeable membrane, permits liquid molecules to pass through it unimpeded, but prevents dissolved molecules from passing through. The membrane is free to slide to the left or right. If it is moved to the position shown in Fig. 7.1b, a force toward the left must be applied to hold it in place. This force equals the net osmotic pressure force acting to the right. The latter can be thought of as a pressure created by the high concentration of dissolved molecules on the left side of the membrane or a pressure caused by the liquid trying to get to the region of low liquid concentration (due to the high concentration of dissolved molecules) and equalize the concentration. The force is directly proportional to the difference in concentration on either side of the membrane. These concepts apply equally well to particles suspended in gases and to dissolved molecules in liquids. The osmotic pressure p_o is given by van't Hoff's law for n suspended particles (or molecules) per unit volume,

$$p_o = kTn \tag{7.3}$$

where k is Boltzmann's constant and T is the absolute temperature.

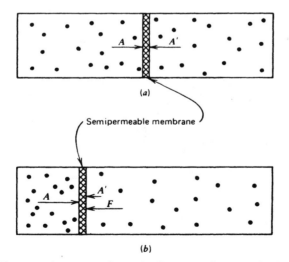

FIGURE 7.1 The osmotic pressure force. (a) Force A = force A'. (b) Force A > force A'. An additional force F must be applied to keep the membrane in position.

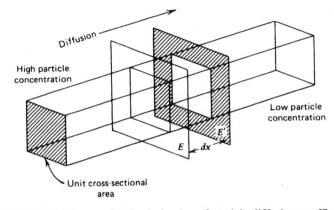

FIGURE 7.2   Diagram for the derivation of particle diffusion coefficient.

Consider the situation shown in Fig. 7.2. Diffusion of aerosol particles is taking place from left to right through surfaces E and E'. Because there is a concentration gradient from left to right, the osmotic pressures at E and E' differ slightly, as predicted by Eq. 7.3. The *net* osmotic pressure creates the diffusion force F_{diff}, which pushes the particles in the volume between E and E' to the right.

$$F_{diff} = kTn_E - kTn_{E'} = -kTdn \tag{7.4}$$

We consider here only a unit cross-sectional area of the surfaces E and E', as shown in Fig. 7.2. The volume of gas between E and E' is equal to dx, and the number of particles in that volume is $n\,dx$. Therefore, the diffusion force acting on *each* particle in the volume dx is

$$F_{diff} = -\frac{kT}{n}\frac{dn}{dx} \tag{7.5}$$

This diffusion force results in a net particle motion resisted by Stokes drag. Substituting Eq. 7.5 into Eq. 7.2 and rearranging gives

$$nV = -\frac{kTC_c}{3\pi\eta d}\frac{dn}{dx} \tag{7.6}$$

The left-hand side of Eq. 7.6, the product of n, the number of particles per unit volume, and V, the net velocity of particle motion resulting from the diffusion process, is the flux J in number of particles per unit area per second. Therefore, Eq. 7.6 has the same form as Eq. 7.1, with

$$D = \frac{kTC_c}{3\pi\eta d} \tag{7.7}$$

TABLE 7.1 Diffusion-Related Properties of Standard-Density Spheres at 293 K [20°C]

Particle Diameter (μm)	Mobility (m/N · s)	Diffusion Coefficient (m²/s)	Mean Thermal Velocity (m/s)
0.00037[a]	—	2.0×10^{-5}	460[b]
0.01	1.3×10^{13}	5.4×10^{-8}	4.4
0.1	1.7×10^{11}	6.9×10^{-10}	0.14
1.0	6.8×10^{9}	2.7×10^{-11}	0.0044
10	6.0×10^{8}	2.4×10^{-12}	0.00014

[a]Diameter of an "air molecule."
[b]Calculated by Eq. 2.22.

Equation 7.7 is called the Stokes–Einstein equation. It gives the diffusion coefficient of an aerosol particle. It is commonly written in terms of particle mobility, Eq. 3.16 multiplied by slip correction, as

$$D = kTB \qquad (7.8)$$

The diffusion coefficient has units of m²/s [cm²/s]. It increases with temperature, as expected from the similar relationship for gases. For large particles in which slip correction can be neglected, D is inversely proportional to the particle size. For small particles with large slip correction factors, D is approximately proportional to d^{-2}, as expected from Eq. 2.35 for gas molecules. The effect of particle size on the diffusion coefficient is shown in Table 7.1. Even a 0.01-μm particle has a diffusion coefficient that is nearly three orders of magnitude less than that of an "air molecule." Not only does the diffusion coefficient of a particle characterize the intensity of its Brownian motion, but it is also equal to the rate of particle transport in a unit concentration gradient. Thus, a 0.01-μm particle will be transported by diffusion 20,000 times faster than a 10-μm particle.

EXAMPLE

Calculate the diffusion coefficient for a 0.1-μm particle at standard conditions.

$$D = kTB = \frac{kTC_c}{3\pi\eta d} = \frac{1.38 \times 10^{-23}(293)2.93}{3\pi(1.81 \times 10^{-5})(0.1 \times 10^{-6})} = 6.9 \times 10^{-10} \ \text{m}^2/\text{s}$$

$$\left[D = \frac{1.38 \times 10^{-16}(293)2.93}{3\pi(1.81 \times 10^{-4})(0.1 \times 10^{-4})} = 6.9 \times 10^{-6} \ \text{cm}^2/\text{s} \right]$$

Because aerosol particles are exchanging energy with the surrounding gas molecules, both the particles and the gas molecules have the same average kinetic energy, equal to $(3/2)kT$ (Eq. 2.19). This equation can be solved for the rms forward velocity of a spherical particle of mass m and diameter d:

$$c_{rms} = \left(\frac{3kT}{m}\right)^{1/2} = \left(\frac{18kT}{\pi\rho_p d^3}\right)^{1/2} \tag{7.9}$$

Similarly, the mean thermal velocity is

$$\bar{c} = \left(\frac{8kT}{\pi m}\right)^{1/2} = \left(\frac{48kT}{\pi^2\rho_p d^3}\right)^{1/2} \tag{7.10}$$

While the diffusion coefficient does not depend on particle density, the thermal velocity does. This rather unintuitive result can be seen from the derivation of these two quantities: Neither the osmotic pressure force nor the Stokes drag depends on particle density, whereas kinetic energy, by definition, depends on mass. During diffusion the net motion of a particle is the statistical combination of many bits of wiggling motion, some in the plus direction and some in the minus direction. The mean thermal velocity, on the other hand, is the average forward velocity of a particle during each step. Note also that the aerodynamic diameter is inappropriate to characterize particle diffusion. Instead, the physical diameter should be used, because net transport is independent of ρ_p.

7.2 PARTICLE MEAN FREE PATH

In Chapter 2, the small-scale motion of a gas molecule was characterized in terms of the average velocity and mean free path of the molecule. It is useful to make a similar characterization of the Brownian motion of an aerosol particle. The average thermal velocity of a particle is given by Eq. 7.10, the same equation as that for a gas molecule. The motion of an aerosol particle is quite different from that of a gas molecule, however, as shown in Fig. 7.3, a scale drawing of the motion of an "air molecule" and the center of a 0.1-μm-diameter sphere. The trajectory of the gas molecule consists of straight-line segments between collisions. Particle motion, by contrast, is changed imperceptibly by each collision with a gas molecule. The cumulative effect of billions of collisions is to cause the particle to follow the meandering path shown. It is useful to characterize the scale of a particle's motion, as shown in Fig. 7.3, by a measure similar to the mean free path of a gas molecule. The particle mean free path λ_p is defined as the average distance from any origin that the center of a particle travels before it completely changes direction. As shown in Fig. 7.4, this distance is equal to the average straight-line distance in a given direction that the particle travels before it has lost all velocity in that direction—that is, before it has turned 90° from its original direction. On average, a particle trav-

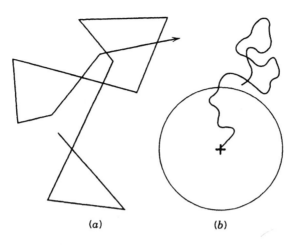

(a)   (b)

FIGURE 7.3 A two-dimensional projection (magnified 330,000 times) of the path of (a) an "air molecule" ($\lambda = 0.066$ μm and elapsed time = 1.6 ns) and (b) the center of a 0.1-μm particle ($\lambda_p = 0.012$ μm and elapsed time = 1.7 μs).

els at its mean thermal velocity (Eq. 7.10), so λ_p is simply the stopping distance or "persistence" of the particle at that velocity.

$$\lambda_p = \tau \bar{c} \qquad (7.11)$$

Table 7.2 gives τ, $\bar{c}$, and the particle mean free path for particles of various sizes. While τ and $\bar{c}$ change significantly over the range of sizes given, their product λ_p changes by less than a factor of four for a thousand-fold change in particle size. This is one of very few aerosol properties that are not strongly dependent on particle size. Larger particles have greater inertia than smaller particles, but this is offset by a lower average thermal velocity, so that the scale of Brownian motion for aerosol particles is approximately the same for all particle sizes: about one-third of the gas mean free path. The particle mean free path increases with increasing particle density.

The diffusion coefficient of a particle can be expressed in terms of its mean velocity and mean free path by an equation analogous to Eq. 2.34 for gas molecules. Combining Eqs. 5.3, 7.8, 7.10, and 7.11 gives

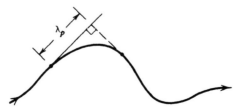

FIGURE 7.4 Trajectory of a particle, showing its apparent mean free path.

TABLE 7.2   Particle Mean Free Path for Standard-Density Spheres

d (μm)	τ (s)	$\bar{c}$ (m/s)	λ_p (μm)
0.00037[a]	—	460	0.066
0.01	6.8×10^{-9}	4.4	0.030
0.1	8.8×10^{-8}	0.14	0.012
1.0	3.6×10^{-6}	0.0044	0.016
10	3.1×10^{-4}	0.00014	0.044

[a]Diameter of an "air molecule."

$$D = \frac{\pi}{8} \lambda_p \bar{c} \qquad (7.12)$$

Since λ_p varies little with particle size, Eq. 7.12 shows the close relationship between $\bar{c}$ and D.

7.3   BROWNIAN DISPLACEMENT

The particle mean free path characterizes the small-scale motion of an aerosol particle undergoing Brownian motion. We now consider the net displacement of a particle during its wandering motion. At one moment the particle is moving in a certain direction, and a moment later it is moving in the opposite direction. The net displacement over a time that is long compared with the particle's relaxation time depends on the statistical combination of many such small-scale motions.

Consider a process of diffusion from left to right along the x-axis as shown in Fig. 7.5, a side view of a cylinder having unit cross-sectional area. We would like to determine the net flux J of particles from left to right through the imaginary surface S. During a brief time t greater than τ, each particle will have moved a net distance x_i. Consider only motion along the x-axis. Approximately half the displace-

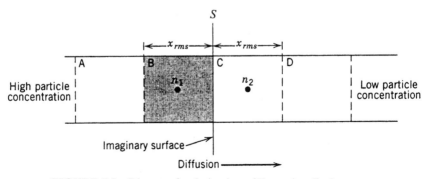

FIGURE 7.5   Diagram for derivation of Brownian displacement.

ments will be to the right and the other half to the left. For simplicity, we replace the individual displacements with their rms average value, x_{rms}. Now consider only those particles within a distance x_{rms} of surface S—that is, the particles in volume B (and C). During time t, half the particles in B move to the right a distance x_{rms} and, having crossed surface S, now occupy volume C. The other half have moved to the left and now occupy volume A. Similarly, half the particles in C have migrated to B and half to D.

Let n_1 and n_2 represent the number concentration of particles at the midpoints of volumes B and C. From the definition of a gradient,

$$n_2 = n_1 + x_{rms} \frac{dn}{dx} \tag{7.13}$$

where dn/dx is negative for concentration decreasing to the right—the direction of increasing x. The net transfer rate J through surface S from left to right equals the number of particles crossing from volume B in unit time minus the number crossing from volume C; that is,

$$J = \frac{x_{rms}}{2t} n_1 - \frac{x_{rms}}{2t} n_2 = \frac{x_{rms}}{2t} (n_1 - n_2) \tag{7.14}$$

where x_{rms} is numerically equal to volume B (and C) for a unit cross-sectional area. Substituting Eq. (7.13) for n_2 into Eq. (7.14) gives

$$J = \frac{x_{rms}}{2t} \left(n_1 - n_1 - x_{rms} \frac{dn}{dx} \right) \tag{7.15}$$

Rearranging yields

$$J = - \frac{(x_{rms})^2}{2t} \frac{dn}{dx} \tag{7.16}$$

which is another statement of Fick's first law (Eq. 7.1), with the diffusion coefficient given by

$$D = \frac{(x_{rms})^2}{2t} \tag{7.17}$$

Thus, the root-mean-square net displacement along any axis during a time t is

$$x_{rms} = \sqrt{2Dt} \tag{7.18}$$

The preceding derivation shows the link between Brownian motion and diffusion. At any point particles are undergoing Brownian motion, but because of the particle concentration gradient, there are always more particles coming past a point from the left (in Fig. 7.5), where the concentration is higher, than from the right, where the concentration is lower. That is, there is always a net mass transfer in the direction of decreasing particle concentration. If there is no gradient, transport due

TABLE 7.3   Net Displacement in 1 Second Due to Brownian Motion and Gravity for Standard-Density Spheres at Standard Conditions

Particle Diameter (μm)	rms Brownian Displacement in 1 s (x_{BM}) (m)	Settling in 1 s (x_{grav}) (m)	x_{BM}/x_{grav}
0.01	3.3×10^{-4}	6.9×10^{-8}	4800
0.1	3.7×10^{-5}	8.8×10^{-7}	42
1.0	7.4×10^{-6}	3.5×10^{-6}	0.21
10	2.2×10^{-6}	3.1×10^{-3}	7.1×10^{-4}

to Brownian motion is the same in both directions, so there is no *net* mass transfer. Values of x_{rms} in 1 s are compared with V_{TS}, the gravitational displacement in 1s, in Table 7.3. It is apparent that diffusion and Brownian motion are the most important mechanisms for transporting particles with diameters less than 0.1 μm through a gas.

Another way to consider Brownian motion is to imagine the situation shown in Fig. 7.6a. At time $t = 0$, a large number of particles are at $x = 0$ and nowhere else. If no external forces are present, Brownian motion will cause the particles to spread out with time as shown in Fig. 7.6b. Equal numbers of particles are to the left and right of the origin. The distribution of particles, shown in Fig. 7.6c, is a normal distribution with a mean of zero and a standard deviation equal to the root-mean-square displacement, x_{rms} (Eq. 7.18).

$$\sigma = x_{rms} = \sqrt{2Dt} \tag{7.19}$$

The fraction of the total number of particles originally released that lie between x and $x + dx$ at time t is given by

$$\frac{dn(x,t)}{n_0} = \frac{1}{(4\pi Dt)^{1/2}} \exp\left(\frac{-x^2}{4Dt}\right) dx \tag{7.20}$$

where n_0 is the number released at time $t = 0$. Equation 7.20 also expresses the probability that a single particle starting at the origin at time $t = 0$ will be between x and $x + dx$ at time t. Table 7.4 gives the probability that a particle released from the origin at $t = 0$ will be 1 mm or more away from the origin along a given axis at various times following release.

Thermal diffusion is most important for small particles, with $d < 0.1\mu$m, in situations where distances are short and times are relatively long. The Brownian displacements described here and given in Table 7.4 are extremely small compared to the natural convective motion of the atmosphere around us. In outdoor convective diffusion, aerosols are transported by the wind and by eddies that are superimposed on the average wind velocity. The size and velocity of the eddies depend on the wind

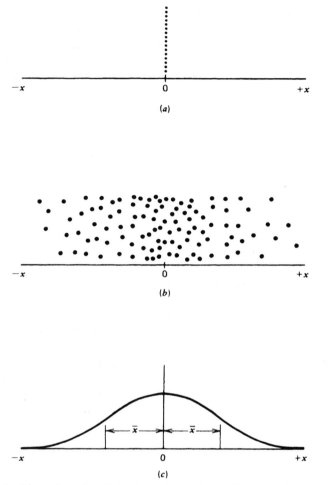

FIGURE 7.6 Dispersion of particles due to Brownian motion. (*a*) Initial condition ($t = 0$). (*b*) Distributions of particles at time $t > 0$. (*c*) Frequency distribution at time $t > 0$.

TABLE 7.4   Probability of Finding a Particle Greater than 1 mm from the Origin along a Given Axis due to Brownian Motion for Various Times Following Release

Diameter	Time (s)			
(μm)	1	10	100	1000
0.001	0.759	0.923	0.976	0.992
0.01	0.002	0.338	0.762	0.924
0.1	0	0	0.007	0.396
1.0	0	0	0	2×10^{-5}

velocity and the stability of the atmosphere. Indoors, there is usually little average wind velocity. The air motion is dominated by eddies that serve to mix and transport aerosols. The concepts of thermal diffusion can be applied to eddy diffusion by defining an eddy diffusion coefficient analogous to the thermal diffusion coefficient. Convective diffusion for aerosol flow in a tube is covered in Sections 7.4 and 7.5; the general topic of convective diffusion of aerosols is reviewed by Friedlander (1977) and Reist (1993).

Particles also have a rotational Brownian motion (Fuchs, 1964), rotating first one way and then another. This motion is of little practical importance for spheres, but it is a reason for assuming random orientation for irregular particles in the absence of an aligning force.

7.4 DEPOSITION BY DIFFUSION

Gas molecules rebound when they collide with a surface. This transfer of momentum is the mechanism by which gas pressure is transferred to the walls of a container. Unlike gas molecules, aerosol particles adhere when they collide with a surface. This means that the aerosol concentration at the surface is zero and that a gradient is established in the region near the surface. This concentration gradient causes a continuous diffusion of aerosol particles to the surface, which leads to a gradual decay in concentration.

The simplest case is a plane vertical surface in an infinitely large volume of aerosol that has a uniform initial concentration n_0. We assume initially that there is no gas velocity near the surface. The first objective is to determine the rate at which particles are removed from the aerosol by deposition onto the surface. Let x be the horizontal distance from the surface. Then the particle concentration at x at any time t, $n(x, t)$, must satisfy Fick's second law of diffusion,

$$\frac{dn}{dt} = D\frac{d^2n}{dx^2} \quad \begin{cases} n(x,0) = n_0 & \text{for } x > 0 \\ n(0,t) = 0 & \text{for } t > 0 \end{cases} \tag{7.21}$$

The solution of this equation is

$$n(x,t) = \frac{n_0}{(\pi Dt)^{1/2}} \int_0^x \exp\left(\frac{-p^2}{4Dt}\right) dp \tag{7.22}$$

where p is a dummy variable that goes from 0 to x. Figure 7.7 shows the concentration profiles for 0.05-μm aerosol particles near a surface at various times, and Fig. 7.8 gives the general form for any size of particle, distance from the wall, or time. As time passes, the concentration gradient becomes less steep, but extends farther and farther away from the surface. Even after 16 min, this process has not affected the aerosol concentration beyond 6 mm from the wall. Thus, the process will not rapidly deplete aerosol concentration, except where the dimension scale is on the order of 1 mm or smaller or the particle size is significantly less than 0.05

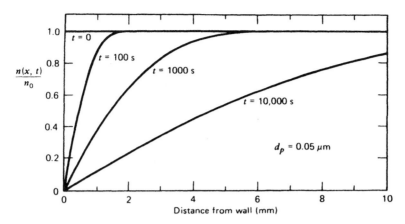

FIGURE 7.7   Concentration profiles for a stagnant aerosol of 0.05-μm particles near a wall. Parameter is the time after initial mixing.

μm. Although the derivation of Eq. 7.22 assumes an infinite aerosol volume, it can be used for finite volumes, provided that the gradients established at other walls do not interfere.

We can determine the rate of particle deposition per unit area of surface by evaluating the concentration gradient at the surface and applying Fick's first law. The concentration gradient at the surface, dn/dx at $x = 0$, is given by Eq. 7.22 as

$$\frac{dn}{dx} = \frac{n_0}{(\pi D t)^{1/2}} \quad \text{for } x = 0 \qquad (7.23)$$

Substituting Eq. 7.23 into Eq. 7.1 gives the rate of deposition of particles onto a unit area of surface at any time t.

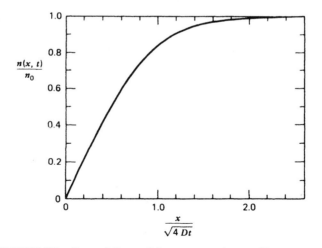

FIGURE 7.8   General form of the concentration profile near a wall.

$$J = n_0 \left(\frac{D}{\pi t} \right)^{1/2} \tag{7.24}$$

It is useful to integrate Eq. 7.24 to get the cumulative number of particles deposited per unit area of surface during a time t, $N(t)$.

$$N(t) = \int_0^t n_0 \left(\frac{D}{\pi t} \right)^{1/2} dt \tag{7.25}$$

$$N(t) = 2 n_0 \left(\frac{Dt}{\pi} \right)^{1/2} \tag{7.26}$$

Although Eq. 7.26 applies only to an infinite volume of stagnant aerosol maintained at a concentration of n_0 outside the gradient region, it is useful for predicting the upper limit of losses to the walls of a finite container. Table 7.5 compares the deposition of particles onto a horizontal surface by diffusion with that by gravitational settling. For a period of 100 s, deposition rates for the two mechanisms are equal for particles of standard density that are about 0.2 µm in diameter.

The deposition rate given by Eq. 7.24 can also be expressed in terms of a *deposition velocity* V_{dep}, defined as the deposition flux divided by the undisturbed concentration:

$$V_{dep} = \frac{J}{n_0} = \frac{\text{number deposited/m}^2 \cdot \text{s}}{\text{number/m}^3} = \text{m/s} \tag{7.27}$$

The deposition velocity is the effective velocity with which particles migrate to a surface and is analogous to the settling velocity for deposition by settling.

A problem of great practical importance is the diffusion of aerosol particles to the walls of a tube as they flow through it. Although this situation is more compli-

TABLE 7.5   Cumulative Deposition of Standard-Density Particles onto a Horizontal Surface from Unit Aerosol Concentration[a] during 100 Seconds by Diffusion and by Gravitational Settling

Diameter (µm)	Cumulative Deposition		Ratio Diffusion/Settling
	Diffusion (number/m^2)	Settling (number/m^2)	
0.001	2.6×10^4	0.68	3.8×10^4
0.01	2.6×10^3	6.9	380
0.1	3.0×10^2	88	3.4
1.0	59	3500	1.7×10^{-2}
10	17	3.1×10^5	5.5×10^{-5}
100	5.5	2.5×10^7	2.2×10^{-7}

[a]Assuming an aerosol concentration of 1 particle/cm^3 beyond the gradient region.

cated than the stagnant one just considered, mathematical solutions have been obtained for a few simple cases, such as fully developed laminar flow through a tube having a circular cross section. This solution gives the penetration (the fraction of entering particles that exit) through a tube of diameter d_t as a function of the dimensionless deposition parameter, μ,

$$\mu = \frac{4DL}{\pi d_t^2 \overline{U}} = \frac{DL}{Q} \tag{7.28}$$

where D is the diffusion coefficient of the particles, L is the length of the tube, $\overline{U} = Q/A$ is the average flow velocity through the tube, and Q is the volume flow rate through the tube. That the loss in a tube does not depend on the tube diameter for a given volumetric flow rate is a surprising result. The extra distance particles must diffuse in a wider tube is just offset by the longer time for diffusion permitted by the wider tube. Equation 7.29, a simplified version of more complicated and more accurate expressions, gives the penetration **P** as a function of μ with an accuracy of 1% for all values of μ.

$$\mathbf{P} = \frac{n_{out}}{n_{in}} = 1 - 5.50 \, \mu^{2/3} + 3.77 \, \mu \quad \text{for } \mu < 0.009$$

$$\tag{7.29}$$

$$\mathbf{P} = 0.819 \exp(-11.5 \, \mu) + 0.0975 \exp(-70.1 \, \mu) \quad \text{for } \mu \geq 0.009$$

Figure 7.9 is a graph of **P** versus μ for tubes, calculated by Eq. 7.29, and for rectangular channels, calculated by Eq. 7.34. For $\mu < 0.001$, particle losses by diffusion to the walls are small. For $\mu > 0.3$, nearly all particles are lost to the tube walls.

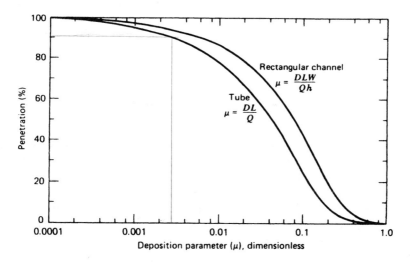

FIGURE 7.9 Penetration versus deposition parameter for circular tubes and for channels with a rectangular cross section.

TABLE 7.6   Fractional Loss to the Walls by
Diffusion for an Aerosol Flowing through a Tube 1
Meter Long at Three Laminar Flow Rates[a]

Particle Diameter (μm)	Flow Rate (L/min)		
	0.1	1.0	10
0.001	1.000	0.978	0.422
0.01	0.428	0.108	0.025
0.1	0.029	0.006	0.001
1.0	0.003	0.0008	0.0002

[a]Tube can be any diameter, provided that the flow is
laminar.

Table 7.6 compares the fractional loss in a tube 1 meter long for three aerosol flow rates.

EXAMPLE

An aerosol of 0.03-μm particles is sampled through a 5-mm-diameter tube 1.2 m long at 8×10^{-7} m³/s [0.8 cm³/s]. What fraction of the entering particles will pass through the tube?

$$\mu = \frac{DL}{Q} = \frac{6.39 \times 10^{-9} \times 1.2}{8.0 \times 10^{-7}} = 0.0096 = \left[\frac{6.39 \times 10^{-5} \times 120}{0.8} = 0.0096 \right]$$

From Fig. 7.9, **P** = 78%. Alternatively, **P** can be calculated by Eq. 7.29:

$$P = 0.819e^{-11.5(0.0096)} + 0.0975e^{-70.1(0.0096)}$$

$$= 0.733 + 0.0497 = 78\%$$

Deposition onto surfaces *by diffusion* from turbulent flow is more complicated and equations describing it cannot be solved explicitly. It is customary to assume that the turbulent flow provides a constant concentration n_0 everywhere beyond a thin boundary layer next to a surface, which may be the walls of a tube or the Atlantic Ocean. In the boundary layer, flow is laminar, and the concentration is assumed to decrease linearly from n_0 to zero at the surface. Under these conditions, the deposition velocity is given by

$$V_{dep} = \frac{D}{\delta} \tag{7.30}$$

where D is the diffusion coefficient of the particles and δ is the thickness of the boundary layer. The difficulty in applying Eq. 7.30 lies in determining the proper value of δ, which depends on flow mechanics, the nature of the velocity boundary layer, and the size of the particles. Wells and Chamberlain (1967) give the diffusive deposition velocity through the laminar boundary layer for turbulent flow in a tube as

$$V_{dep} = \frac{0.04\overline{U}}{Re^{1/4}}\left(\frac{\rho_g D}{\eta}\right)^{2/3} \tag{7.31}$$

where $\overline{U}$ is the average velocity in the tube, Re is the flow Reynolds number, and D is the particle diffusion coefficient. Equation 7.31 can be combined with Eq. 7.27 to get the deposition flux J or with Eq. 7.32 to get the penetration through a tube.

For particles greater than 1 μm, the dominant mechanism for deposition from turbulent flow is particle inertia. *Inertial deposition* occurs when a particle near the wall in the turbulent region is given sufficient lateral velocity toward the wall to be thrown through the laminar layer and deposit on the tube wall. Equations for deposition velocity due to this mechanism are given by Friedlander and Johnstone (1957) and reviewed by Davies (1966) and Lee and Gieseke (1994). Lee and Gieseke (1994) also present an empirical equation for the transition region between diffusive and inertial deposition. The overall penetration through a tube of length L subject to losses to the walls by diffusion or inertia from turbulent flow is

$$\mathbf{P} = \frac{n_{out}}{n_{in}} = \exp\left(\frac{-4V_{dep}L}{d_t\overline{U}}\right) \tag{7.32}$$

7.5 DIFFUSION BATTERIES

Diffusion batteries are devices that rely on diffusion to remove small particles from a laminar-flow aerosol stream. The extent of particle loss for a given particle size in a diffusion battery can be predicted using the relationships for diffusion losses described in this and the preceding section. In practice, one measures the concentration of particles entering and exiting a tube or some other similar conduit. With this information, the average value of μ and the diffusion coefficient of the aerosol particles flowing through a tube diffusion battery are obtained by inverse application of Eq. 7.29. Diffusion batteries are applicable to particle sizes from 0.002 to 0.2 μm. Multiple units can be operated in series or parallel to get information about the particle size distribution.

The three principal types of diffusion batteries, shown in Fig. 7.10, are the tube bundle, parallel-plate, and screen batteries. An examination of Eq. 7.28 shows that the same value of μ is obtained by using a single long tube or by cutting it into a large number of parallel short tubes, each carrying $1/n$ of the total flow. Since μ does not depend on the tube diameter, a diffusion battery equivalent to a long tube can be constructed in a compact shape using a bundle of tubes, as shown in

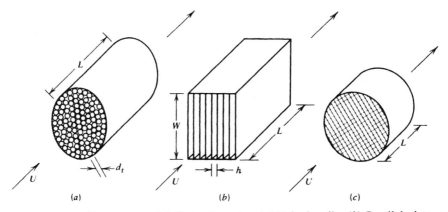

FIGURE 7.10 Three types of diffusion batteries. (*a*) Tube bundle. (*b*) Parallel plate, or rectangular channel. (*c*) Screen.

Fig. 7.10*a*. Once μ is deduced from Eq. 7.29 and the penetration measurement, Eq. 7.28 is used to determine the diffusion coefficient and the size of the flowing particles.

Mathematical solutions also exist for determining the penetration of an aerosol flowing between closely spaced parallel plates. These rectangular channels can be stacked together, as shown in Fig. 7.10*b*, provided that the plate width W is much greater than the plate spacing h. In this case,

$$\mu = \frac{DLn'W}{Qh} \tag{7.33}$$

where L is the length of the channel in the direction of flow, n' is the number of channels, W is the width of the channels, and h is the separation between the plates. The following are simplified equations for penetration, accurate to with 1% for all values of μ:

$$P = \frac{n_{\text{out}}}{n_{\text{in}}} = 1 - 2.96\,\mu^{2/3} + 0.4\,\mu \quad \text{for } \mu < 0.005$$
$$\tag{7.34}$$
$$P = 0.910\,\exp(-7.54\,\mu) + 0.0531\,\exp(-85.7\,\mu) \quad \text{for } \mu \geq 0.005$$

The parallel-plate diffusion battery is simpler to construct and usually more compact than the equivalent tube bundle type, but it must be constructed precisely to perform as predicted by Eq. 7.34.

The third type of diffusion battery uses multiple layers of very fine mesh wire cloth. The flow through the gaps in the weave is equivalent to flow through a large number of small pores, as in the tube bundles just described. For 635-mesh screen, there are 62,000 holes/cm², each approximately 20 μm square. The mechanism of collection is slightly different from that occurring in a tube. The penetration is given by (Cheng and Yeh, 1980)

$$\mathbf{P} = \frac{n_{\text{out}}}{n_{\text{in}}} = \exp\left(\frac{-10.8\alpha L D^{2/3}}{\pi(1-\alpha)d_w^{5/3}\overline{U}^{2/3}}\right) \qquad (7.35)$$

where α is 1 minus the screen porosity, d_w is the wire diameter, and L is the depth of the holes. Equation 7.35 agrees (within 2%) with more detailed equations by Cheng et al. (1985) for representative screens (145 to 635 mesh, $U > 0.02$ m/s [2 cm/s], and $0.002 < d_p < 0.5$ μm). For 635-mesh stainless-steel screen, $\alpha = 0.345$, $d_w = 20$ μm, and $L = 50$ μm per layer.

Measurement of the particle size of a monodisperse aerosol with a diffusion battery is a straightforward inversion of the appropriate penetration equation. Such measurement may be facilitated by a plot of penetration versus μ, such as Fig. 7.9. As shown in Fig. 7.11, a diffusion battery does not produce a sharp cutoff, as an impactor does, but rather, it has a gradual cutoff spanning more than an order of magnitude of particle size. A single measurement of a polydisperse aerosol with a diffusion battery yields an average diffusion coefficient that can be converted to a diffusion average diameter, also called a *thermodynamic diameter* or diffusion equivalent diameter. The diffusion average diameter cannot be converted to another type of average diameter, unless additional information about the distribution is known. For polydisperse aerosols, a set of diffusion batteries with different effective lengths, or the same battery at different flow rates, can be used to get a series of penetration measurements. Because of the gradual, and frequently overlapping cutoffs, the inversion of diffusion battery data is more complicated than that for cascade impactor data. Cheng (1993) reviews graphical and computer inversion methods for determining the size distribution of polydisperse aerosols from diffusion battery data.

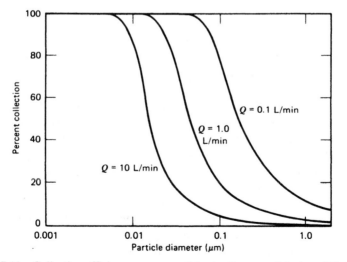

FIGURE 7.11 Collection efficiency versus particle size for a parallel-plate diffusion battery at three flow rates. Thirty channels, each $8 \times 0.1 \times 20$ cm.

A related device, the *denuder* or stripper, removes gases or vapors from an aerosol stream without affecting the particles. Two geometries are commonly used: cylindrical and annular tubes. For both types, the inner walls of the flow path are coated with a material that absorbs a particular gas or vapor. Penetration of the gas or vapor through a cylindrical tube denuder can be predicted by Eq. 7.29 using the diffusion coefficient for the gas or vapor in Eq. 7.28. By carefully selecting the flow rate and geometry, one can achieve a nearly complete capture of the gas or vapor, with a high penetration for the aerosol particles.

In an annular denuder, the aerosol flows axially through the annular space between two concentric tubes. For laminar flow with $d_2 - d_1 << d_1$, penetration can be estimated by Eq. 7.34 with

$$\mu = \frac{\pi DL(d_1 + d_2)}{Q(d_2 - d_1)} \tag{7.36}$$

where d_1 and d_2 are the inner and outer diameters of the annular flow path, respectively. Typically, an annular denuder can be operated at a much higher flow rate, for the same separation of gases and particles, than a cylindrical denuder of the same length.

PROBLEMS

7.1 A hundred particles of equal size and density are released simultaneously and allowed to settle in still air for 5 min. Because of Brownian motion along the vertical axis, some particles will progress farther than others. If the particles are 0.6-μm spheres of standard density, what are the mean and standard deviation of their displacements from origin after 5 min?

ANSWER: 4.1 mm, 0.17 mm.

7.2 Derive an expression giving the diameter of the spherical particle that has an rms displacement in 1 second due to Brownian motion equal to its terminal settling velocity. Neglect slip correction.

7.3 In 1914, Fletcher (*Phys. Rev.* **4**, 440) published a paper describing his experimental verification of diffusion coefficients. His measurements were based on the distribution of settling times for a particle settling over a fixed vertical distance. Fletcher was able to make 6000 replications on a single particle by raising the particle back up to the starting point with an electric field. If his particle was 0.36 μm in diameter and the settling distance was 0.74 mm, what were the theoretical mean and standard deviation of the settling times? Assume that there was no experimental error and that the standard deviation was much less than the mean. $\rho_p = 1000$ kg/m³ [1.0 g/cm3]. Approximately how long did it take to conduct this experiment? Assume that 25 s was required to reposition the particle after each measurement.

ANSWER: 131 s, 28 s, 10.8 days (nonstop!).

7.4 How long, *on average*, would you have to wait to have a 32% probability of observing a 0.1-μm-diameter aerosol particle at a distance greater than 1.0 m (in either direction along a given axis) from its position at time $t = 0$? Assume ambient conditions, and that the only motion is Brownian motion. The slip correction factor for a 0.1-μm particle is 2.93.
ANSWER: 23 yrs.

7.5 A tube 5 mm long and 3 mm in diameter joins two chambers. The concentration of 0.01-μm particles in one chamber is 10,000/cm^3. The other chamber has a concentration of 0/cm^3. Calculate the diffusion coefficient for the particles. At what rate will the particles diffuse from one chamber to the other? Assume that there is no airflow through the tube.
ANSWER: 5.4×10^{-8} m^2/s [5.4×10^{-4} cm^2/s], 0.76 particles/s.

7.6 An aerosol sample is taken at a rate of 0.2 L/min through a tube 1 m [100 cm] long with an inside diameter of 0.4 cm. Determine the loss by diffusion to the walls of the sampling tube for 0.01-μm particles.
ANSWER: 29%.

7.7 A sphere 32.5 mm in diameter is placed inside a chamber containing a radioactive aerosol of 0.1-μm diameter particles for 5 min. The sphere is removed, and the number of particles attached to its surface is determined by radioactive counting to be 8×10^7. What is the number concentration of particles in the chamber?
ANSWER: 4.7×10^{13}/m^3 [4.7×10^7/cm^3].

7.8 At a Reynolds number of 3000, flow in a tube could be laminar or turbulent. A tube 5 cm in diameter and 3 m long is used to sample an aerosol at a Reynolds number of 3000. What is the ratio of loss to the tube walls by diffusion for turbulent flow to that for laminar flow for 0.1-μm particles?
ANSWER: 2.7.

REFERENCES

Chang, Y. S., Yeh, H. C., and Brinsko, K. J., *Aerosol Sci.Tech.*, **4**, 165–174 (1985).

Cheng, Y. S., and Yeh, H. C., *J. Aerosol Sci.*, **11**, 313–320 (1980).

Davies, C. N., "Deposition from Moving Aerosols," in Davies, C. N. (Ed.), *Aerosol Science*, Academic Press, London, 1966.

Einstein, A., "On the Kinetic Molecular Theory of Thermal Movements of Particles Suspended in a Quiescent Fluid," *Ann. Physik*, **17**, 549–60 (1905). English translation, Einstein, A., *Investigations on the Theory of Brownian Movement*, R. Furth (Ed.), translated by A. D. Cowper, Dover, New York, 1956.

Friedlander, S. K., and Johnstone, H. F., *Ind. Engr. Chem.*, **49**, 1151 (1957).

Friedlander, S. K., *Smoke, Dust and Haze*, Wiley, New York (1977).

Fuchs, N. A., *The Mechanics of Aerosols*, Pergamon, Oxford, U.K., 1964.

Lee, K. W., and Gieseke, J. A., *J. Aerosol Sci.*, **25**, 699–704 (1994).

Reist, P. C., *Aerosol Science and Technology*, McGraw-Hill, New York, 1993.

Wells, A. C., and Chamberlain, A. C., *Brit. J. Appl. Phys.*, **18**, 1793 (1967).

Cheng Y-S., "Condensation Detection and Diffusion Size Separation Techniques," in Willeke, K., and Baron, P. A. (Eds.), *Aerosol Measurement*, Van Nostrand Reinhold, New York, 1993.

8 Thermal and Radiometric Forces

A thermal force arises from asymmetrical interactions of a particle with the surrounding gas molecules in a temperature gradient. A similar force, called the radiometric force, arises when a temperature gradient is established in a particle by a light beam. A concentration gradient in the surrounding gas can also produce a force on a particle. The particle motion produced by these forces is called *thermophoresis*, *photophoresis*, or *diffusiphoresis*, depending on the type of gradient. Closely related to these forces are forces arising from *radiation pressure* and *Stefan flow*. All these forces are very weak and cause significant aerosol motion only because of the extremely high mobility of aerosol particles. Such forces are insignificant for macroscopic objects, and we have no intuitive understanding of them from our everyday experience. There has been considerable scientific interest in these exotic forces over the past century, but their practical significance and application have been modest. Thermophoresis is the basis for the thermal precipitator, a device used for aerosol sampling.

8.1 THERMOPHORESIS

When a temperature gradient is established in a gas, an aerosol particle in that gas experience a force in the direction of decreasing temperature. The movement of the particle that results from this force is called thermophoresis. The magnitude of the thermal force depends on gas and particle properties, as well as the temperature gradient. The earliest studies of thermophoresis were empirical investigations of the dust-free layer observed around a heated object, such as a metal rod immersed in smoke. The smoke particles appear to be repelled by the heated object and form a particle-free layer usually less than 1 mm thick, with a well-defined boundary, as shown in Fig. 8.1. Measurements of the thickness of this layer established that it is independent of the particle material and proportional to the square root of the difference in temperature between the object and the gas. Later investigations established that the particle-free layer is a manifestation of thermal forces in the temperature gradient near the surface of the heated object.

The thermal force and the aerosol particle motion are always in the direction of decreasing temperature. When a cold surface is proximate to a warm gas, thermophoresis causes particles in the gas to be deposited onto the surface. This may occur for a hot gas flowing through a metal tube or a warm flow of air over a cold window.

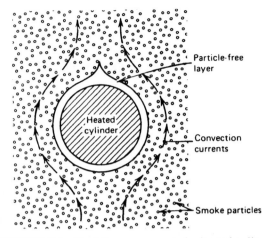

FIGURE 8.1   Dust-free space around a heated cylinder.

For a small particle ($d < \lambda$), the thermal force is a result of a greater transfer of momentum from the gas molecules on the hot side of the particles, relative to those on the cold side. As shown in Fig. 8.2, the gas molecules coming from the left side have a greater velocity than those coming from the right side because of the difference in temperature between the gas on the left and right sides of the particle. The greater momentum received from the left causes a net force—the thermal force—to the right, in the direction of cooler temperature. The exact description of this force is more complicated than that given here, because it depends on the nature of molecular reflection at the particle surface (Waldmann and Schmitt, 1966). Theoretically the thermal force on a particle of diameter d is

$$F_{\text{th}} = \frac{-p\lambda d^2 \nabla T}{T} \quad \text{for } d < \lambda \tag{8.1}$$

where p is the gas pressure, λ is the gas mean free path, ∇T is the temperature gradient, and T is the absolute temperature of the particle. The minus sign is required because the force is in the direction of decreasing temperature. The velocity of thermophoresis is given by Waldmann and Schmitt (1966) as

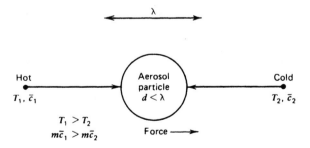

FIGURE 8.2   Molecular impacts on a particle ($d < \lambda$) in a temperature gradient.

$$V_{th} = \frac{-0.55\eta\nabla T}{\rho_g T} \quad \text{for } d < \lambda \tag{8.2}$$

V_{th} is independent of particle size and directly proportional to the temperature gradient. For air at standard pressure and the particle at 293 K [20°C], the thermophoretic velocity for $d < 0.1\,\mu m$ is

$$V_{th} = -2.8 \times 10^{-8}\nabla T \quad \text{for } V_{th} \text{ in m/s and } \nabla T \text{ in K/m}$$

$$[V_{th} = -2.8 \times 10^{-4}\nabla T \quad \text{for } V_{th} \text{ in cm/s and } \nabla T \text{ in K/cm}] \tag{8.3}$$

For large particles ($d > \lambda$), the mechanism is more complicated because a temperature gradient is established in the particle. This gradient affects the temperature gradient in the gas immediately surrounding the particle. The net result is that the particle still receives more momentum from the gas molecules on the hot side than on the cold side, and the net force is in the same direction as for smaller particles: toward the colder region. In this case, the thermal force is influenced by the thermal conductivity of the particle, k_p, relative to that of the air, k_a. The thermal force for $d > \lambda$ is given by

$$F_{th} = \frac{-9\pi d\eta^2 H\nabla T}{2\rho_g T} \quad \text{for } d > \lambda \tag{8.4}$$

where the coefficient H includes the effect of the temperature gradient inside the particle and ∇T is the overall gradient in the gas without the particle present. Using the molecular accommodation coefficients suggested by Brock (1962), the equation for H is

$$H \cong \left(\frac{1}{1 + 6\lambda/d}\right)\left(\frac{k_a/k_p + 4.4\lambda/d}{1 + 2k_a/k_p + 8.8\lambda/d}\right) \tag{8.5}$$

The thermal conductivities of air and various aerosol materials are given in Table 8.1. The thermophoretic velocity for $d > \lambda$ is obtained by equating Eq. 8.4 to the Stokes drag force, to get

$$V_{th} = \frac{-3\eta C_c H\nabla T}{2\rho_g T} \quad \text{for } d > \lambda \tag{8.6}$$

As shown in Fig. 8.3, the thermophoretic velocity of large particles ($d > \lambda$), in contrast to that of small particles ($d < \lambda$), depends on the particle's size and thermal conductivity. Brownian rotation of the particle does not prevent the establishment of a temperature gradient inside the particle, because the time required for significant rotation is greater than that required for temperature adjustment (Fuchs, 1964).

Table 8.2 compares the terminal settling and thermophoretic velocities in a temperature gradient of 100 K/m [1 K/cm]. Even in such a weak gradient, V_{th} exceeds

TABLE 8.1   Thermal Conductivity of Several Materials[a]

Material	Thermal Conductivity	
	W/m · K	(cal/cm · s · K)[b]
Air at 20°C[c]	0.026	0.000062
Castor oil	0.18	0.00043
Glycerol	0.27	0.00064
Mercury	8.4	0.02
Paraffin oil	0.13	0.00030
Water	0.59	0.0014
Asbestos	0.079	0.00019
Carbon	4.2	0.01
Clay	0.71	0.0017
Fused silica	1.0	0.0024
Glass	0.84	0.002
Granite	2.1	0.005
Iron	66.9	0.16
Magnesium oxide	0.13	0.0003
Quartz	9.6	0.023
Sodium chloride	6.6	0.016
Stearic acid	0.13	0.0003

[a]After Mercer (1973).
[b](k in cal/cm · s · K) = 0.00239 × (k in W/m · K).
[c]$k_a = (0.00265 \times T^{1.5}) \times (T + 245 \times 10^{-(12/T)})^{-1}$ W/m · K.

V_{TS} for particles less than 0.1 μm. Table 8.3 shows that deposition onto surfaces by thermophoresis during a 100-s period in this temperature gradient exceeds deposition by diffusion for particles larger than 0.2 μm.

EXAMPLE

A sodium-chloride particle 3.0 μm in diameter flows between two vertical parallel plates 1 mm apart maintained at 40°C and 0°C. What is the thermophoretic velocity toward the cooler plate?

$$k_a/k_p = 0.026/6.6 = 0.0039 \quad \lambda/d = 0.066/3 = 0.022$$

From Eq. 8.5,

$$H = \left(\frac{1}{1 + 6(0.022)}\right)\left(\frac{0.0039 + 4.4(0.022)}{1 + 2(0.0039) + 8.8(0.022)}\right) = 0.074$$

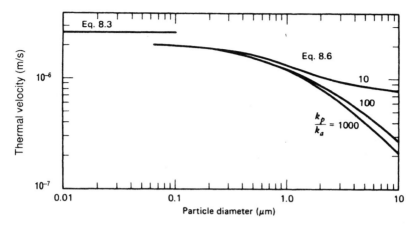

FIGURE 8.3   Thermophoretic velocity in a temperature gradient of 100 K/m [1.0 °C/cm] versus particle size at standard conditions. Parameter is k_p/k_a.

TABLE 8.2   Terminal Settling and Thermophoretic Velocities in a Temperature Gradient of 100 K/m [1 °C/cm] at 293 K [20°C]

Particle Diameter (μm)	Terminal Settling Velocity (m/s)	Thermophoretic Velocity[a] (m/s)
0.01	6.7×10^{-8}	2.8×10^{-6}
0.1	8.6×10^{-7}	2.0×10^{-6}
1.0	3.5×10^{-5}	1.3×10^{-6}
10.0	3.1×10^{-3}	7.8×10^{-7}

[a] $k_p = 10k_a$.

TABLE 8.3   Cumulative Deposition during 100 Seconds by Diffusion and Thermophoresis from Unit Aerosol Concentration in a Temperature Gradient of 100 K/m [1°C/cm] at 293 K [20°C]

Particle Diameter (μm)	Cumulative Deposition	
	Difffusion (number/m²)	Thermophoresis[a] (number/m²)
0.01	2600	280
0.1	290	200
1.0	59	130
10.0	17	78

[a] $k_p = 10k_a$.

From Eq. 8.6,

$$V_{th} = \frac{-3 \times 1.81 \times 10^{-5} \times 0.074((0 - 40)/0.001)}{2 \times 1.2 \times 293} = 2.3 \times 10^{-4} \text{ m/s}$$

$$\left[V_{th} = \frac{-3 \times 1.81 \times 10^{-4} \times 0.074((0 - 40)/0.1)}{2 \times 1.2 \times 10^{-3} \times 293} = 0.023 \text{ cm/s} \right]$$

8.2 THERMAL PRECIPITATORS

Instruments that collect aerosol particles using thermophoresis to deposit the particles onto a surface are called *thermal precipitators*. These devices employ a heated element, such as a wire, ribbon, or plate, and pass the aerosol between the heated element and an ambient temperature surface onto which the particles deposit. Because the thermophoretic velocity does not decrease with particle size, these devices are very efficient collectors of small particles. For properly designed precipitators, the collection efficiency is virtually 100% for particles less than 5 or 10 µm. Typically, thermal precipitators have very low volumetric flow rates, about a few cubic centimeters per minute. These qualities make thermal precipitators particularly well suited for collecting small quantities of particles for observation in an optical or electron microscope.

Thermal precipitators create a difference in temperature of 50–200 K between a heated element and the collection surface. By using a small spacing between the heated element and the collection surface—usually less than 1 mm—these temperature differences give large temperature gradients, 10^5–10^6 K/m [1000–10,000 K/cm]. The best known thermal precipitator is the wire-and-plate design of Green and Watson (1935) shown in Fig. 8.4. For many years, this was the reference sampler for measuring dust concentrations in British mines. The device has an electrically heated Nichrome wire, 0.25 mm in diameter, maintained at 120°C and positioned midway between two glass microscope cover slips 0.5 mm apart. The cover slips are held to brass cylinders that act as heat sinks and maintain the collection surface at ambient temperature to give a maximum temperature gradient of 8×10^5 K/m [8000 K/cm]. The sampling flow rate is 120 mm³/s [7.2 cm³/min]. Collected particles are analyzed by optical microscopy.

Because small particles have a greater thermophoretic velocity than large ones, there is some size segregation of the collected particles in a wire-and-plate thermal precipitator. The smallest particles are deposited first, and larger particles are deposited farther downstream. This problem can be minimized by using a heated plate instead of a heated wire or by slowly moving the deposition surface relative to the wire. For a heated plate, the temperature gradient does not need to be as great, because the particle is exposed to the thermal force for a longer time.

There are no large-scale air cleaners that use thermal precipitation as the primary mechanism of collection, although thermophoresis contributes to the collection of

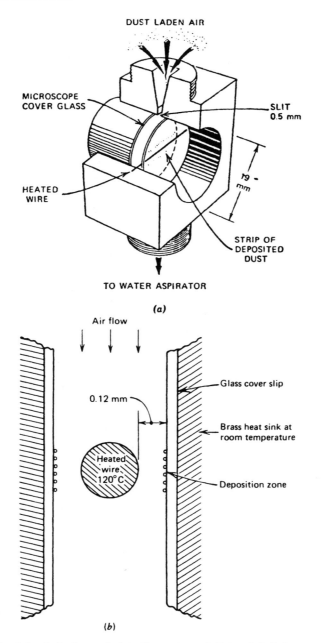

FIGURE 8.4 A heated wire-and-plate thermal precipitator. (*a*) Cutaway view. British Crown copyright. Reproduced with the permission of the Controller of Her Britannic Majesty's Stationery Office. (*b*) Cross-sectional view.

fine particles in air cleaners having significant internal temperature gradients. Thermophoresis is used to reduce particle deposition onto surfaces in the manufacture of microelectronics by maintaining the product at a higher temperature than that of the surrounding air and surfaces.

8.3 RADIOMETRIC AND CONCENTRATION GRADIENT FORCES

Two types of aerosol particle motion can be induced by illumination: photophoresis and motion due to radiation pressure. *Photophoresis* is a special case of thermophoresis in which the absorption of light by the particle creates a temperature gradient in the particle. The gas immediately around the particle takes on the same gradient and establishes a radiometric force, as described in Section 8.1. The mechanism is the same as that for thermophoresis, but there is no overall temperature gradient in the gas. It is much more difficult to determine the temperature gradient in the particle for photophoresis than for thermophoresis. The photophoretic force depends on the intensity and wavelength of the illumination; the size, shape, and material of the particle; and the gas pressure. The force reaches a maximum when $\lambda \approx d$. Unlike radiation pressure, described shortly, photophoresis is a result of an interaction between the particle and the surrounding gas and will not occur in a vacuum.

For photophoresis to occur, the particles must absorb some of the light incident on them. If they are strongly absorbing, the particles will equilibrate with the hot side facing the light source, and the photophoretic force will be away from the source. Weakly absorbing particles that meet certain requirements of size and absorption coefficient will focus the incident light so that the side of the particle away from the light source becomes hot and the particle experiences a radiometric force toward the source of illumination. This is called *reverse photophoresis*. Photophoresis, which has been reviewed by Preining (1966), Kerker (1974), and Mackowski (1989), may have significance in the movement of submicrometer particles in the upper atmosphere, but it has no practical application.

Radiation pressure, or light pressure, results from a direct transfer of momentum by the deflection and absorption of the light itself. The light can be thought of as a stream of photons whose momentum is altered as they are deflected (or absorbed) by the particle, causing a force that is directed away from the source of light.

The radiation pressure force on a particle is

$$F_{\text{rp}} = \frac{I_0 \pi d^2 Q_{\text{rp}}}{4c} \tag{8.7}$$

where I_0 is the incident intensity, c is the velocity of light, and Q_{rp} is an efficiency factor that defines the fraction of the light geometrically incident on the particle that is effective in transferring momentum to the particle in the direction away from the source. Q_{rp} is a complicated function of the particle size, wavelength, index of refraction, and absorption coefficient (Kerker, 1969). The radiation pressure due to visible light decreases rapidly with decreasing particle size below 0.3 μm.

Because radiation pressure is a direct effect of electromagnetic radiation, the radiation pressure force is the same whether the particle is in air or a vacuum. It is this mechanism that is responsible for pushing the tails of comets in a direction away from the sun. The tails are composed of microscopic particles that are shed by the comet and may be millions of miles in length. High-intensity laser beams have been used to lift particles (1–100 μm) against the force of gravity (Ashkin, 1980).

When a gas or vapor diffuses through air, particles suspended in the mixture may experience a *diffusiophoretic force* due to the unequal transfer of momentum from the two species of gas. In this situation complementary gradients are established, with the vapor diffusing one way and the air diffusing the other way to maintain a constant total pressure. A particle in this region is bombarded unequally by the molecular traffic coming from both sides. If the vapor molecules are heavier than the air molecules, they will impart a greater momentum to the particle from the side of higher vapor concentration. This action results in a net force on the particle in the direction of the diffusion of the heavier molecules. The magnitude of the force depends on the molecular weights and diffusion coefficients of the gases and on the concentration gradient; it is approximately independent of the particle size.

A more complicated situation exists near an evaporating or condensing liquid surface. Besides the binary concentration gradients and diffusiophoresis, there exists an aerodynamic flow of air, called *Stefan flow*, that exerts Stokes drag on a particle near the evaporating or condensing surface. Stefan flow, which is directed away from an evaporating surface and toward a condensing surface, causes a thin, dust-free layer similar to that observed around a heated object (see Section 8.1) to form near the evaporating surface under isothermal conditions. This phenomenon was first observed more than 100 years ago.

To maintain a constant total pressure near an evaporating surface, the concentration gradient of vapor is balanced by an equal and opposite concentration gradient of air. The latter causes a continuous diffusion of air molecules toward the surface, but since the air cannot accumulate at the surface, there must be an aerodynamic flow of air away from the evaporating surface to compensate for the diffusion of air to the surface. The transfer of air away from the surface cannot be by diffusion, because it would be opposed by the air concentration gradient toward the surface. There would be no Stefan flow if the air mysteriously condensed when it reached the evaporating surface, but since it does not at normal temperatures, this air must flow aerodynamically away from the surface. When condensation takes place at the surface, the situation is reversed and the Stefan flow is toward the surface.

In most situations, both diffusiophoresis and Stefan flow will be acting on a particle near an evaporating or condensing surface. These forces may be in the same direction or may oppose each other. If the molecular weights of the gas and vapor are the same, there will be no diffusiophoresis but the Stefan flow will remain. For water vapor in air at standard conditions, Goldsmith and May (1966) give an empirical expression for the velocity resulting from combined diffusiophoresis and Stefan flow.

$$V_{dsf} = -1.9 \times 10^{-7} \frac{dp}{dx} \tag{8.8}$$

where V_{dsf} is in m/s and dp/dx *is* the partial pressure gradient of the vapor in kPa/m. For cgs units, V_{dsf} is in cm/s, dp/dx in kPa/cm, and the coefficient is -1.9×10^{-3}. The velocity predicted by Eq. 8.8 is away from an evaporating droplet and toward a condensing droplet. The latter situation is exploited in air-cleaning equipment, such as venturi scrubbers, in which condensing conditions are maintained and the resulting Stefan flow carries fine particles onto condensing droplets and surfaces.

PROBLEMS

8.1 A paraffin oil droplet 2 μm in diameter (density = 900 kg/m³ [0.9 g/cm³]) flows across a horizontal surface. What temperature gradient is required to prevent the particle from settling onto the surface? $T = 20°C$.
ANSWER: 8300 K/m [83 K/cm].

8.2 Estimate the maximum thermophoretic velocity of a fused silica particle 1.0 μm in diameter in the thermal precipitator shown in Fig. 8.4. Assume a uniform temperature gradient between the wire and the plate, a wire temperature of 120°C, and a plate temperature of 20°C. [*Hint*: ρ_g, η, λ, and C_c should be evaluated at the midpoint temperature.]
ANSWER: 12 mm/s [1.2 cm/s].

8.3 Calculate the force due to radiation pressure on a 0.5-μm diameter stearic acid particle in direct sunlight at standard conditions. What is the velocity of the particle? The intensity of sunlight is approximately 1200 W/m². Use $Q_{rp} = 1.0$.
ANSWER: 7.8×10^{-19} N, 1.2×10^{-8} m/s [7.8×10^{-14} dyn, 1.2×10^{-6} cm/s].

8.4 What size paraffin oil droplet ($\rho_p = 900$ kg/m³ [0.9 g/cm³]) will not settle in a vertical temperature gradient of −2000 K/m [−20 K/cm]? (The negative sign indicates that the temperature decreases in the upward direction.) Assume that $T = 20°C$ and $H = 0.17$.
ANSWER: 0.98 μm

REFERENCES

Ashkin, A., "Applications of Laser Radiation Pressure," *Science*, **210**, 1081–1088 (1980).

Brock, J. R., "On the Theory of Thermal Forces Acting on Aerosol Particles," *J. Colloid Sci.*, **17**, 768–780 (1962).

Fuchs, N. A., *The Mechanics of Aerosols*, Pergamon, Oxford, U.K., 1964.

Goldsmith, P., and May, F. G., in Davies, C. N. (Ed.), *Aerosol Science*, Academic Press, London, 1966.

Green, H. L., and Watson, H. H., *Med. Res. Council, Spec. Rept.* No. 199, His Majesty's Stationery Office, London, 1935.

Kerker, M., "Movement of Small Particles by Light," *Am. Sci.*, **62**, 92–98 (1974).

Kerker, M., and Cooke, D. C., "Photophoretic Force on Aerosol Particles in the Free-molecule Regime," *J. Opt. Soc. Am.*, **72**, 1267–1272 (1982).

Kerker, M., *The Scattering of Light*, Academic Press, New York, 1969.

Mackowski, D. W., "Photophoresis of Aerosol Particles in the Free-Molecule and Slip-Flow Regimes," *Int. J. Heat Mass Trans.*, **32**, 843–854 (1989).

Mercer, T. T., *Aerosol Technology in Hazard Evaluation*, Academic Press, New York, 1973.

Preining, O., "Photophoresis," in Davies, C. N. (Ed.), *Aerosol Science*, Academic Press, London, 1966.

Waldmann, L., and Schmitt, K. H., "Thermophoresis and Diffusiophoresis of Aerosols," in in Davies, C. N. (Ed.), *Aerosol Science*, Academic Press, London, 1966.

9 Filtration

The capture of aerosol particles by filtration is the most common method of aerosol sampling and is a widely used method for air cleaning. Filtration is a simple, versatile, and economical means for collecting samples of aerosol particles. At low dust concentrations, fibrous filters are the most economical means for achieving high-efficiency collection of submicrometer particles. Aerosol filtration is used in diverse applications, such as respiratory protection, air cleaning of smelter effluent, processing of nuclear and hazardous materials, and clean rooms.

The process of filtration is complicated, and although the general principles are well known there is still a gap between theory and experiment. Nevertheless, filtration is an active area for theoretical and experimental research, and there is an extensive scientific literature on the subject.

In this chapter, we review filtration in order to provide an understanding of the properties of fibrous and porous membrane filters, the mechanisms of collection, and how collection efficiency and resistance to airflow change with filter properties and particle size. The discussion of filtration that follows allows one to use and integrate nearly all the concepts covered in the preceding chapters.

9.1 MACROSCOPIC PROPERTIES OF FILTERS

The most important types of filters for aerosol sampling are fibrous and porous membrane filters. Fibrous filters consist of a mat of fine fibers arranged so that most are perpendicular to the direction of airflow. As shown in Fig. 9.1, these filters are mostly air, having porosities from 70 to greater than 99%. The fibers range in size from submicrometer to 100 μm. Cellulose fibers (wood fibers), glass fibers, and plastic fibers are the most common types. The air velocity through high-efficiency filters is usually quite low, about 0.1 m/s [10 cm/s]. Because of the low velocities used, it is often necessary to pleat the filter material to obtain a large filter area in an element of convenient size.

A common misconception is that aerosol filters work like microscopic sieves in which only particles smaller than the holes can get through. This view may be appropriate for the liquid filtration of solid particles, but it is not how *aerosol* filtration works. As will be described, particles are removed by a fibrous filter when they collide and attach to the surface of the fibers.

As Fig. 9.2 shows, porous membrane filters have a structure different from that of fibrous filters, with less porosity—50 to 90%—than fibrous filters. The gas flow

FIGURE 9.1   Scanning electron microscope photograph of a high-efficiency glass fiber filter. Magnification of (*a*) 4150× and (*b*) 800×.

through the filter follows an irregular path through the complex pore structure. Particles are lost from the gas stream as they deposit on the structural elements that form the pores. Membrane filters have high efficiency and a greater resistance to airflow, pressure drop, than do other types of filters. The high collection efficiency of porous membrane filters extends to aerosol particles much smaller than the manufacturer's stated pore size, which is based on liquid filtration. Porous membrane filters are made from cellulose esters, sintered metals, polyvinyl chloride, Teflon™, and other plastics.

The capillary pore membrane filter (Nuclepore®), shown in Fig. 9.3, has an array of microsopic cylindrical holes of uniform diameter, approximately perpendicular to the surface of the filter. These filters are constructed of 10-μm-thick polycarbonate film that is subject to bombardment by fission fragments followed by etching

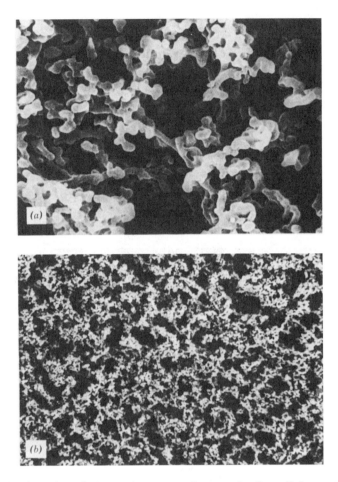

FIGURE 9.2 Scanning electron microscope photograph of a cellulose ester porous membrane filter with a pore size of 0.8 μm. Magnification of (*a*) 4150× and (*b*) 800×.

to produce the cylindrical pores. The efficiency for particles smaller than the pore size is not as good as that of porous membrane filters. Because of their smooth surfaces, capillary pore membrane filters are particularly useful for collecting particles for observation in a scanning electron microscope.

Fabric filtration, often confused with fibrous filtration, is used in industrial air cleaning for high-efficiency filtration at high dust concentrations. These applications are frequently large installations containing thousands of fabric bags operating in parallel. Each bag may be 0.12 - 0.4 m [12–40 cm] in diameter and 3–10 m in height. The woven or felted fabric has a low initial collection efficiency, but becomes highly efficient when a dust layer builds up on the fabric. It is this porous dust layer supported by the fabric that is responsible for the method's high filtration efficiency. Home vacuum cleaners use a simplified version of this type of fil-

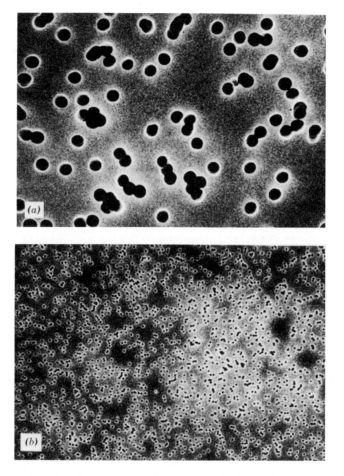

FIGURE 9.3   Scanning electron microscope photograph of a capillary pore membrane filter with a pore size of 0.8 μm. Magnification of (*a*) 4150× and (*b*) 800×.

tration. In industrial operations, the bags are shaken periodically to reduce the thickness of the accumulated dust layer and facilitate the flow of air. These filters operate at a higher pressure drop than other types of filters. Special attention has been given in recent years to fabrics that can operate at high temperatures to clean flue gases.

In granular-bed filtration, aerosol particles are collected as they pass through a bed (horizontal layer) of fine granules. Collection occurs by the same mechanisms that operate in a fibrous filter. Granular-bed filtration is used primarily for air cleaning, especially for corrosive aerosols and aerosols at high temperatures. It is not covered in this text. [See Tien (1989).]

When filters are used for sampling, their ability to collect particles is usually characterized by their efficiency of collection—the fraction of entering particles retained by the filter. This efficiency can be expressed either in terms of a particle number collection efficiency $\mathbf{E}$ or a mass collection efficiency $\mathbf{E}_m$. The latter refers to the fraction of the entering particulate mass retained by the filter. Generally, mass efficiency is higher than count efficiency.

$$\mathbf{E} = \frac{N_{in} - N_{out}}{N_{in}} \tag{9.1}$$

$$\mathbf{E}_m = \frac{C_{in} - C_{out}}{C_{in}} \tag{9.2}$$

where N and C refer to the number and mass concentration, respectively, of particles entering and leaving the filter.

Air-cleaning equipment, on the other hand, is usually characterized in terms of its penetration $\mathbf{P}$—the fraction of entering particles that exit or penetrate the filter.

$$\mathbf{P} = \frac{N_{out}}{N_{in}} = 1 - \mathbf{E} \tag{9.3}$$

$$\mathbf{P}_m = \frac{C_{out}}{C_{in}} = 1 - \mathbf{E}_m \tag{9.4}$$

At high efficiencies, large changes in penetration are associated with small changes in efficiency. For example, penetration changes by a factor of 10 when efficiency changes by less than 1%, from 99 to 99.9%.

The velocity of the air at the face of a filter, just before the air enters, is called the *face velocity*, U_0.

$$U_0 = \frac{Q}{A} \tag{9.5}$$

where Q is the volumetric flow rate through the filter and A is the cross-sectional area of the filter exposed to the entering airstream. Inside a filter, the air velocity U is slightly greater than U_0, because the volume for air passage is reduced by the volume of fibers (or membrane or granules), so that

$$U = \frac{Q}{A(1 - \alpha)} \tag{9.6}$$

where α is the volume fraction of fibers, called the *packing density* or *solidity*. For fibrous filters, α, defined as follows, is typically between 0.01 and 0.3:

$$\alpha = \frac{\text{fiber volume}}{\text{total volume}} = 1 - \text{porosity} \tag{9.7}$$

For the collection of particles by sieves, a single ideal sieve collects 100% of particles larger than the sieve openings, and 100% of those smaller pass through.

In contrast to sieves, fibrous filters can be thought of as many thin layers of filters, each having a certain probability of collecting particles of a given size. Thus, the filtration efficiency for monodisperse aerosols increases as the thickness of a filter is increased. If γ represents the fractional capture per unit thickness for a differentially thin layer dt, then the number of particles captured n_c when a unit volume of aerosol passes through the layer is

$$n_c = N\gamma dt \qquad (9.8)$$

where N is the particle number concentration entering the layer. The decrease in concentration of an aerosol passing through the layer is equal to n_c, the number of particles captured per unit volume of aerosol passing through the layer; that is

$$dN = -n_c = -N\gamma dt \qquad (9.9)$$

The combined effect of all the layers is obtained by integrating Eq. 9.9 over the entire thickness t:

$$\int_{N_{in}}^{N_{out}} \frac{dN}{N} = \int_0^t (-\gamma)dt \qquad (9.10)$$

to get

$$\ln\left(\frac{N_{out}}{N_{in}}\right) = -\gamma t$$

$$\mathbf{P} = e^{-\gamma t} \qquad (9.11)$$

Thus, particle penetration decreases exponentially with increasing filter thickness. The value of γ, covered in the next section, depends on particle size, face velocity, solidity, and fiber size. Equation 9.11 holds for monodisperse aerosols and for a particular particle size of a polydisperse aerosol, but does not apply to the overall penetration of a polydisperse aerosol. The easy-to-collect particles are removed primarily in the first few layers, while hard-to-collect particles pass through these layers nearly undiminished. Thus, the particle size distribution changes as the aerosol passes through the filter, and γ, which depends on the particle size, also varies with position in the filter for polydisperse aerosols.

For a given fibrous filter, there is a particle size, usually between 0.05 and 0.5 μm, that has the minimum collection efficiency; that is, all particles, larger *or* smaller than this size, are collected with greater efficiency. For a given size particle, there is also a velocity for minimum collection efficiency. These effects and the mechanisms that cause them are described in detail in the next two sections.

The structure of a filter creates a resistance to the air flowing through it, called the *pressure drop*, Δp. At a given face velocity, the pressure drop of a filter is directly proportional to the thickness of the filter. This should be obvious when one considers that the effect of doubling the thickness is equivalent to operating two identical filters in series. The airflow and resistance will be the same for both filters, and since they are in series, the resistances must be added. As will be discussed, the flow inside most filters is laminar; consequently, the pressure drop is directly proportional to the flow rate.

A change in any property of a filter, such as its fiber or pore size, α, U_0, or t, causes a change in both the efficiency for a given particle size and the pressure drop. The best filter is the one that gives the highest collection efficiency with the least pressure drop. A useful criterion for comparing different types of filters and filters of different thickness is the filter quality q_F, the ratio of γ to the pressure drop per unit thickness, $\Delta p/t$. Since $\gamma = -\ln(\mathbf{P})/t = \ln(1/\mathbf{P})/t$,

$$q_F = \frac{\gamma t}{\Delta p} = \frac{\ln(1/\mathbf{P})}{\Delta p} \tag{9.12}$$

The greater the value of q_F, the better the filter. Comparisons of q_F must be made for the same face velocity and test aerosol particle size. Properties of some commercially available filter materials are given in Table 9.1.

In fibrous filters, both efficiency and pressure drop increase with dust loading, the accumulation of collected particles in the filter. Initially, this is beneficial; that is, the filter quality improves. But eventually, the pressure drop becomes excessive and the filter is said to be clogged. Filters with a small value of α can accommodate the greatest dust loading without clogging. Unlike the collection of vapor on charcoal, there is usually no problem with breakthrough of dust particles in fibrous filters. At very high velocities, filters may shed droplets of accumulated liquid, and mechanical flexing of heavily loaded filters may release dust agglomerates.

Besides collection efficiency and pressure drop, the selection of a filter for aerosol sampling may depend strongly on the analytical method to be used. Gravimetric analysis requires that a filter's weight be stable with age, handling, and changes in temperature and relative humidity. (See Section 10.4.) Under practical weighing and storage conditions, it is difficult to achieve a filter weight stability of less than 0.2%, equivalent to 100 μg for a 37-mm-diameter membrane filter. Cellulose fiber filters are quite hygroscopic and generally are not suitable for gravimetric analysis. Glass fiber and cellulose ester filters are much less affected by moisture and age; polycarbonate, polyvinyl chloride, and Teflon® filters are affected the least.

If the aerosol particles are to be analyzed by chemical methods, one must be concerned with interferences caused by the filter material or contaminants in the filter. For example, some kinds of glass fiber filters contain organic binder, up to 5% by weight, that interferes with the analysis of certain organic compounds. These filters may also suffer from the formation of artifacts; specifically, the glass fibers have a sightly alkaline surface that reacts with SO_2 gas to form sulfates on the fiber surface. Special organic-free "microquartz" filters have a low sulfate artifact formation and are widely used for the analysis of trace organic compounds. For analytical methods that require ashing by incineration or dissolution in acid, consideration must be given to the amount and type of filter residue. The extraction of soluble aerosol particles with a solvent requires solvent-stable filter material. The depth at which particles are captured is an important consideration for radioassay of alpha-emitting particles, because of alpha absorption by the filter material. Fil-

TABLE 9.1 Characteristics of Some Common Aerosol-Sampling Filter Materials

Filter	Type	Material	Thickness (mm)	Fiber Diameter or Pore Diameter (μm)	Solidity	Pressure Drop[a] (kPa)	Efficiency[a,b]	Filter Quality[a,b] (kPa^{-1})
Whatman 41	Fiber	Cellulose	0.19	3–20	0.35	2.5[c]	72[c]	0.52
Nuclepore	CPM[d]	Polycarbonate	0.01	0.8	0.85	5.9	90[e]	0.39
Microsorban	Fiber	Polystyrene	1.5	0.7	0.04	2.9	99.5	1.10
MSA 1106B	Fiber	Glass	0.23	0.1–4	0.10	2.0[c]	99.93[c]	3.70
Millipore AA	Membrane	Cellulose ester	0.15	0.8	0.19	9.5[c]	99.98[c]	0.93

[a] At U_0 = 0.27 m/s [27 cm/s].
[b] For d_p = 0.3 μm.
[c] From Lippmann (1995).
[d] CPM = capillary pore membrane.
[e] Estimated.

ter sampling for microscopic analysis is covered briefly in Sections 10.6, 20.3, and 20.5 and for bioaerosols in Section 19.2.

9.2 SINGLE-FIBER EFFICIENCY

Fibrous filtration is a complex process, and only by analyzing this process at its most elementary level, the collection of a particle by an individual fiber, does the influence of various parameters such as d_p, α, U_0, and d_f become apparent. The approach is to consider a single fiber, positioned with its axis perpendicular to the airflow in the middle of a filter, and analyze the several mechanisms by which particles can be collected on that fiber. We assume that a particle sticks if it contacts the fiber and is permanently removed from the aerosol stream.

The classical theory of filtration considers only collection on isolated fibers, as if the other fibers did not exist. An examination of the Reynolds number Re_f that characterizes the flow around a fiber having a diameter d_f reveals that, under most conditions, the flow inside a filter will be laminar.

$$Re_f = \frac{\rho_g d_f U}{\eta} \qquad (9.13)$$

In laminar flow, the distortion of the flow caused by a fiber influences the flow around neighboring fibers, even if they are many fiber diameters distant. The more modern single-fiber theory follows the same approach as the isolated fiber theory, but takes into account the effect of neighboring fibers.

The efficiency with which a fiber removes particles from an aerosol stream is defined in terms of a single-fiber efficiency E_Σ, a dimensionless particle deposition rate on a unit length of fiber. As shown in Fig. 9.4, E_Σ is the fraction of particles approaching the fiber in the region defined by the projected area of the fiber that are ultimately collected on the fiber.

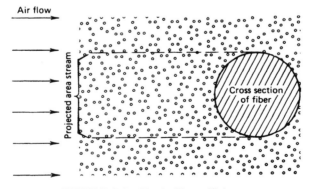

FIGURE 9.4 Single-fiber efficiency.

$$E_\Sigma = \frac{\text{number collected on unit length}}{\text{number geometrically incident on unit length}} \qquad (9.14)$$

Put another way, E_Σ is the ratio of the number of particles actually collected by a fiber in one second to the number that would have passed through an imaginary outline of the fiber in one second.

Clearly, the overall efficiency of a filter, **E**, is a function of the single-fiber efficiency, E_Σ. Assuming that all fibers in a filter have the same diameter, the total length L of fiber in a unit volume of the filter can be obtained from the definition of α, Eq. 9.7.

$$L = \frac{4\alpha}{\pi d_f^2} \qquad (9.15)$$

From the definition of E_Σ, the number of particles collected when a unit volume of aerosol passes through an element of unit cross-sectional area and thickness dt is

$$n_c = N E_\Sigma d_f L \, dt \qquad (9.16)$$

Equation 9.16 is the same as Eq. 9.8 with

$$\gamma = E_\Sigma d_f L \qquad (9.17)$$

Combining Eqs. 9.15 and 9.17 gives

$$\gamma = \frac{4\alpha E_\Sigma}{\pi d_f} \qquad (9.18)$$

and Eq. 9.11 becomes

$$\mathbf{P} = 1 - \mathbf{E} = e^{-\gamma} = \exp\left(\frac{-4\alpha E_\Sigma t}{\pi d_f}\right) \qquad (9.19)$$

Equation 9.19 relates the macroscopic property of filter penetration to the microscopic property of single-fiber efficiency E_Σ. The difficulty in applying Eq. 9.19 is in determining the value of E_Σ. Sections 9.3 and 9.4 deals with this problem.

9.3 DEPOSITION MECHANISMS

There are five basic mechanisms by which an aerosol particle can be deposited onto a fiber in a filter:

1 Interception
2 Inertial impaction
3 Diffusion

4 Gravitational settling

5 Electrostatic attraction

These five deposition mechanisms form the basic set of mechanisms for all types of aerosol particle deposition, including deposition in a lung, in a sampling tube, or in an air cleaner. The method of analysis and prediction is different for each situation, but the deposition mechanisms are the same. The first four mechanisms are called *mechanical collection mechanisms*. Each of the five deposition mechanisms is described below, along with equations that predict the single-fiber efficiency due to that mechanism. The theoretical analysis is complex, and only simplified equations are presented. Still, these equations are accurate enough to show the trend of collection efficiency with filter parameters. Wherever possible, the equations are based on experimentally verified theory and except where noted, are valid for standard conditions and $0.005 < \alpha < 0.2$, $0.001 < U_0 < 2$ m/s [0.1–200 cm/s], $0.1 < d_f < 50$ µm, and $Re_f < 1$.

Collection by *interception* occurs when a particle follows a gas streamline that happens to come within one particle radius of the surface of a fiber, as shown in Fig. 9.5. The particle hits the fiber and is captured because of its finite size. Thus, for a given size particle, certain streamlines will result in capture of the particle while other streamlines will not. For pure interception, it is assumed that the particles follow the streamlines perfectly; that is, they have negligible inertia, settling, and Brownian motion. Interception is the only mechanism that is not a result of a particle departing from its original gas streamline.

The single-fiber efficiency due to interception depends on the dimensionless parameter R, where

$$R = \frac{d_p}{d_f} \qquad (9.20)$$

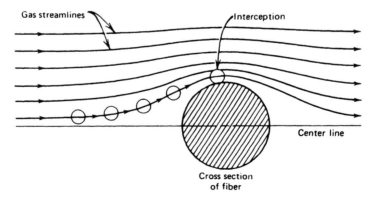

FIGURE 9.5 Single-fiber collection by interception.

The single-fiber efficiency for interception, E_R, is given by Lee and Ramamurthi (1993) as

$$E_R = \frac{(1 - \alpha)R^2}{\text{Ku}(1 + R)} \tag{9.21}$$

where Ku is the Kuwabara hydrodynamic factor, a dimensionless factor that compensates for the effect of distortion of the flow field around a fiber because of its proximity to other fibers. Ku depends only on the solidity α,

$$\text{Ku} = -\frac{\ln \alpha}{2} - \frac{3}{4} + \alpha - \frac{\alpha^2}{4} \tag{9.22}$$

Ku ranges from 1.9 for $\alpha = 0.005$ to 0.25 for $\alpha = 0.2$. For $d_f < 2$ μm, a slip-corrected version of Eq. 9.22 must be used (Yeh and Liu, 1974). Adding $2\lambda/d_f$ to the right side of Eq. 9.22 provides a satisfactory approximation (Kirsch and Stechkina, 1978). E_R increases with increasing R, but cannot exceed the maximum theoretical value of $1 + R$ based on the definition of single-fiber efficiency. Interception is an important collection mechanism in the particle size range of minimum efficiency and is the only mechanism that does not depend on flow velocity U_0.

Inertial impaction of a particle on a fiber occurs when the particle, because of its inertia, is unable to adjust quickly enough to the abruptly changing streamlines near the fiber and crosses those streamlines to hit the fiber, as shown in Fig. 9.6. The parameter that governs this mechanism is the Stokes number, defined in Eq. 5.23 as the ratio of particle stopping distance to fiber diameter.

$$\text{Stk} = \frac{\tau U_0}{d_f} = \frac{\rho_p d_p^2 C_c U_0}{18 \eta d_f} \tag{9.23}$$

This equation represents the ratio of the "persistence" of a particle to the size of the target. Single-fiber efficiency for impaction increases with an increasing value

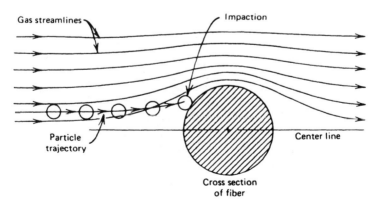

FIGURE 9.6   Single-fiber collection by impaction.

of Stokes number, because of (1) a greater particle inertia (greater d_p or ρ_p), (2) a greater particle velocity, or (3) a more abrupt curvature of streamlines, caused by a smaller fiber size. The single-fiber efficiency for impaction E_I is given by (Yeh and Liu, 1974) as,

$$E_I = \frac{(\text{Stk})J}{2\text{Ku}^2} \tag{9.24}$$

where

$$J = (29.6 - 28\alpha^{0.62})R^2 - 27.5R^{2.8} \quad \text{for } R < 0.4 \tag{9.25}$$

There is no simple equation for J when $R > 0.4$. For approximate analysis, a value of $J = 2.0$ for $R > 0.4$ can be used, as is done in the next section. As expected, impaction is the most important mechanism for large particles, but such particles usually have significant collection by interception as well. The sum of E_I and E_R cannot exceed the theoretical maximum of $1 + R$.

The Brownian motion of small particles is sufficient to greatly enhance the probability of their hitting a fiber while traveling past it on a nonintercepting streamline. Figure 9.7 shows the trajectory of one such particle. This is a special case of diffusion to a surface, covered in Chapter 7. The single-fiber efficiency due to *diffusion*, E_D, is a function of the dimensionless Peclet number, Pe,

$$\text{Pe} = \frac{d_f U_0}{D} \tag{9.26}$$

where D is the particle diffusion coefficient. The single-fiber efficiency due to diffusion E_D is

$$E_D = 2\text{Pe}^{-2/3} \tag{9.27}$$

Equation 9.27 is based on experimental measurements of filter efficiencies, and the coefficient 2 is an empirically determined number (Kirsch and Fuchs, 1968).

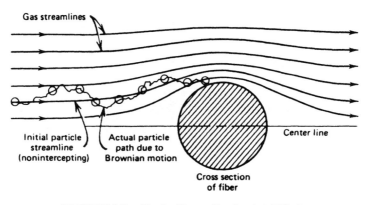

FIGURE 9.7 Single-fiber collection by diffusion.

Single-fiber efficiency increases as Pe and particle size decrease. E_D is the only deposition mechanism that increases as d_p decreases. Theoretical expressions for E_D include the effect of Ku. [See Davies (1973) and Brown (1993).]

In estimating the overall single-fiber collection efficiency near the size of minimum efficiency, it is necessary to include an interaction term to account for enhanced collection due to interception of the diffusing particles.

$$E_{DR} = \frac{1.24 R^{2/3}}{(Ku\ Pe)^{1/2}} \tag{9.28}$$

The dimensionless number that controls deposition due to *gravitational settling* is G.

$$G = \frac{V_{TS}}{U_0} = \frac{\rho_g d_p^2 C_c g}{18\eta U_0} \tag{9.29}$$

When U_0 and V_{TS} are in the same direction, downward airflow, the single-fiber efficiency for settling, E_G, is

$$E_G \approx G(1 + R) \tag{9.30}$$

For gas flow in the direction opposite to V_{TS},

$$E_G \approx -G(1 + R) \tag{9.31}$$

and E_G decreases overall single-fiber efficiency. When flow is horizontal, E_G is much less—on the order of G^2. Generally, E_G is small compared with other single-fiber mechanisms, unless the particle size is large and U_0 is low. When U_0 is greater than about 0.1 m/s [10 cm/s], impaction is more important than settling.

The remaining deposition mechanism, *electrostatic deposition*, can be extremely important, but is difficult to quantify because it requires knowing the charge on the particles and on the fibers. Electrostatic collection is often neglected, unless the particles or fibers have been charged in some quantifiable way. Increasing the charge on either the particles or the fibers, or reducing the face velocity, increases the collection efficiency. The theory of particle collection by charged fibers, charged particles, or both is reviewed by Brown (1993). Charged particles are attracted to oppositely charged fibers by coulombic attraction. A neutral particle can also be attracted to a charged fiber: The electric field created by the charged fiber induces a dipole, or charge separation, in the particle. In the nonuniform field around the fiber, the near side of the particle experiences an attractive force that is greater than the repulsive force on the far side; hence, a net force exists in the direction of the fiber, and the particle migrates in that direction. Finally, a charged particle can be attracted to a neutral fiber at close range by image forces. The charged particle induces an equal and opposite charge in the fiber surface and thus creates its own field for attraction. Image forces are weaker than coulombic forces.

To use equations involving the charge on filter fibers requires knowing the distribution of charge at a microscopic level, something that is usually not known and that may be very different from the macroscopic charge state. As discussed in Chapter 15, the charge on a particle can be estimated from the conditions under which the charge was acquired. Brown (1993) gives the single-fiber efficiency for electrostatic image forces, for a neutral fiber and a particle with charge q, based on experimental measurements with glass fiber filters, as

$$E_q = 1.5\left[\frac{(\varepsilon_f - 1)}{(\varepsilon_f + 1)} \frac{q^2}{12\pi^2\eta U_0\epsilon_0 d_p d_f^2}\right]^{1/2} \tag{9.32}$$

where ε_f is the relative permittivity (dielectric constant) of the fiber, q is the charge on the particle, and ϵ_0 is the permittivity of a vacuum.

Although difficult to quantify at the microscopic level, charged fibers can greatly enhance filter collection. This characteristic is used for filters that require high efficiency and low pressure drop, such as respirator filters. [See Brown (1993).] The oldest type of charged-fiber filter is the resin–wool, or Hansen, filter, made of wool fiber impregnated with insulating resin particles about 1 μm in diameter. The mechanical action of carding the felt causes the resin particles to become highly charged, and they retain their charge for years under favorable conditions. The presence of these highly charged particles in the filter greatly enhances its collection efficiency without increasing its resistance. Unfortunately, charged fiber filters lose their charge and their effectiveness when exposed to ionizing radiation, high temperature, high humidity, or organic liquid aerosols, such as oil or di(2-ethylhexyl) phthalate (DOP). Also, accumulated dust can mask the charge and reduce its effectiveness.

Another type of charged fiber is the electret fiber. This fiber is made from thin sheets of insulating plastic, such as polypropylene, that are corona charged so that one side is positive and the other negative in a more or less permanent configuration. The sheets are split into fibers and incorporated into fibrous filters. Electret fiber filters have advantages and limitations similar to those of resin–wool filters.

9.4 FILTER EFFICIENCY

The overall efficiency of a filter can be determined by Eq. 9.19 if the total single-fiber efficiency E_Σ is known. The mechanical single-fiber efficiencies are correctly combined by Eq. 9.33a as long as each acts independently and is less than 1.0.

$$E_\Sigma = 1 - (1 - E_R)(1 - E_I)(1 - E_D)(1 - E_{DR})(1 - E_G) \tag{9.33a}$$

$$E_\Sigma \approx E_R + E_I + E_D + E_{DR} + E_G \tag{9.33b}$$

where the single-fiber efficiencies are given by Eqs. 9.21, 9.24, 9.27, 9.28, and 9.30. Equation 9.33b is simpler and often used although theoretically incorrect. It may overestimate collection, because different mechanisms are competing for the same particle and its capture could be counted twice. Often one mechanism will predominate, and the overall efficiency can be assumed to depend only on that mechanism. As will be shown, later in this section, the mechanisms of impaction, diffusion, and gravity predominate for different particle sizes and face velocities. Equation 9.33b agrees with Eq. 9.33a within 5% when only one mechanism predominates and the others are less than 0.01. The situation is usually more complicated than is suggested by this discussion, because of the range of fiber diameters and particle sizes present. Integrating Eq. 9.19 over the particle size distribution correctly deals with the effect of the particle size range. This approach is not correct for the distribution of fiber sizes, because the flow field and collection efficiency associated with each fiber size are influenced by the presence of fibers of other sizes. As a practical matter, the effective fiber diameter, based on pressure drop measurements (see Section 9.5), is a reasonable approximation for d_f in these calculations. Equations 9.19 and 9.33 tend to overestimate efficiency because the fibers are not all perpendicular to the airflow, they may be clumped together, and α is not uniform. [See the inhomogeneity factor in Davies (1973).] The use of the effective fiber diameter (Section 9.5) in Eqs. 9.19 and 9.33 avoids this problem.

Table 9.2 gives the single-fiber efficiency for each collection mechanism, E_Σ, and the overall efficiency, calculated using Eqs. 9.19 and 9.33b, for a filter having a thickness of 1 mm, a solidity α of 0.05, and a fiber diameter of 2 μm, operating at a face velocity of 0.1 m/s [10 cm/s]. Figure 9.8 shows the total filter efficiency and the efficiency due to each of the single-fiber mechanisms calculated by Eq. 9.19; data are from Table 9.2. For this filter and face velocity, interception and impaction are negligible for small particles, but increase rapidly for particles larger than 0.3 μm. Diffusion is the only important mechanism for particles below 0.2 μm, but is of decreasing importance for particles above that size. For all particle sizes, gravity

TABLE 9.2   Single-Fiber and Total Efficiency for a Filter Having t = 1 mm, α = 0.05, d_f = 2 μm, and U_0 = 0.1 m/s [10 cm/s]

Particle Diameter (μm)	Single-Fiber Efficiency[a]						Overall Filter Efficiency[b] (%)
	E_R	E_I	E_D	E_{DR}	E_G	E_Σ	
0.01	0.000	0.000	0.840	0.020	0.000	0.861	100.0
0.02	0.000	0.000	0.339	0.016	0.000	0.356	100.0
0.05	0.001	0.000	0.106	0.013	0.000	0.119	97.7
0.1	0.003	0.000	0.046	0.011	0.000	0.059	84.9
0.2	0.010	0.002	0.021	0.010	0.000	0.043	74.3
0.5	0.055	0.034	0.009	0.009	0.000	0.108	96.8
1.0	0.183	0.238	0.005	0.010	0.001	0.437	100.0
2.0	0.550	0.887	0.003	0.011	0.003	1.454	100.0
5.0	1.965	3.500	0.002	0.012	0.027	3.500	100.0
10.0	4.585	6.000	0.001	0.014	0.183	6.000	100.0

[a]Calculated by Eq. 9.33b, assuming spheres of standard density at standard conditions.
[b]Calculated by Eq. 9.19.

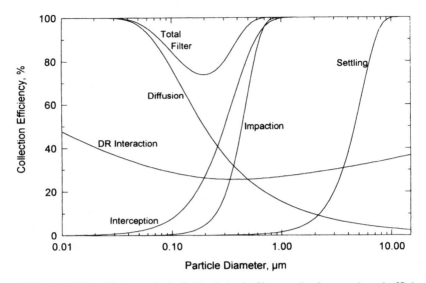

FIGURE 9.8 Filter efficiency for individual single-fiber mechanisms and total efficiency; $t = 1$ mm, $\alpha = 0.05$, $d_f = 2$ μm, and $U_0 = 0.10$ m/s. [10 cm/s].

settling is small compared with the other mechanisms. The collection of particles larger than 0.5 μm is governed by mechanisms that depend on the particle's aerodynamic diameter, but for particles less than 0.5 μm, collection is governed by mechanisms that depend on physical diameter. Figure 9.9 shows the effect of face velocity on filter efficiency as a function of particle size. Generally, decreasing face velocity increases efficiency for particle sizes close to the minimum efficiency.

The particle size that gives the minimum efficiency, about 0.2 μm in Figs. 9.8 and 9.9, is an in-between size that is too large for diffusion to be effective and too small for impaction or interception to be effective. Because these competing mechanisms are most effective in different size ranges, all filters have a particle size that gives minimum efficiency, usually in the range 0.05–0.5 μm. A standard test for the collection efficiency of high-efficiency filters uses 0.3-μm DOP particles, on the assumption that this size is near the minimum efficiency point and efficiency will be greater for all other sizes or size distributions. Lee and Liu (1980) derived the following equations to predict the particle size of minimum efficiency, $\hat{d}_p$, and the minimum single-fiber efficiency $\hat{E}_\Sigma$, based on the assumption that the only important mechanisms near the minimum efficiency are interception and diffusion.

$$\hat{d}_p = 0.885\left[\left(\frac{\mathrm{Ku}}{1-\alpha}\right)\left(\frac{\sqrt{\lambda}\,kT}{\eta}\right)\left(\frac{d_f^2}{U_0}\right)\right]^{2/9} \tag{9.34}$$

$$\hat{E}_\Sigma = 1.44\left[\left(\frac{1-\alpha}{\mathrm{Ku}}\right)^5\left(\frac{\sqrt{\lambda}\,kT}{\eta}\right)^4\left(\frac{1}{U_0^4 d_f^{10}}\right)\right]^{1/9} \tag{9.35}$$

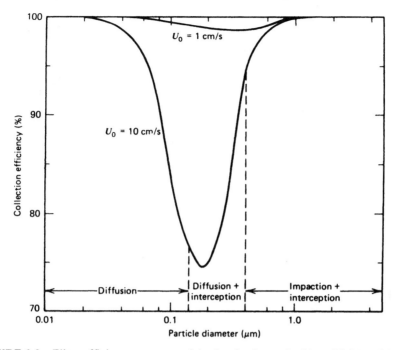

FIGURE 9.9   Filter efficiency versus particle size for face velocities of 0.01 and 0.1 m/s [1 and 10 cm/s]; $t = 1$ mm, $\alpha = 0.05$, and $d_f = 2$ μm.

These equations are valid for $0.075 < \lambda/d_p < 1.3$, where λ is the mean free path of the gas, and $R < 1$. Table 9.2 and Fig. 9.8 support the assumption that interception and diffusion are the only important mechanisms operating when efficiency is near the minimum. The particle size for minimum efficiency is proportional to $(d_f)^{4/9}$, which suggests that $\hat{d}_p$ will decrease as fiber size is decreased. Furthermore, by Eq. 9.35, the minimum efficiency becomes greater as fiber size is decreased. Lee (1981) gives equations similar to Eqs. 9.34 and 9.35 for granular-bed filters.

As shown in Fig. 9.10, there is a face velocity that gives minimum efficiency for a given particle size and filter. As the velocity is increased, collection by diffusion is reduced, but collection by impaction rises. The velocity for minimum efficiency increases with decreasing particle size. Measuring the efficiency of a filter at several velocities will suggest which collection mechanisms are most important for the particle size tested.

Figure 9.11a summarizes the range of particle sizes and face velocities that are of importance for the various filtration mechanisms. The filter conditions used are the same as those for Table 9.2 and Figs. 9.8–9.10. Regions are defined where a particular mechanism contributes more than 20% to the total single-fiber efficiency based on Eq. 9.33b. Horizontal lines on Fig. 9.11 at 1 and 10 cm/s correspond to the lines portrayed in Fig. 9.9. Similarly, vertical lines at 0.2, 0.4, 0.5, and 1 μm, correspond to those in Fig. 9.10. For this filter, diffusion is the only important mechanism for particles smaller than 0.1 μm. Interception contributes significantly

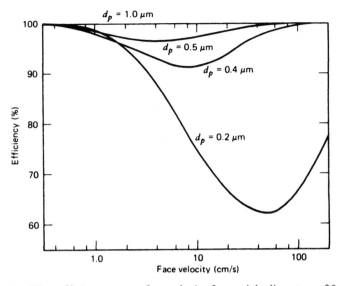

FIGURE 9.10   Filter efficiency versus face velocity for particle diameters of 0.2, 0.4, 0.5, and 1 μm; $t = 1$ mm, $\alpha = 0.05$, and $d_f = 2$ μm.

to the collection of particles larger than 0.1 μm over a wide region. Impaction is important for larger particles and face velocities greater than 1 cm/s. Figure 9.11b shows the overall filter efficiency contours on the same axes as in Fig. 9.11a. The region of minimum efficiency occurs where diffusion and interception are the primary collection mechanisms.

As the fibers in a filter collect dust particles, the particles protrude from the surface of the fiber into the airstream, like short, small-diameter fibers. This situation improves the collection efficiency, but also causes an increase in the resistance to airflow. Eventually, clogging occurs with an excessive increase in resistance. This subject is reviewed by Brown (1993). Preferential collection of particles on previously collected particles has been observed, leading to the development of chains of particles, or *dendrites*, extending from the fiber surface. Liquid particles that spread out on the fiber surface do not show an enhancement in collection efficiency. A small value of solidity has the obvious advantage of providing lots of room for particle collection before clogging occurs. At high face velocities (> 500 cm/s), solid particles may hit the fibers with sufficient force to bounce and continue through the filter. This is not a problem for the usual case of $U_0 < 1.0$ m/s [100 cm/s].

9.5   PRESSURE DROP

In a fibrous filter, the resistance to airflow, or *pressure drop*, across a filter is caused by the combined effect of each fiber resisting the flow of air past it. The pressure

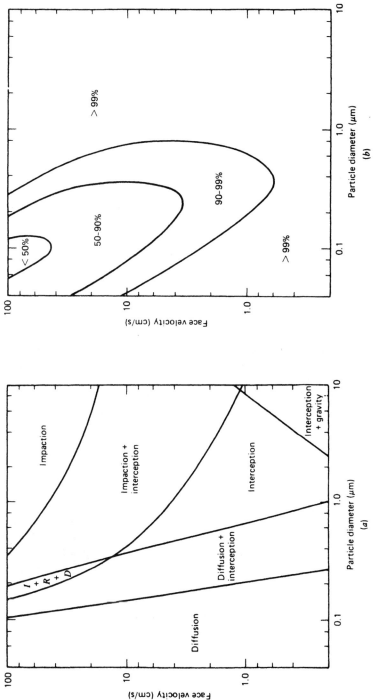

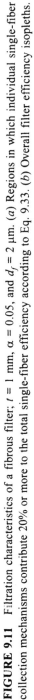

FIGURE 9.11 Filtration characteristics of a fibrous filter; $t = 1$ mm, $\alpha = 0.05$, and $d_f = 2$ μm. (*a*) Regions in which individual single-fiber collection mechanisms contribute 20% or more to the total single-fiber efficiency according to Eq. 9.33. (*b*) Overall filter efficiency isopleths.

drop represents the total drag force of all the fibers. Davies (1973) gives the pressure drop as

$$\Delta p = \frac{\eta t U_0 f(\alpha)}{d_f^2} \qquad (9.36)$$

where

$$f(\alpha) = 64\alpha^{1.5}(1 + 56\alpha^3) \quad \text{for } 0.006 < \alpha < 0.3 \qquad (9.37)$$

Pressure drop is thus directly proportional to t and inversely proportional to d_f^2. Pressure drop is also directly proportional to face velocity, as expected for laminar flow inside the filter. The pressure drop for filters having fiber sizes less than 1 μm is slightly less than that predicted by Eq. 9.36, because of the effect of gas slip at the fiber surface. For example, a filter having a d_f of 0.1 μm and α of 0.05 has a Δp that is 70% of that predicted by Eq. 9.36.

Equation 9.37 is based on an empirical correlation, so it includes nonideal effects, such as a nonperpendicular alignment of the fibers to the flow. Measurements of Δp, t, and α can be used in Eq. 9.36 to determine an effective fiber diameter, which in turn can be used to estimate efficiency by the methods of Sections 9.2–9.4.

Fig. 9.12 illustrates the effect of fiber size on collection efficiency. In the figure, solidity is held constant, and the thickness of each filter is adjusted, so that all of the filters have the same pressure drop. In this situation, an increase in efficiency for a given particle size is equivalent to an increase in filter quality. As the fiber size is decreased, the particle size for minimum efficiency decreases and the minimum efficiency increases. Thus, filter quality increases with decreasing fiber size for $d_p > 0.2$ μm.

9.6 MEMBRANE FILTERS

As mentioned in Section 9.1, porous membrane filters are important for aerosol sampling. These filters come in pore sizes of 0.01–10 μm, pore densities of 10^4–10^6/mm² [10^6–10^8/cm²], and solidities of 0.15–0.30. Their efficiency is high for all particle sizes. The most common material for such filters is cellulose ester, but other materials, such as polyvinyl chloride, nylon, polytetrafluoroethylene (PTFE), and sintered metals, also are commercially available.

Experimental evaluation of porous membrane filters has demonstrated that porous membrane filtration is equivalent to fibrous filtration for filters having the same thickness and solidity, and an effective fiber diameter that is slightly less than the pore size (Rubow, 1981). For example, a membrane filter with a pore size of 0.8 μm has an effective fiber diameter of 0.55 μm. The structural elements between the pores, shown in Fig. 9.2, act like fibers and have dimensions that are approximately that of their effective fiber diameter. The mechanisms of collection and their region

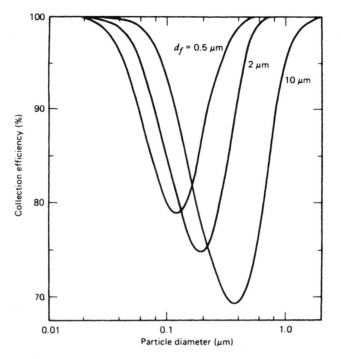

FIGURE 9.12 The effect of fiber size on filter efficiency as a function of particle size; $\alpha = 0.05$ and $U_0 = 0.1$ m/s [10 cm/s]. Filter thickness has been adjusted so that all three filters have the same pressure drop, as calculated by Eq. 9.36. The parameter is the fiber diameter.

of application are the same as for fibrous filters with the same properties. Measured pressure drops for porous membrane filters agree with those calculated by Eq. 9.36 using the effective fiber diameters for these filters.

Capillary pore membrane filters (Fig. 9.3) are thin (6–11 μm) polycarbonate films having 1000–10^7 pores/mm² [10^5–10^9 pores/cm²]. The pores are uniform cylindrical holes that are approximately perpendicular to the filter surface. Pore sizes range from 0.015 to 14 μm in diameter. The smooth surface of the polycarbonate film makes it an ideal sample collector for optical and scanning electron microscopy and surface analytical methods such as X-ray fluorescence. Because of the simple geometry of these filters, the collection efficiency due to diffusion to the pore walls (neglecting inlet effects) can be calculated by Eq. 7.29. Gravity settling is the same as in tubes. Impaction and interception occur primarily at the inlet to the pores, where streamlines pass close to the rim of each pore. The overall efficiency of capillary pore membrane filters is less than 100%—often considerably less—for particles smaller than the pore size and larger than about 0.01 μm. Minimum efficiencies are exhibited at about 0.05 μm. For larger pore sizes, liquid particles, show higher efficiency than solid particles, due to bounce at the pore rim.

The selection of a filter medium for a specific sampling application depends on the collection efficiency for the particle sizes of interest, flow rate versus pressure drop, and the method of analysis of the collected particles. The latter, which includes gravimetric, microscopic, and chemical analysis, is covered in Section 10.4.

PROBLEMS

9.1 A filter has a fiber diameter of 10 μm and a solidity of 1%. What face velocity will give a particle size of minimum efficiency of 0.5 μm?

ANSWER: 0.12 m/s [12 cm/s].

9.2 What minimum efficiency is predicted by Eq. 9.35 for a filter with $t = 1$ mm, $\alpha = 0.05$, $d_f = 2$ μm, and $U_0 = 0.1$ m/s [10 cm/s]?

ANSWER: 56%

9.3 Determine the relative importance of single-fiber efficiency due to interception, impaction, and diffusion in collecting standard-density particles 0.5 μm in diameter. The filter has an effective fiber diameter of 1.0 μm and a solidity of 0.02. $U_0 = 0.5$ m/s [50 cm/s].

ANSWER: 28%, 71%, 1%.

9.4 What face velocity is required to achieve 50% collection of 0.2-μm-aerodynamic diameter particles by *impaction?* Assume that the fiber diameter is 4 μm, the filter solidity is 0.05, and the filter thickness is 2 mm.

ANSWER: 8.5 m/s [850 cm/s].

9.5 What is the effective fiber diameter of a membrane filter with a pore size of 0.45-μm? The solidity is 0.20, the thickness is 0.15 mm, and the pressure drop is 1 cm Hg at $U_0 = 0.015$ m/s [1.5 cm/s].

ANSWER: 0.50 μm.

REFERENCES

Brown, R.C., *Air Filtration: An Integrated Approach to the Theory and Applications of Fibrous Filters*, Pergamon, Oxford, U.K., 1993.

Davies, C. N. (Ed.), *Air Filtration*, Academic Press, London, 1973.

Kirsch, A .A., and Fuchs, N. A., "Studies of Fibrous Filters—III: Diffusional Deposition of Aerosols in Fibrous Filters," *Ann. Occup. Hyg.*, 11, 299–304 (1968).

Kirsch, A. A., and Stechkina, I. B., "The Theory of Aerosol Filtration with Fibrous Filters," in Shaw, D. T. (Ed.), *Fundamentals of Aerosol Science*, Wiley, New York, 1978.

Lee, K. W., "Maximum Penetration of Aerosol Particles in Granular Bed Filters," *J. Aerosol Sci.*, 12, 79–87 (1981).

Lee, K. W., and Ramamurthi, M., "Filter Collection," in Willeke, K. and Baron, P. A. (Eds.), *Aerosol Measurement: Principles, Techniques, and Applications*, Van Nostrand Reinhold, New York, 1993.

Lee, K. W., and Liu, B. Y. H., "On the Minimum Efficiency and Most Penetrating Particle Size for Fibrous Filters," *J. Air Poll. Control Assoc.*, **30**, 377–381 (1980).

Lippman, M., "Filters and Filter Holders," in *Air Sampling Instruments for Evaluation of Atmospheric Contaminants*, 8th ed., American Conference of Governmental Industrial Hygienists, Cincinnati, 1995.

Spurny, K. R., *Advances in Aerosol Filtration*, CRC/Lewis, Boca Raton, FL, 1998.

Rubow, K., "Submicrometer Aerosol Filtration Characteristics of Membrane Filters," Ph.D. thesis, University of Minnesota, 1981.

Tien, C., *Granular Filtration of Aerosols and Hydrosols*, Butterworth, Boston, 1989.

Yeh, H. C., and Liu, B. Y. H., "Aerosol Filtration by Fibrous Filters," *J. Aerosol Sci.*, **5**, 191–217 (1974).

10 Sampling and Measurement of Concentration

An essential element in the study of aerosols is the ability to collect representative samples for analysis. These samples must accurately reflect the airborne particles in both concentration and size distribution. In this chapter, we address this problem and the sampling-related aspects of determining mass and number concentration.

10.1 ISOKINETIC SAMPLING

Isokinetic sampling is a procedure to ensure that a representative sample of aerosol enters the inlet of a sampling tube when sampling from a moving aerosol stream. The sampling may be extractive—for example, when a probe is used to obtain a sample from a duct or stack—or direct—for instance, when one takes samples in a windy environment. Sampling is isokinetic when the inlet axis of the sampler, a thin-walled tube or probe, is aligned parallel to the gas streamlines and the gas velocity entering the probe equals the free-stream velocity approaching the inlet. As shown in Fig. 10.1, this condition is equivalent to taking a sample so that there is no distortion of the streamlines just upstream of the inlet. If sampling is isokinetic, there is no particle loss at the inlet, regardless of particle size or inertia. Isokinetic sampling in no way ensures that there are no losses between the inlet and the collector; instead, it guarantees only that the concentration and size distribution of the aerosol entering the tube is the same as that in the flowing stream.

Anisokinetic sampling, a failure to sample isokinetically, may result in a distortion of the size distribution and a biased estimate of the concentration. These effects arise because of particle inertia in the region of curved streamlines near the inlet. Depending on the conditions, the sample may contain an excess or a deficiency of large particles. Figure 10.2 shows three conditions of anisokinetic sampling. In Fig. 10.2a, the probe is not aligned with the gas flow streamlines; in Fig. 10.2b, the velocity in the probe exceeds the stream velocity, *superisokinetic* sampling; and in Fig. 10.2c, the stream velocity exceeds the velocity in the probe, *subisokinetic* sampling.

If sampling is not done isokinetically, there is no way to determine the true concentration unless the original particle size distribution is known or can be estimated. The errors resulting from anisokinetic sampling for a given particle size can be

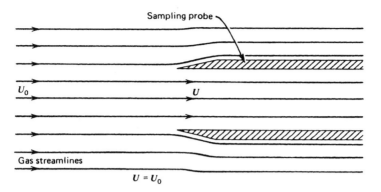

FIGURE 10.1 Isokinetic sampling.

estimated by the equations that follow in this section; losses in sampling tubes are covered in Section 10.3.

For a free-stream velocity U_0 and a gas velocity U in the probe, the isokinetic condition for a properly aligned probe is

$$U = U_0 \qquad (10.1)$$

To meet this condition, the flow rates in the duct and probe must be proportional to their respective cross-sectional areas. If a sampling probe of diameter D_s is used at a sampling flow rate Q_s in a circular duct of diameter D_0 carrying a flow rate Q_0, Eq. 10.1 becomes

$$\frac{Q_s}{Q_0} = \left(\frac{D_s}{D_0}\right)^2 \qquad (10.2)$$

EXAMPLE

What is the isokinetic sampling flow rate for a 20-mm-diameter probe in a 0.3-m [30-cm] duct carrying 1 m³/s?

$$Q_s = 1 \times \left(\frac{0.02}{0.3}\right)^2 = 0.0044 \text{ m}^3/\text{s} = \left[10^6\left(\frac{2}{30}\right)^2 = 4400 \text{ cm}^3/\text{s}\right]$$

In some situations, such as sampling with a cascade impactor, the sample flow rate is fixed, and the diameter of the sampling probe must be selected to ensure that the sampling is done isokinetically. Sampling probes are available with a set of interchangeable inlets of different diameters to enable the user to match the sampling velocity to the stream velocity. The *maximum* sampling error occurs when the

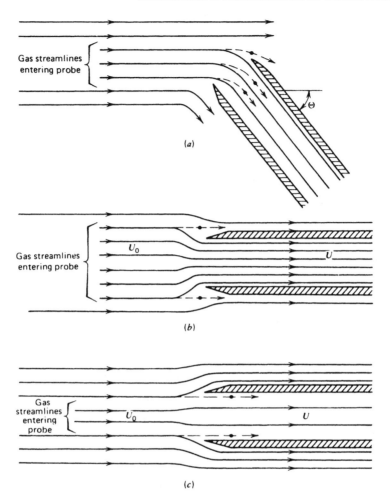

FIGURE 10.2   Anisokinetic sampling. (a) Misalignment, $\Theta \neq 0$. (b) Superisokinetic sampling, $U > U_0$. (c) Subisokinetic sampling, $U < U_0$.

particles have so much inertia that they continue in a straight line as the gas curves into the inlet. At the other extreme, particles with negligible inertia have no sampling error, because they follow the gas streamlines perfectly.

When the sampling flow rate is isokinetic, but the probe is misaligned, the concentration will be underestimated. As shown in Fig. 10.2a, particles with high inertia originally in the volume of air that is sampled will be unable to turn quickly enough to enter the inlet and will travel beyond the inlet and not be included in the sample.

If Θ is the angle of misalignment of the probe, as shown in Fig. 10.2a, then the *maximum* error in concentration that can result from sampling at the isokinetic flow rate with a misaligned probe is

$$\frac{C}{C_0} = \cos \Theta, \quad \text{for } 0° < \Theta < 90° \text{ and Stk} > 6 \tag{10.3}$$

where C is the concentration in the probe, C_0 is the concentration in the free stream, and Stk is the Stokes number for the inlet.

$$\text{Stk} = \frac{\tau U_0}{D_s} \tag{10.4}$$

When Stk < 0.01, particle inertia is negligible and the concentration ratio $C/C_0 = 1$. For sampling conditions such that $0.01 < \text{Stk} < 6$, the situation is more complicated. Durham and Lundgren (1980) give the following empirical equation for probes that are misaligned ($0° \leq \Theta \leq 90°$), but are operated at the correct (isokinetic) flow rate,

$$\frac{C}{C_0} = 1 + (\cos \Theta - 1)\left(1 - \frac{1}{1 + 0.55(\text{Stk}') \exp(0.25\text{Stk}')}\right) \tag{10.5}$$

where

$$\text{Stk}' = \text{Stk} \exp(0.022 \Theta)$$

for Θ in degrees. Equation 10.5 gives a value of 1 for C/C_0 when $\Theta = 0$ or Stk = 0. Figure 10.3 shows the effect, calculated by Eq. 10.5, of the Stokes number and mis-

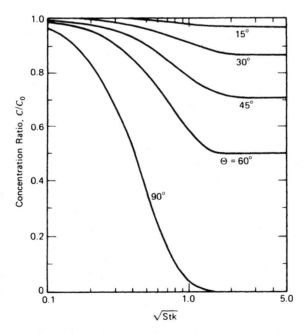

FIGURE 10.3 Effect of misalignment of the probe on concentration ratio for $U = U_0$.

alignment angle on the concentration ratio C/C_0 when $U = U_0$. As was done for cascade impactors in Chapter 5, the concentration ratio is plotted against the square root of the Stokes number, because this quantity is directly proportional to particle size. The particle size corresponding to a particular value of Stk depends on the actual sampling conditions. For example, referring to Fig. 10.3, we see that values of Stk of 0.1, 1.0, and 5.0 correspond to aerodynamic diameters of 1.8, 18, and 90 μm, respectively, for a 1-cm-diameter probe sampling from a stream moving at 1000 cm/s. A misalignment of less than 15° produces a small error in measured concentration for all particle sizes. Since this amount of misalignment is readily discernible, positioning of a sampling probe by eye is adequate for situations in which the direction of the streamlines is known.

When the probe is correctly aligned, but the probe inlet velocity exceeds the stream velocity (Fig. 10.2b), some particles with high inertia originally in the volume sampled cannot follow the converging streamlines to enter the probe and are lost from the sample. Thus, a sampling flow rate that is greater than the isokinetic sampling flow rate (super isokinetic) underestimates the true concentration.

As with misalignment, the *maximum* sampling error occurs for large, heavy particles that continue in a straight line. For this situation, the concentration ratio equals the cross-sectional area of the cylinder of particles that actually enter the probe, divided by the cross-sectional area of the cylinder of sampled gas approaching the probe at a velocity U_0. From Figs. 10.2b (and c) and Eq. 2.46,

$$\frac{C}{C_0} = \frac{(\pi/4)D_s^2}{(U/U_0)(\pi/4)D_s^2} = \frac{U_0}{U}, \quad \text{for Stk} > 6 \qquad (10.6)$$

For a properly aligned sampling probe, the concentration ratio is given by Belyaev and Levin (1974) as

$$\frac{C}{C_0} = 1 + \left(\frac{U_0}{U} - 1\right)\left(1 - \frac{1}{1 + (2 + 0.62\,U/U_0)\text{Stk}}\right) \qquad (10.7)$$

Equation 10.7 yields $C/C_0 = 1$ when $U_0/U = 1$ or when Stk = 0.

When the velocity entering the sampling probe is less than that passing by the probe, the streamlines diverge at the inlet, as shown in Fig. 10.2c, and some large particles that were not originally in the volume of gas that is sampled will travel into the probe and be included in the sample. In this situation, with the sampling flow rate less than isokinetic, the concentration will be overestimated (assuming, of course, that the actual sampling flow rate is known and is used in the calculation of concentration). The effect of this subisokinetic sampling situation on the concentration ratio C/C_0 is also given by Eqs. 10.6 and 10.7.

Figure 10.4 shows the effect of velocity mismatch on the concentration ratio as a function of the square root of the Stokes number. Figure 10.5 gives the same information, but concentration ratio is plotted against velocity ratio for various values of the Stokes number. When sampling conditions are such that Stk < 0.01 and $0.2 < U_0/U < 5$, the anisokinetic loss is negligible and $C/C_0 \approx 1$.

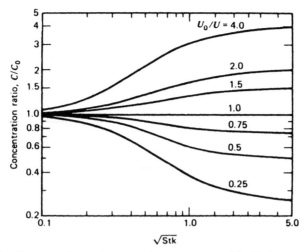

FIGURE 10.4 Concentration ratio versus the square root of the Stokes number for several values of the velocity ratio ($\Theta = 0°$).

EXAMPLE

What is the concentration ratio for 20-μm particles of standard density sampled at 8 L/min by a 10-mm-diameter thin-wall probe from a duct having a flow velocity of 10 m/s [1000 cm/s]? Assume that the probe is correctly aligned.

$$U = \frac{0.008}{60(\pi/4)0.01^2} = 1.70 \text{ m/s} \quad \left[\frac{8000}{60(\pi/4)1^2} = 170 \text{ cm/s} \right]$$

$$U_0/U = 10/1.70 = 5.88, \quad U/U_0 = 0.170$$

$$\text{Stk} = \frac{\tau U_0}{D_s} = \frac{1.24 \times 10^{-3} \times 10}{0.01} = 1.24 \quad \left[\frac{1.24 \times 10^{-3} \times 1000}{1} = 1.24 \right]$$

$$\frac{C}{C_0} = 1 + (5.88 - 1)\left(1 - \frac{1}{1 + (2 + 0.62(0.170))1.24}\right)$$

$$= 1 + 4.88(1 + 0.28) = 4.5$$

When *both* a velocity mismatch and a misalignment of the probe are present, the situation is slightly more complicated. [See Brockmann (1993).] The limiting concentration ratio is then

$$\frac{C}{C_0} = \frac{U_0}{U} \cos \Theta \quad \text{for } 0° \leq \Theta \leq 90° \text{ and Stk} > 6 \tag{10.8}$$

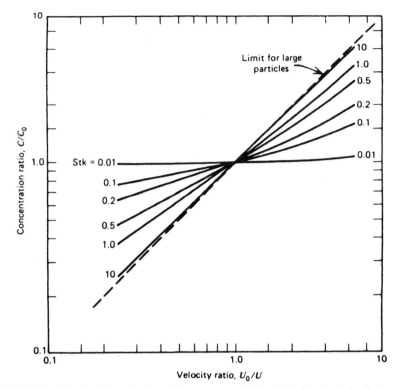

FIGURE 10.5   Effect of the velocity ratio on concentration ratio for several values of the Stokes number ($\Theta = 0°$).

Equation 10.8 indicates that, for a given misalignment, there is a velocity mismatch that will just compensate to give a concentration ratio of unity.

   The preceding analysis assumes that the probe is a thin-walled tube with a sharp-edged inlet. Blunt samplers are considerably more complicated to analyze. [See Vincent (1989).] The analysis also assumes that the flow is laminar, but, as a practical matter, the relationships hold for turbulent flow as well, because the instantaneous fluctuations in velocity rarely exceed ±10% of the mean velocity and average to zero. Isokinetic sampling from stacks is further complicated by the fact that the sample flow is usually measured after the gas has cooled and changed composition. This change in volumetric flow must be accounted for in the calculation of the correct isokinetic sampling rate.

   To take accurate samples in locations that are less than 5–10 duct diameters downstream from bends or obstructions in a duct requires multiple samples from different regions of the duct cross section. This measurement is usually facilitated by dividing the cross-sectional area of the duct into segments of equal area, determining U_0, and sampling with the correct velocity for each segment. Some isokinetic probes combine a pitot tube for measuring velocity and a sampling probe in one unit, to provide continuous measurement of velocity at the probe. Automated versions continuously sense velocity and adjust the sampling flow rate to maintain isokinetic conditions.

10.2 SAMPLING FROM STILL AIR

Closely related to isokinetic sampling is the possibility of a bias when one samples from still or nearly still air. Clearly, the isokinetic criteria are of no value here because they require a zero sampling flow rate in still air. There are two sources of error associated with sampling in still air, one due to the settling velocity of the particles and one due to particle inertia. Davies (1968) established criteria that ensure a negligible sampling error from these two sources of bias.

A bias due to particle settling velocity arises when the inlet velocity is low and the inlet tube faces upwards. Additional particles that were not in the original volume of sampled air will settle into the inlet and cause an overestimation of the concentration. In the extreme case when the sampling flow rate is zero, the collection of settling particles will lead to an infinite sampling error. When the sampling tube is facing down, there will be a corresponding underestimation of concentration. When the axis of the sampling tube is horizontal, there will be no sampling bias due to particle settling velocity.

The criterion for negligible sampling bias due to particle settling for sampling probes in any orientation is given by Davies (1968) as,

$$U \geq 25 V_{TS} \qquad (10.9)$$

where U is the air velocity in the probe. Using Eq. 5.4, we can express Eq. 10.9 in terms of the sample flow rate Q, probe diameter D_s, and particle relaxation time τ as,

$$D_s \leq \frac{2}{5} \left(\frac{Q}{\pi \tau g} \right)^{1/2} \qquad (10.10)$$

For standard conditions, neglecting slip correction, Eq. 10.10 can be written as,

$$D_s \leq 680 \left(\frac{Q^{1/2}}{d_a} \right) \qquad (10.11)$$

for D_s in mm, Q in m³/h, and d_a (the aerodynamic diameter) in μm. For D_s in cm and Q in cm³/s, the coefficient is 4.1.

The inertia of a particle also can introduce a bias. As a particle approaches an inlet along a curved streamline, its velocity increases, and if it has enough inertia, its stopping (or starting) distance will be large compared with the dimensions of the inlet. Particles undergoing such motion may be unable to enter, or may be thrown across or away from, the inlet and escape collection. This bias will become greater for larger particles and higher inlet velocities.

Using an assumed flow field at the probe inlet, Davies (1968) derived the following criterion to ensure negligible sampling loss due to particle inertia:

$$D_s \geq 10\left(\frac{Q\tau}{4\pi}\right)^{1/3} \qquad (10.12)$$

This inequality can be rewritten for standard conditions and negligible slip correction as,

$$D_s \geq 4.05Q^{1/3}d_a^{2/3} \qquad (10.13)$$

where D is in mm, Q is in m^3/h, and d_a is in μm. For D_s in cm and Q in cm^3/s, the coefficient is 0.062.

Equations 10.10 and 10.12 respectively set upper and lower limits on the probe size that can be used to sample a given particle size at a given flow rate. Figure 10.6 gives the range of allowable probe sizes for a specified sampling flow rate and aerodynamic particle size, based on these criteria. The set of lines increasing to the right

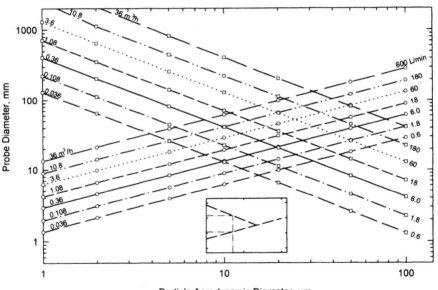

FIGURE 10.6 Probe diameter versus particle size for unbiased sampling in still air at seven flow rates from 0.036 to 36 m^3/h [0.6 to 600 L/min]. Lines with circle symbols represent minimum probe sizes (inertial sampling criterion). Those with square symbols represent maximum probe sizes (settling velocity criterion). The two lines for each flow rate define the range of acceptable probe diameters for particle sizes to the left of their intersection. (See inset.) For particle sizes to the right of the intersection point, the probe size should exceed the upper line (circle symbols) and the probe should be positioned with its axis horizontal.

represents minimum probe sizes necessary to conform with the inertial sampling criterion, Eq. 10.12, for seven sampling flow rates. The set of lines decreasing to the right represents maximum probe sizes necessary to conform to the settling velocity criterion, Eq. 10.10, for the same seven sampling flow rates. The triangular region to the left of the intersection point for the two lines associated with a given flow rate defines a range of acceptable probe diameters for the indicated particle size. To the right of the intersection point for a given flow rate, both criteria cannot be met simultaneously; the maximum conforming probe size is smaller than the minimum conforming probe size. The maximum probe size is defined by the settling velocity criterion, Eq. 10.10, for the worst case, a sampling probe facing upward. For a sampling probe with its axis horizontal, there is no settling velocity bias and the probe size need not be bound by the indicated maximum. Thus, unbiased samples can be obtained by using a horizontal-axis probe whose diameter exceeds the minimum size given by Eq. 10.12 or Fig. 10.6.

Equations 10.9–10.13 and Fig. 10.6 apply to sampling in still air, which, for these criteria, is defined as

$$U_0 \leq \frac{1}{5}\left(\frac{Q}{4\pi\tau^2}\right)^{1/3} \tag{10.14}$$

The maximum air velocity U_0 for which the still-air sampling criteria can be used is given as a function of the sampling rate and d_a in Fig. 10.7. For air velocities greater than those given by the Figure, isokinetic sampling criteria should be used.

Davies's criteria—particularly the inertial criterion (Eq. 10.12)—are considered by some to be too restrictive; nevertheless, they define conditions that ensures unbiased sampling. Agarwal and Liu (1980) used a mathematical simulation of the flow field around a sampling inlet in still air to calculate that there will be less than a 10% error if the size of an upward-facing probe meets the criterion

$$D_s \geq 20\tau^2 g \tag{10.15}$$

This criterion has the advantage of being relatively simple and independent of the inlet velocity, or sample flow rate. It is much less restrictive than Davies' criteria and suggests that there are no practical restrictions on still-air sampling for particles less than 100 μm in aerodynamic diameter.

It must be emphasized that the foregoing criteria apply only to the efficient entry of particles into a sampling probe and do not reflect losses elsewhere in the sampling system, see Section 10.3.

EXAMPLE

In sampling 5-μm particles from still air at 0.36 m³/h [6 L/min], what range of probe diameters will meet Davies's criteria? Assume standard conditions.
Using Eq. 10.11 and 10.13,

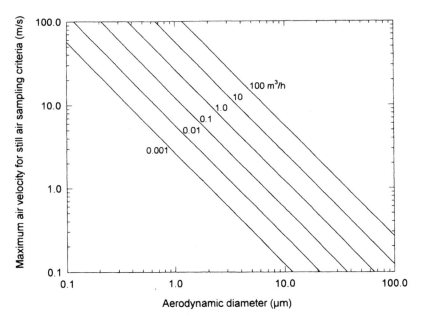

FIGURE 10.7 Maximum ambient air velocity for use with still-air sampling criteria versus aerodynamic diameter for sampling flow rates of 0.001 to 100 m³/h. Isokinetic sampling should be used if the air velocity is greater than that indicated by the graph (1 m³/h = 16.7 L/min).

$$D_s \le \frac{680(0.36)^{1/2}}{5} = 82 \text{ mm} \qquad \left[\le \frac{4.1(6000/60)^{1/2}}{5} = 8.2 \text{ cm} \right]$$

$$D_s \ge 4.05(0.36)^{1.3} \times 5^{2/3} = 8.4 \text{ mm} \qquad [\ge 0.062(6000/60)^{1/3} \times 5^{2/3} = 0.84 \text{ cm}]$$

These results agree with results obtained from Figure 10.6, see inset.

10.3 TRANSPORT LOSSES

Even with efficient entry of particles into a sampler inlet, particles may be lost in the tubing and fittings between the inlet and the collection or measuring device. Losses can occur on the inside of the sampling probe, right near the inlet, because of the curving streamlines entering the inlet, or along the conducting tubing. Both types of losses are reviewed by Brockmann (1993).

Losses in sample lines are a result of the five deposition mechanisms covered in Section 9.3. Each mechanism can be important, although interception is usually small compared with the others. In addition, thermophoresis can be important when one samples a hot aerosol stream through cooler tubing. For a given particle size, losses due to the different mechanisms can be combined in a manner analogous to that for the single-fiber mechanisms in filters (Eq. 9.33). In general, losses are reduced by making the transport path as short and straight as possible and using the same diameter tubing throughout. For some sampling situations, it is desirable to dilute the aerosol stream with clean air to reduce coagulation or temperature or to prevent condensation.

For laminar flow in tubes with a circular cross section, particle loss by diffusion to the tube walls can be calculated by Eq. 7.29. For turbulent flow, loss to the tube walls also can occur by diffusion (Eqs. 7.31 and 7.32) or by inertial projection through the boundary layer. [See the discussion at the end of Section 7.4.] Settling loss in rectangular channels with uniform flow is equivalent to that in the horizontal elutriator, covered in Section 3.9. For laminar flow in tubes of circular cross section, Fuchs (1964) gives the loss due to settling as

$$\text{Settling loss} = \frac{2}{\pi}\left(2k_1 k_2 - k_1^{1/3} k_2 + \arcsin(k_1^{1/3})\right) \qquad (10.16)$$

where

$$k_1 = \left(\frac{3LV_{TS}}{4D_s U}\right)\cos\theta \qquad k_2 = (1 - k_1^{2/3})^{1/2}$$

and L is the tube length, θ is the inclination of the tube from the horizontal, and $\arcsin\theta$ is in radians. For laminar flow, the inertial deposition of particles in a tube bend is given by the empirical equation (Crane and Evans, 1977)

$$\text{Bend loss} = (\text{Stk})\,\phi \qquad (10.17)$$

where ϕ is the bend angle in radians and $\text{Stk} = \tau U/D_s$. For turbulent flow in a tube bend, an empirical equation by Pui et al. (1987) gives

$$\text{Bend loss} = 1 - \exp(-2.88(\text{Stk})\,\phi) \qquad (10.18)$$

Electrostatic and thermophoretic depositions are more difficult to predict. Electrostatic velocity towards the wall can be calculated by Eq. 15.15 when the particle charge and field strength are known. Similarly, thermophoretic velocity can be calculated by Eqs. 8.2 and 8.6 if the temperature gradients are known. For turbulent flow, loss by these mechanisms can be estimated using Eq. 15.38 and the calculated value of electrostatic or thermophoretic velocity.

10.4   MEASUREMENT OF MASS CONCENTRATION

Direct measurement of mass concentration is based on sensing particle mass or inertia, for example, by gravimetric analysis. Indirect measurement relies on measuring another property, such as light scattering, that can be related to particle mass (See Section 16.5.) *Direct-reading* instruments provide near-real-time concentration measurements, with reading intervals from less than a second to several minutes. (See Section 10.5.) Filter samples integrate particle mass to give the average aerosol concentration over a period that may range from minutes to days. In the general atmosphere and many indoor environments, the aerosol concentration fluctuates strongly in space and time.

The most accurate and direct way to determine aerosol mass concentration is to pass a known volume of the aerosol through a filter and determine the increase in mass of the filter due to the aerosol particles collected. It is nearly universal to use high-efficiency filters with a collection efficiency of nearly 100% for all particle sizes. For such filters, the collected particle mass is divided by the sample volume to get the mass concentration. This is called "*total*" mass sampling to distinguish it from mass sampling of a subset of all the particle sizes present, which is called *inhalable*, or *respirable*, sampling (see Sections 11.4 and 11.5) for workplace sampling and *PM-10* or *PM-2.5* for ambient air sampling. Particles collected by any of these filtering methods may be further analyzed to determine the mass of any specific elements or compounds present.

Gravimetric analysis of particles collected on a filter is a simple, low-cost, accurate, and widely used method for determining mass concentration. It requires the ability to accurately weigh a filter before and after sampling and accurately measure the sampling flow rate and sampling time. The measurement of flow rate is discussed in Section 2.6.

Usually, an analytical balance or an Electrobalance® (Cahn Instruments, Inc., Cerritos, CA) is used. An analytical balance is a mechanical beam balance that has a sensitivity of 0.1 mg (0.01 mg on specially equipped models) and a capacity of more than 100 g. The Electrobalance is more sensitive, using the torque on a coil in a magnetic field to balance a mechanical beam. Direct measurement of the current in the coil is employed to determine the sample weight. Sensitivity ranges from 0.1 μg on the most sensitive scale (0–25 mg) to 10 μg on the least sensitive scale (0–1250 mg). For weighing most air-sampling filters, 10 μg is a practical limit of sensitivity.

Common filter sizes are 13, 25, 37, and 47 mm in diameter for circular disks and 200 × 250-mm (8 × 10-in.) rectangular sheets for high-volume samplers. It is advantageous to use a filter with a low tare weight because the collected particles represent a larger proportion of the total weight, and therefore, the relative effect of a change in filter weight caused by moisture or temperature is less. Table 10.1 gives information on weight and weight stability for several common types of filter. The accumulated dust on a filter must be about 0.5 mg for a concentration measurement not to be unduly affected by the weight stability of the filter. The sample volume must be adjusted, by adjusting the time or flow rate, so that the collected mass is large compared with the variability in filter weight given in the table.

The ideal way to weigh filters is to locate the balance in a chamber or room with controlled temperature (±1°C) and humidity (±5% for a relative humidity between 30 and 50%) and permit the filters to equilibrate for at least 1 h (24 h is often used) in this environment before each weighing. Any change in humidity between initial and final weighings results in an error in the net weight. Charell and Hawley (1981) found that at 20°C, 47-mm cellulose ester membrane filters gain 28–35 μg and PVC

TABLE 10.1 Weight Stability Characteristics of Some Common Types of Filters[a]

Type of Filter	Average Weight 37 mm diameter (mg)	Weight Stability at Standard Conditions[b] (mg)	Average Change in Weight Following 24 hr in a Dessiccator[c] (mg)	Average Change in Weight following 24 hr at 80% RH[c] (mg)
Gelman glass filter type A/E (without organic binder)	86	0.14	−0.02	+0.01
MSA #457193; 5-μm pore-size PVC membrane with fibrous backup filter	243	0.04	−0.04	0
Millipore AA; 0.8-μm pore-size cellulose ester membrane	53	0.13	0	+0.35
Millipore Teflon; 5-μm pore-size Teflon membrane	108	0.02	+0.01	0
Selas Flotronics FM-37; 0.8-μm pore-size silver membrane	460	0.03	0	+0.01

[a]From Lowry and Tillery (1979)
[b]95% confidence interval based on three measurements a day for six filters of each type over 30 days.
[c]Filters were equilibrated at room conditions for 24 hr, weighed, desiccated or humidified for 24 hr, equilibrated at room conditions for 24 hr, and reweighed.

filters gain about 0.7 μg for each percent increase in relative humidity. Three blank (control) filters and a reference mass of approximately the same weight as the filters should be weighed each time the sample filters are weighed.

Static charge can accumulate on filters and cause an error in weighing. It is common practice to hold the filter near an alpha radiation source, such as polonium-210, to neutralize the accumulated charge before each weighing.

When a sample is taken to determine whether the concentration is below a certain concentration standard C'_m, the sample volume should be sufficiently large that the collected mass (if the concentration is at the standard) will be 10 times the sensitivity of the analytical method. For gravimetric analysis, this condition is equivalent to

$$\text{sample volume} = \frac{10S}{C'_m} \qquad (10.19)$$

where S is the sensitivity of the balance. This requirement assumes perfect sampling conditions and does not address the statistical problems associated with varying concentrations.

Filter holders for common filter sizes may be open-face or in-line, as shown in Fig. 10.8. The former must be used to collect samples for microscopic analysis because they ensure that the particles are uniformly distributed over the filter. Generally, open-face filter holders reduce the possibility of losses in the inlet section. The in-line, or closed-face, type is used when the samples are extracted by means of a probe or in air supply lines to ensure a particle-free air supply for equipment.

Personal sampling is commonly used in the field of occupational hygiene to evaluate workers' exposure to aerosols by measuring their concentration in the worker's breathing zone, which is defined as the region within 0.3 m [30 cm] of the worker's mouth. Personal sampling is facilitated by lightweight, battery-powered pumps (see Section 10.7) and plastic filter holders (Fig. 10.8). In-line 37-mm filter holders are recommended by NIOSH for most personal sampling because they protect the filter from accidental contamination and mechanical abuse (Doemeny and Smith, 1981). Personal sampling is necessary whenever local sources cause a

FIGURE 10.8 Open-face and in-line 47-mm filter holders and 37-mm disposable filter cassettes.

large spatial variation in concentration. When the concentration is reasonably uniform, *area* or *fixed sampling* can be used. Filter holders are mounted at breathing zone height on a stand or tripod. Fairchild et al. (1980) found greater sampling errors for the open-face type than the in-line type. Both types may have significant wall deposition for particles larger than a few micrometers.

It is important that the rim of the filter be well sealed to eliminate any possibility of leakage from the outside or bypassing around the filter. A screen or support pad is required to prevent glass fiber or membrane filters from rupturing. Open-face or in-line disposable plastic filter holders are available in 25 and 37 mm sizes. Commercially available filter holders are reviewed by Lippmann (1995).

The high-volume sampler with a PM-10 inlet is the most widely used device for sampling ambient air to evaluate air quality. The inlet (see Section 11.5) is omnidirectional and uses a combination of inlet losses, impaction, and elutriation to achieve the required PM-10 cutoff. This type of sampler operates at 0.188 m³/s [40 ft³/min, 1130 L/min]. The PM-10 dichotomous (dichot) sampler uses a PM-10 inlet and a virtual impactor to separate the sampled aerosol—the PM-10 subfraction—into fractions greater and less than 2.5 μm. Its sampling flow rate is 1.0 m³/hr [16.7 L/min].

Stack sampling, also called stationary-source sampling, for particulates is usually conducted following EPA Method 5 or 17 (U.S. Environmental Protection Agency, 1992). The setup for Method 5 is shown in Fig. 10.9. The isokinetic sam-

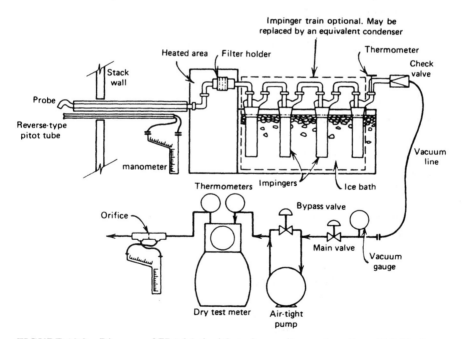

FIGURE 10.9 Diagram of EPA Method 5 stack-sampling system. From U.S. Environmental Protection Agency (1992).

pling probe has an integral pitot tube for monitoring velocity at the sampling location. The flow path from the inlet to the filter is lined with glass and heated to prevent condensation. The filter holder is in a heated box. Downstream of the filter holder, the stream is cooled to remove condensable gases and then goes to the pump and flow measuring units. The simpler Method 17 can be used when the particulate mass concentration is independent of temperature. This method is the same as Method 5, except that the filter holder is positioned in the stack a few centimeters downstream from the inlet. The heated box and tube are not needed. The rest of the system is the same as Method 5.

10.5   DIRECT-READING INSTRUMENTS

Gravimetric analysis of filter samples requires that the filters be weighed before and after sampling, usually at a location remote from the sampling site. This process may involve substantial delay, so there is a need, particularly in occupational hygiene work, for portable direct-reading instruments (survey instruments) that can measure mass concentration in a few minutes at the sampling site. In general, direct-reading instruments are less accurate, but more convenient, than gravimetric filter measurements. Their speed of operation permits assessment of concentrations that are changing in space or time.

The direct-reading mass concentration measuring methods described in this section are covered in detail by Williams et al. (1993). In addition to these instruments, condensation nuclei counters are covered in Chapter 13, electrical mobility analyzers in Chapter 15, and photometers, nephelometers, and optical particle counters in Chapter 16.

Direct-reading instruments that measure mass concentration require a means of collecting particles and a sensitive means of determining their mass. The quartz crystal microbalance collects aerosol particles by electrostatic precipitation or impaction onto the surface of a piezoelectric quartz crystal. The crystal is oscillated at its resonant frequency (5–10 MHz), which decreases linearly with the mass deposited on the surface of the crystal. For one version of this type of instrument, the Piezobalance® (TSI, Inc., St. Paul, MN), shown schematically in Fig 10.10, an increase in mass of only 0.005 µg changes the resonant frequency by 1 Hz. The instrument measures the net change in resonant frequency during a 24- or 120-s sampling period, converts the value obtained to mass concentration, and displays the result on a digital indicator. A reference crystal is used to compensate for changes in temperature. The instrument is suitable for measuring concentrations of 0.05–5 mg/m^3 with an accuracy of ±15% and a precision of ±10%. It weighs 4 kg and has a sampling flow rate of 1 L/min. The portable model can be used with an upstream impactor having a cutoff size of 0.5–10 µm for size-selective or respirable mass sampling. A direct-reading cascade impactor (QCM,® California Measurements, Inc., Sierra Madre, CA) uses this sensing method for each impactor stage.

Although the collection efficiency is high for all particle sizes, solid particles larger than about 10 µm do not completely couple to the crystal surface. When such particles are collected, an underestimation of mass concentration results. Changes

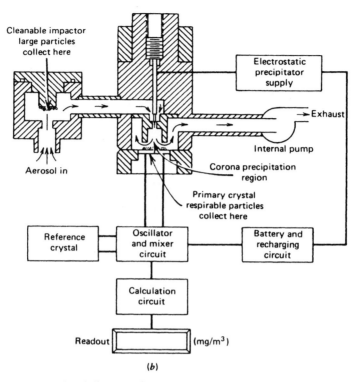

FIGURE 10.10   Functional diagram of the TSI Piezobalance respirable aerosol mass monitor model 3500. Courtesy of TSI, Incorporated, St. Paul, MN.

in air temperature and relative humidity *during* a measurement can cause errors. When the loading on the crystal is heavy enough, several layers of particles (about 5–10 μg), additional particles do not couple well to the crystal. The crystal can be cleaned by careful wiping with moistened sponges. In the portable version of this instrument, this is done semiautomatically with built-in sponges.

Another type of vibrating mass sensor is the Tapered Element Oscillating Microbalance (TEOM®), manufactured by Rupprecht & Patashnick Co., Inc., Albany, NY. The vibrating element is a hollow, tapered glass stalk 100–150 mm long with a 13-mm filter and filter holder attached to the narrow end, as shown in Fig. 10.11. The wide end of the stalk is fixed and the filter end vibrates transversely at a few hundred hertz. Aerosol is drawn through the filter and along the hollow stalk at 0.5 to 5 L/min. The accumulated particle mass collected on the filter decreases the resonant frequency of the element. This change in resonant frequency is sensed electronically, converted to mass concentration, and displayed. The increase in collected mass, Δm, is related to the initial and final frequencies, f_i and f_f, by

$$\Delta m = K_0 \left(\frac{1}{f_f^2} - \frac{1}{f_i^2} \right) \qquad (10.20)$$

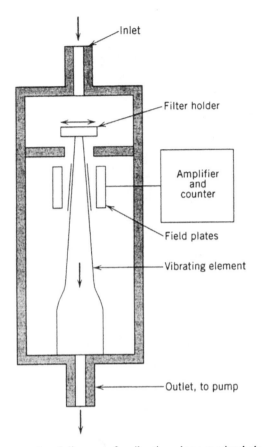

FIGURE 10.11 Cross-sectional diagram of a vibrating-element microbalance for measuring collected aerosol mass.

where K_0 is a calibration constant, specific to each element. The 1400A instrument provides an ambient concentration measurement in $\mu g/m^3$ with an averaging time of 10 min to 24 hr. To maximize the effect of a unit mass of aerosol particles, it is desirable to keep the filters and holder as small and light as possible. The instrument can be used with PM-10, PM-2.5, and PM 1.0 preselector inlets. Its calibration can be checked by adding a known mass to the filter.

Another type of portable survey instrument for measuring mass concentration uses the beta gauge method. Particles are collected on a filter or impacted on a thin Mylar™ film located between a beta source and a beta detector, as shown in Fig. 10.12. The mass of the collected particles is determined by their ability to attenuate beta radiation. The detector measures the radiation from the source before and after the sample is collected. The difference in these measurements is proportional to the deposited particulate mass and is converted to mass concentration and displayed digitally. Most beta gauge instruments have substrates with surface density—that is, mass per unit area—of 100–1000 g/m^2 [10–100 mg/cm^2] and respond well

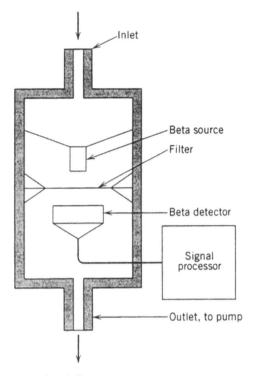

FIGURE 10.12   Cross-sectional diagram of a beta gauge for measuring collected aerosol mass.

to aerosol deposits of 0.2–5 g/m^2 [10–500 mg/cm^2]. Longer sampling times are required for lower concentrations and less narrowly focused particle deposits. Because impactors can deposit particles onto a small spot, they can produce sufficient beta attenuation in 1 min for aerosol concentrations of 1–25 mg/m^3 sampled at 2 L/min. Filter-type instruments with a 1-cm-diameter filter typically require 0.1 to 4 hr to make concentration measurements for this concentration range and flow rate. Measurement accuracy is ±25%. These instruments are not suitable for particles greater than about 10 μm, which completely absorb all the beta radiation incident on them. Either type of instrument can be preceded by a PM-10 or respirable preselector. Impactor type instruments do not collect particles smaller than the impactor cutoff size and this may cause a measurement bias. Beta attenuation is somewhat dependent on the aerosol material, but the range of this effect is only 35% for common aerosol materials.

10.6   MEASUREMENT OF NUMBER CONCENTRATION

Particle number concentration can be determined by sampling particles on a membrane filter (such as a Millipore® AA or HA) and examining the filter with an optical microscope. Cellulose ester membrane filters have two advantages in this application: They collect all particles close to their surface because of their very high efficiency, and they can be made transparent by saturating them with an immersion oil of appropriate refractive index. Counting of the particles collected follows the procedure for microscopic sizing discussed in Chapter 20. The microscope is calibrated so that known areas (fields) are counted. The total number of particles collected on the filter is calculated from the count and the ratio of the filter area to the area of the fields counted. The number concentration is obtained by dividing the total number of particles collected by the total volume of gas sampled. For 47-mm filters, a total collection of about 10^7 particles provides a convenient sample density for microscopic counting.

The standard deviation of the count, σ, is given by

$$\sigma = \sqrt{n} \tag{10.21}$$

where n is the number of particles counted and a Poisson distribution of counts is assumed.

EXAMPLE

A 5-min sample taken at 10 L/min on a 47-mm filter (active diameter = 41 mm) yields a total count of 441 from five microscope fields, each 0.30 mm in diameter. What is the aerosol number concentration?

The total number of particles on the filter is the count scaled up by the ratio of areas.

$$N = 441\left(\frac{41^2}{5 \times 0.30^2}\right) = 1.65 \times 10^6$$

The sample volume is $5 \times 10 = 50$ L $= 0.05$ m^3 = [50,000 cm^3]. The number concentration is

$$\frac{1.65 \times 10^6}{0.05} = 3.3 \times 10^7/\text{m}^3 \quad \left[\frac{1.65 \times 10^6}{50,000} = 33/\text{cm}^3\right]$$

Between 1925 and 1950, the standard method in the United States for measuring number concentration was to collect a sample with an impinger and use a dust-counting cell to determine the number of particles collected. Measurements made in this way were found to correlate with the incidence of respiratory disease observed in mines and other dusty trades. Many of the American Conference of Governmental Industrial Hygienists (ACGIH) threshold limit values for mineral dusts are based on data obtained by this method, and it continued to be recommended by the ACGIH until 1984. The impinger, shown in Fig. 10.13, operates much like an impactor, except that the jet is immersed in water or alcohol. Particles

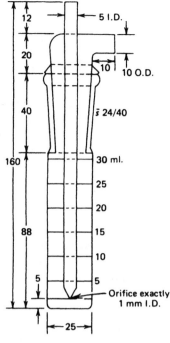

Dimensions (mm)

FIGURE 10.13 Diagram of an all-glass midget impinger. Reprinted with permission from *Ind. Eng. Chem. (Anal. Ed.)*, **16**, 346 (1944). Copyright 1944, American Chemical Society.

larger than about 1 μm are captured by inertial mechanisms and end up suspended in the liquid. The collection efficiency drops off rapidly for particles less than 1 μm, as might be expected from the calculated d_{50} of 0.7 μm for the midget impinger. (See Section 5.5.) Characteristics of the standard (modified Greenberg–Smith) and midget impingers are given in Table 10.2.

After an air sample of known volume is taken, the liquid is brought back up to the standard volume and an aliquot of the liquid is examined in a dust-counting cell. The cell immobilizes a 0. l-mm-thick layer of liquid between glass surfaces to al-

TABLE 10.2 Characteristics of Impingers

	Standard	Midget
Flow rate	28.3 L/min	2.83 L/min
Jet velocity	120 m/s	60 m/s
Nozzle diameter	2.3 mm	1.0 mm
Impingement distance	5 mm	5 mm
Liquid volume	75 mL	10 mL
Pressure drop	9.6 kPa [72 mm Hg]	3.6 kPa [27 mm Hg]
Collection efficiency	high for $d_a > 1$ μm	high for $d_a > 1$ μm

low microscopic examination. The number of particles in a known liquid volume is determined by counting the particles in an area defined by an eyepiece graticule or by a grid etched on the cell base. The total number of collected particles is calculated on the basis of the ratio of the liquid volume counted to the total volume of liquid. The number concentration is calculated by dividing the total number of particles collected by the volume of air sampled, as for filter samples. This method of sampling has been largely replaced by respirable mass sampling. (See Section 11.5.) Particle-counting techniques continue to be used for evaluating bioaerosols (Chapter 18) and asbestos concentrations (Section 20.5). Condensation and optical instruments for measuring number concentration are covered in Sections 13.6 and 16.5, respectively.

10.7   SAMPLING PUMPS

Rubow (1995) provides a review of pumps and air movers suitable for aerosol sampling. The characteristics of four common pumps, shown in Fig. 10.14, that span the range of aerosol-sampling pumps are given in Table 10.3 and Fig. 10.15.

Personal sampling pumps are lightweight, portable, battery-powered pumps designed to be worn on a person's belt. They are used to sample the atmosphere to which a worker is exposed during a full 8-h shift at 1–4 L/min. They have built-in rotameters and flow control valves or constant-flow-rate circuitry. In addition to the personal sampling pumps described in Table 10.3, at least five other, similar pumps are commercially available. Most use a diaphragm-type pump in which a motor drives a diaphragm back and forth to pump the air. They operate relatively quietly. Internal dampers are used to reduce pulsations in the flow rate. Advanced versions provide constant-flow-rate sampling over a wide range of flow rates (e.g., 5–5000 cm^3/min) and programmable control for timed start and stop and intermittent sampling. They also display the sampling time.

FIGURE 10.14   Aerosol-sampling pumps: MSA personal pump, Gast rotary-vane pump, Gast diaphragm vacuum pump, and GMW high-volume sampler.

TABLE 10.3 Characteristics of Some Sampling Pumps

Type of Pump	Power Requirement (W (Hp))	Weight (kg)	Usual Filter Size (mm)	Maximum Flow Rate (L/min)	Maximum Vacuum (kPa (cm Hg))
MSA Monitaire[a] personal sampler, diaphragm	0.7 (0.001)	0.5	37	3.2	12 (9)
SKC 224-PC488[b] personal sampler, diaphragm	1.5 (0.002)	0.9	37	5.0	40 (30)
GAST DOL-101-DB[c] diaphragm vacuum pump	90 (0.12)	8	47	31	80 (60)
GAST 0522-V103-G18DX[c] rotary-vane vacuum pump	190 (0.25)	14	47	110	93 (70)
GMWL-2000[d] high-volume sampler, two-stage, high-speed blower	420 (0.56)	6	200 × 250 (8 in. × 10 in.)	2700[e]	21 (16)[e]

[a]Mine Safety Appliances Co., Pittsburgh, PA.
[b]SKC, Inc., Eighty Four, PA.
[c]Gast Manufacturing Corp., Benton Harbor, MI.
[d]General Motor Works, Inc., Cleves, OH.
[e]Estimated.

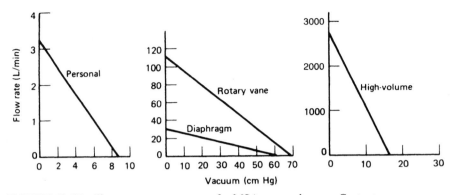

FIGURE 10.15 Flow rate versus vacuum for MSA personal pump, Gast rotary-vane pump, Gast diaphragm vacuum pump, and GMW high-volume sampler.

A larger version of this pump is the diaphragm vacuum pump, which operates on line voltage, 120 or 240 V. It, too, is relatively quiet and has considerable pulsation in the flow. It can also be used to supply air under pressure. Pressure gauges, vacuum gauges, and regulators can be mounted directly on the pump. It has stainless-steel valves that occasionally may have to be cleaned.

The rotary-vane pump does not have valves. Pumping is done by sliding carbon vanes mounted in a hub that rotates in an eccentric chamber. These pumps are noisier than all of the previously described pumps. Lubrication and wear of the vanes generates a carbon and oil mist aerosol in the outlet stream. Rotary-vane pumps are available in free-air capacities of 10–1500 L/min.

The high-volume ("hi-vol") sampler is a two-stage, high-speed blower capable of sampling up to 1700 L/min (60 ft^3/min) through a standard glass fiber high-volume filter (8 × 10 in.). These blowers, mounted in protective shelters, were used for many years as the standard sampler for particulate matter for ambient air quality measurements. The inlet efficiency of the standard high-volume sampler housing with a gable roof depends on wind speed and direction. It is 50% for 30-μm particles at a wind speed of 8 km/h—greater at lower wind speed and less at higher wind speed. These samplers have largely been replaced by PM-10 high-volume samplers operating at 0.19 m^3/s [40 ft^3/min, 1130 L/min]. (See Section 11.5.)

PROBLEMS

10.1 To obtain a sample of highway aerosol, you decide to sample at 40 L/min, using a probe from your car window while driving at 50 mi/hr. You wish to have less than a ±10% error in mass concentration for any size particle due to anisokinetic sampling. What range of probe diameters can be used? Neglect the flow distortion due to flow around the car and assume proper probe alignment.

ANSWER: 5.9–6.5 mm [0.59–0.65 cm].

10.2 What is the maximum particle size that can be sampled at 10 L/min in still air with a 20-mm diameter probe in conformance with Davies's criteria? Neglect slip correction and assume that the probe axis is horizontal.

ANSWER: 14 μm.

10.3 What probe diameter is required to sample isokinetically at 1 ft³/min from a 6-in. diameter duct with a flow rate of 400 ft³/min? Assume proper probe alignment. What error will occur in sampling 20-μm particles at this sampling rate if the probe is misaligned by 30°?

ANSWER: 7.6 mm [0.76 cm], − 11%.

10.4 What range of flow rates will meet Davies's criteria for sampling 18-μm particles in still air with a 33-mm probe?

ANSWER: 0.76–1.7 m³/h [12.6–28 L/min].

10.5 A 30-min sample with a midget impinger yields a total count of 283 particles in 100 small squares of a dust-counting cell. The liquid thickness is 0.1 mm and the small squares are 50 × 50 μm. What is the dust concentration in the sampled air, in number/cm³?

ANSWER: 1330/cm³.

10.6 What concentration error will result if a 1.3-cm probe is used to sample 20-μm particles from a stack at a flow rate of 128 L/min? The velocity in the stack is 25 m/s. Assume correct probe alignment.

ANSWER: +47%.

10.7 A 47-mm-diameter filter holder (active diameter = 40 mm) is used with a Millipore AA filter at a negative pressure of 2.5 cm Hg. What size probe must be connected to this filter holder to sample isokinetically from a gas stream moving at 10 m/s? [Hint: use Table 9.1.]

ANSWER: 3.9 mm [0.39 cm].

10.8 A 10-μm aerosol is sampled isokinetically at 6 L/min through a 10-mm-diameter horizontal tube 0.3 m long. The tube is followed by a 90° bend. What is the particle loss in the straight tube and in the bend? Assume standard density.

ANSWER: 8.8 %, 6.2 %.

REFERENCES

Agarwal, J. K., and Liu, B. Y. H., "A Criterion for Accurate Aerosol Sampling in Calm Air," *Am. Ind. Hyg. Assoc. J.*, **41**, 191–197 (1980).

Belyaev, S. P., and Levin, L. M., "Techniques for collection of representative aerosol samples," *J. Aerosol Sci.*, **5**, 325–338 (1974).

Brockmann, J. E., "Sampling and Transport of Aerosols," in Willeke, K., and Baron P. (Eds.), *Aerosol Measurement*, Van Nostrand Reinhold, New York, 1993.

Charell, P. R., and Hawley, R. E., "Characteristics of Water Adsorption on Air Sampling Filters," *Am. Ind. Hyg. Assoc. J.*, **42**, 353–360 (1981).

Crane, R. I., and Evans, R. L., "Motion of particles in bends of circular pipes," *J. Aerosol Sci.*, **8**, 161–170 (1977).

Davies, C. N., "The Entry of Aerosols into Sampling Tubes and Heads," *Brit. J. Appl. Phys.* (*J. Phys. D*), **1**, 921–932 (1968).

Doemeny, L. J., and Smith J. P., "Letter to the Forum," *Am. Ind. Hyg. Assoc. J.*, **42**, A22 (1981).

Durham, M. D., and Lundgren, D. H., "Evaluation of Aerosol Aspiration Efficiency as a Function of Stokes Number, Velocity Ratio, and Nozzle Angle," *J. Aerosol Sci.*, **11**, 179–188 (1980).

Fairchild, C. I., Tillery, M. I., Smith, J. P., and Valdez, F. O., *Collection Efficiency of Field Sampling Cassettes*, Los Alamos Scientific Laboratory Report No. LA-8646-MS (1980).

Fuchs, N. A., *The Mechanics of Aerosols*, Pergamon, Oxford, 1964.

Fuchs, N. A., "Sampling of Aerosols," *Atmos. Env.*, **9**, 697–707 (1975).

Lippmann, M., "Filters and Filter Holders," in Cohen B. S. and Hering S. V., (Eds.), *Air Sampling Instruments for Evaluation of Atmospheric Contaminants*, 8th ed., ACGIH, Cincinnati, 1995.

Lowry, P. L., and Tillery, M. I., *Filter Weight Stability Evaluation*, Los Alamos Scientific Laboratory Report No. LA-8061-MS (1979).

Pui, D. Y. H., Romay-Novas, F., and Liu, B. Y. H., "Experimental Study of Particle Deposition in Bends of Circular Cross-sections," *Aerosol Sci. Tech.*, **7**, 301–315 (1987).

Rubow, K. "Air Movers and Samplers," in Cohen B. S. and Hering S. V., (Eds.), *Air Sampling Instruments for Evaluation of Atmospheric Contaminants*, 8th ed., ACGIH, Cincinnati, 1995.

U.S. Environmental Protection Agency, *Standards of Performance for New Stationary Sources*, 40CFR60, pp. 195–1142 (Rev. July 1, 1992), U.S. Govt. Printing Office, Washington, DC (1992).

Vincent, J. H., *Aerosol Sampling: Science and Practice*, Wiley, Chichester, U.K., 1989.

Williams, K., Fairchild, C., and Jaklevic, J., "Dynamic Mass Measurement Techniques," in Willeke, K., and Baron P. (Eds.), *Aerosol Measurement*, Van Nostrand Reinhold, New York, 1993.

11 Respiratory Deposition

The hazard caused by inhaled particles depends on their chemical composition and on the site at which they deposit within the respiratory system. Thus, an understanding of how and where particles deposit in our lungs is necessary to evaluate properly the health hazards of aerosols. Such an understanding is also central to the effective administration of pharmaceutical aerosols by inhalation. Humans have evolved effective defense mechanisms against aerosol hazards, and we consider here the first line of defense: mechanisms that restrict access of particles the sensitive regions of the lungs.

The deposition of particles in the lungs relies on the same basic mechanisms that cause collection in a filter, but the relative importance of each mechanism is quite different. While filtration occurs in a fixed system at a steady flow rate, respiratory deposition occurs in a system of changing geometry, with a flow that changes with time and cycles in direction. This added complexity means that predicting deposition from basic theory is much more difficult, and we must rely to a greater extent on experimental data and empirically derived equations. In this chapter, we review the basic mechanisms of particle deposition as they apply to the respiratory system, the characteristics of particle deposition in the lungs, and the entry of particles into the mouth or nose during inhalation. To gain an understanding of these features, it is necessary to review first the characteristics of the human respiratory system.

11.1 THE RESPIRATORY SYSTEM

From the standpoint of respiratory deposition, the respiratory system can be divided into three regions, each covering several anatomical units. These regions differ markedly in structure, airflow patterns, function, retention time, and sensitivity to deposited particles. The first is the *head airways region*, which includes the nose, mouth, pharynx, and larynx. It is also called the extrathoracic or nasopharyngeal region. Inhaled air is warmed and humidified in this region. The second region is the *lung airways* or *tracheobronchial* region, which includes the airways from the trachea to the terminal bronchioles. This region resembles an inverted tree, with a single trunk, the trachea, subdividing into smaller and smaller branches. Finally, beyond the terminal bronchioles is the *pulmonary* or *alveolar* region, where gas exchange takes place.

The respiratory system of a normal adult processes 10–25 m^3 (12–30 kg) of air per day. The surface area for gas exchange is about 75 m^2, half the size of a singles

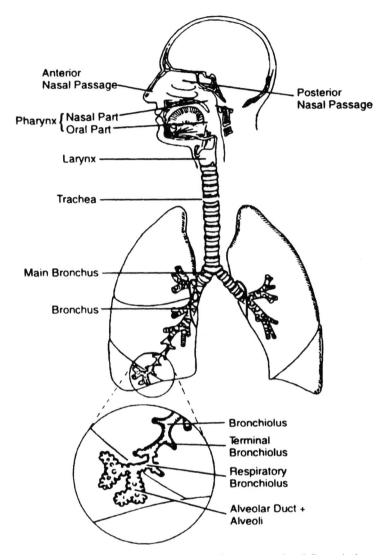

FIGURE 11.1   The respiratory system. Adapted from International Commission on Radiological Protection (1994).

tennis court, and is perfused with more than 2000 km of capillaries. Figure 11.1 shows the parts of the respiratory system. At rest, about 0.5 L of tidal air is inhaled and exhaled with each breath. During heavy work, the tidal volume may exceed three times this amount. A resting adult breathes about 12 times a minute, a rate that triples during heavy work. About 2.4 L of reserve air is not exhaled during normal breathing, but nearly half of this can be exhaled by forced exhalation. Inhaled air follows a flow path that goes through a sequence of 23 airway branchings as it travels from the trachea to the alveolar surfaces. The first 16 branchings take place in the tracheobronchial region and the remainder in the gas exchange region.

Once deposited, particles are retained in the lung for varying times, depending on their physicochemical properties, their location within the lung, and the type of clearance mechanism involved. The airway surfaces of the first two respiratory regions, the head and lung airways, are covered with a layer of mucus that is slowly propelled by ciliary action to the pharynx, where it is subconsciously swallowed to the gastrointestinal tract. This mucociliary escalator transports particles deposited in the airways out of the respiratory system in a matter of hours. This clearance mechanism can be accelerated by low doses of irritating gases or aerosols, or it can be slowed by high doses of such materials or by overloading with particles. Because of its gas exchange function, the alveolar region does not have this protective mucus layer. Hence, insoluble particles deposited in this region are cleared very slowly, over a period of months or years. Dissolved particles pass through the thin alveolar membrane into the bloodstream. Solid particles may dissolve slowly or be engulfed by alveolar macrophages (phagocytic cells) and dissolved or transported to lymph nodes or the mucociliary escalator. Fibrogenic dusts, such as silica, interfere with this clearance mechanism and cause gradual scarring or fibrosis of the alveolar region.

Table 11.1 gives the characteristics of various parts of the lung. During inhalation at a steady rate of 3.6 m^3/hr [1 L/s], the velocity in the airways increases slightly until the air reaches the lobar bronchi. After that, the velocity decreases rapidly during the remaining 70 or 80 mm of transit. This rapid decrease in velocity is a result of the tremendous increase in the total airway cross-sectional area due to the large number of small airways. The total cross-sectional area of the airways expands by a factor of 250 from the lobar bronchi to the respiratory bronchioles, and velocity decreases by the same factor. In normal breathing, freshly inhaled air pushes the residual air ahead of it, so that the fresh air travels only as far as the alveolar ducts. Nevertheless, gas exchange takes place readily by diffusion of O_2 and CO_2 over the very short terminal distances (< 1 mm). In the trachea and main bronchi, the airflow can be turbulent at peak inspiratory and expiratory flow rates for the normal breathing cycle. The remaining airflow is laminar under normal conditions, but, because the airway sections are relatively short compared with their diameters (length $\approx 3\times$ diameter), the airflow in these smaller airways is not fully developed laminar flow. This further complicates mathematical analysis and modeling. The air velocities given in Table 11.1 are based on steady flow, whereas during actual breathing, the airflow rate is continuously changing and reverses direction twice each cycle.

11.2 DEPOSITION

Inhaled particles may deposit in the various regions of the respiratory system by the complex action of the five deposition mechanisms described in Section 9.3 or

TABLE 11.1 Characteristics of Selected Regions of the Lung[a]

Airway	Generation	Number per Generation	Diameter (mm)	Length (mm)	Total Cross Section (cm²)	Velocity[a] (mm/s)	Residence Time[b] (ms)
Trachea	0	1	18	120	2.5	3900	30
Main bronchus	1	2	12	48	2.3	4300	11
Lobar bronchus	2	4	8.3	19	2.1	4600	4.1
Segmental bronchus	4	16	4.5	13	2.5	3900	3.2
Bronchi with cartilage in wall	8	260	1.9	6.4	6.9	1400	4.4
Terminal bronchus	11	2000	1.1	3.9	20	520	7.4
Bronchioles with muscles in wall	14	16,000	0.74	2.3	69	140	16
Terminal bronchiole	16	66,000	0.60	1.6	180	54	31
Respiratory bronchiole	18	0.26×10^6	0.50	1.2	530	19	60
Alveolar duct	21	2×10^6	0.43	0.7	3200	3.2	210
Alveolar sac	23	8×10^6	0.41	0.5	72,000	0.9	550
Alveoli		300×10^6	0.28	0.2			

[a]Based on Weibel's model A; regular dichotomy average adult lung with volume. 0.0048 m³ [4800 cm³] at about three-fourths maximal inflation. Table adapted from Lippmann (1995).

[b]At a flow rate of 3.6 m³/hr [1.0 L/s].

they may be exhaled. The most important of these mechanisms are impaction, settling, and diffusion; interception and electrostatic deposition are only important in certain situations. Particles that contact the airway walls deposit there and are not reentrained. The extent and location of particle deposition depend on particle size, density, and shape; airway geometry; and the individual's breathing pattern.

To describe aerosol deposition within the respiratory system analytically requires a complete solution of the constantly changing hydrodynamic flow field in the respiratory airways and the superposition of particle motion on that flow field. Currently, this is not possible; however, an understanding of the factors involved can be gained by examining the specific mechanisms that cause individual particles to deposit at different locations in the respiratory system. Understanding these mechanisms permits insight into the complex relationship between respiratory deposition and particle size, breathing frequency, flow rate, and tidal volume. The discussion that follows applies to healthy adults. Deposition characteristics may be quite different for others, such as children or people with respiratory disease.

During inhalation, the incoming air must negotiate a series of direction changes as it flows from the nose or mouth down through the branching airway system to the alveolar region. Each time the air changes direction, the suspended particles continue a short distance in their original direction because of their inertia. The net result is that some particles near the airway surfaces deposit there by *inertial impaction*. The effectiveness of this mechanism depends on the particle stopping distance at the airway velocity, which is comparatively low. Consequently, this mechanism is limited to the deposition of large particles that happen to be close to the airway walls. Nonetheless, the mechanism typically causes most aerosol deposition on a mass basis. The greatest deposition by impaction typically occurs at or near the first carina, the dividing point at the tracheal bifurcation, and to a lesser degree at other bifurcations. This is because the streamlines bend most sharply at bifurcations and pass close to the carina. Table 11.2 gives the ratios of particle stopping distances to airway dimensions at velocities associated with a steady inhalation of 3.6 m³/hr [1.0 L/s] for selected airways. The probability of deposition by impaction depends on this ratio and is highest in the bronchial region.

While impaction is of primary concern in the large airways, *settling* is most important in the smaller airways and the alveolar region, where flow velocities are low and airway dimensions are small. Sedimentation has its maximum removal effect in horizontally oriented airways. Table 11.2 gives the ratio of the settling distance (terminal settling velocity × residence time in each airway, at a steady flow of 3.6 m³/hr [1.0 L/s]) to the airway diameter. As can be seen, this mechanism is most important in the distal airways (those farthest from the trachea). Hygroscopic particles grow as they pass through the water-saturated airways, and this increase in size favors deposition by settling and impaction in the distal airways.

The *Brownian motion* of submicrometer-sized particles leads to an increased likelihood that they will deposit on airway walls, especially in the smaller airways, where distances are short and residence times comparatively long. Table 11.2 includes the ratio of the root-mean-square displacement during residence in selected airways to the airway diameter. This ratio determines the relative likelihood of depo-

TABLE 11.2   Relative Importance of Settling, Impaction, and Diffusion Mechanisms for Deposition of Standard Density Particles in Selected Regions of the Lung

Airway	Stopping Distance[a] Airway Diameter (%)			Settling Distance[b] Airway Diameter (%)			Rms Displacement[c] Airway Diameter (%)		
	0.1 μm	1 μm	10 μm	0.1 μm	1 μm	10 μm	0.1 μm	1 μm	10 μm
Trachea	0	0.08	6.8	0	0	0.52	0.04	0.01	0
Main bronchus	0	0.13	10.9	0	0	0.41	0.03	0.01	0
Segmental bronchus	0	0.31	27.2	0	0	0.22	0.05	0.01	0
Terminal bronchus	0	0.17	14.9	0	0.02	2.1	0.29	0.06	0.02
Terminal bronchiole	0	0.03	2.8	0	0.18	15.6	1.1	0.22	0.06
Alveolar duct	0	0	0.23	0.04	1.7	150	3.9	0.79	0.23
Alveolar sac	0	0	0.07	0.12	4.7	410	6.7	1.3	0.40

[a]Stopping distance at airway velocity for a steady flow of 3.6 m^3/hr [1.0 L/s].
[b]Settling distance = settling velocity × residence time in each airway at a steady flow of 3.6 m^3/hr [1.0 L/s].
[c]Rms displacement during residence time in each airway at a steady flow of 3.6 m^3/hr [1.0 L/s].

sition by diffusion. The airway and flow conditions that favor settling—that is, a small diameter and a long residence time—also favor diffusion. Diffusion is the predominant deposition mechanism for particles less than 0.5 μm in diameter and is governed by geometric, rather than aerodynamic, particle size.

Interception is the process by which a particle, without deviating from its gas streamline, contacts the airway surface because of its physical size. The likelihood of interception depends on the proximity of the gas streamline to the airway surface and on the ratio of particle size to airway diameter, which is usually small even in the smallest airways. One exception to this is the case of long fibers, which are large in one dimension, but have small aerodynamic diameters. Long fibers can readily traverse the tortuous path to the small airways, where they have a high likelihood of interceptive deposition.

Highly charged particles are attracted to airway surfaces by the *electrostatic* image charge that the particle induces in the airway surface by its presence. Unipolar charged aerosols with high number concentrations are also deposited because their mutual repulsion drives particles away from each other and toward the airway walls.

Total deposition—the combined deposition of particles in all regions of the respiratory system—is usually determined experimentally by measuring the concentration of inhaled and exhaled monodisperse test aerosols under controlled conditions. A person's breathing frequency, the volume of air he or she inhales, and the length of the pause between inhalation and exhalation all affect the deposition of inhaled particles. For particles with an aerodynamic diameter larger than 0.5 μm, the lower the breathing frequency (number of breaths per minute), the greater the fractional deposition, because there is more time for gravity settling. For particles larger than about 1 μm, deposition increases with the average airflow rate, because of the velocity-dependent mechanism of inertial impaction. A pause in the breathing cycle between inhalation and exhalation increases the deposition of particles for all size ranges, particularly for larger particles and longer pauses. Figure 11.2 shows the total deposition for a wide range of particle sizes, based on the International Commission on Radiological Protection (ICRP) deposition model. (See Section 11.3.)

Although most experimental studies have measured only total deposition, *regional deposition* within the lung is important for assessing the potential hazard of inhaled particles. To evaluate the hazard, the effective dose at the critical site within the lung where injury is initiated must be known. The deposition in any respiratory region depends on deposition in preceding regions, as well as on the deposition efficiency for the region. Generally, deposition in the head and tracheobronchial regions serve to protect the alveolar region of the lungs from irritating or harmful particles. Figure 11.3 shows the total and regional deposition predicted by the ICRP model (see Section 11.3) as a function of particle size, for particles from 0.001 to 100 μm.

Deposition of particles in the *head airways region* is highly variable and depends on several factors, including whether mouth or nose breathing is used, flow rates, and particle size. Air taken in through the nose is warmed and humidified while

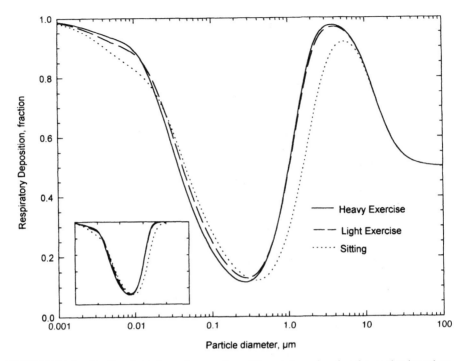

FIGURE 11.2   Predicted total respiratory deposition at three levels of exercise based on ICRP deposition model. Average data for males and females. Inset does not include the effect of inhalability.

passing around shelf-like turbinates in the nasal passage. The largest particles are removed by settling and by impaction on nasal hairs and at bends in the airflow path. Those particles deposited on the ciliated surfaces of the nasal cavity are cleared to the pharynx and swallowed. For mouth breathing at an inspiratory flow rate of 1.8 m³/hr [30 L/min], approximately 20% of 5-µm aerodynamic diameter particles and 70% of 10-µm particles (aerodynamic diameter) are deposited before the inhaled air reaches the larynx. Under conditions of light exercise and nose breathing, 80% of inhaled 5-µm particles and 95% of inhaled 10-µm particles are trapped in the nose. For both mouth and nose breathing, deposition in the head region increases when average inspiratory flow rate increases. Ultrafine particles less than 0.01 µm have significant deposition in the head airways due to their high diffusion coefficients.

At an average inspiratory flow rate of 1.2 m³/hr [20 L/min] or greater, impaction is the dominant mechanism for the deposition of particles larger than 3 µm in the *tracheobronchial region*. For particles 0.5–3 µm or flow rates less than 1.2 m³/hr [20 L/min], settling is the predominant mechanism of deposition, although overall tracheobronchial deposition for particles in this size range is quite small. For conditions of light exercise, particles with aerodynamic diameters of 5 and 10 µm that reach the tracheobronchial region are deposited there with approximately 35 and

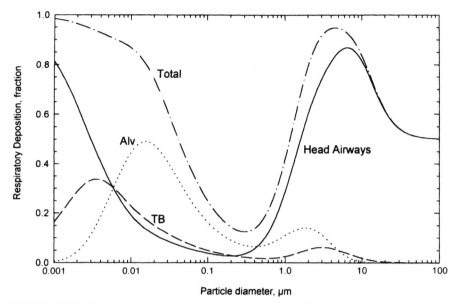

FIGURE 11.3   Predicted total and regional deposition for light exercise (nose breathing) based on ICRP deposition model. Average data for males and females.

90% efficiency, respectively. These efficiencies increase significantly with increased breathing rate. During exercise, there can be significant mixing of inhaled air with the reserve air in the first few sections of the tracheobronchial region due to turbulence in these airways. This mixing transfers particles from tidal to reserve air facilitating subsequent deposition of submicrometer-sized particles in the alveolar region. Ultrafine particles, especially those less than 0.01 μm, have enhanced deposition in the tracheobronchial region due to their rapid Brownian motion.

Alveolar deposition is usually expressed as the fraction of the inhaled particles traversing the head airways region that ultimately deposit in the alveolar region. Because of size-selective particle deposition in the tracheobronchial region, particles larger than 10 μm generally do not reach the alveolar region, and particles in the 2–10 μm range reach the alveolar region in attenuated numbers. Deposition in the alveolar region depends on particle size, breathing frequency, and tidal volume. As shown in Figure 11.3, alveolar deposition is reduced whenever tracheobronchial and head airway deposition is high. Thus, these rapidly cleared regions serve to protect the more vulnerable alveolar region.

Inhaled air penetrates the alveolar region with a thin parabolic velocity profile along the axis of each airway. With normal breathing, the apices of the parabolas do not enter the alveoli and gas exchange takes place by molecular diffusion over the last millimeter. At the low air velocities and small dimensions in this region, gas diffusion is a faster transport mechanism than flow. Inhaled submicrometer-sized particles are not directly deposited in the alveolar region, because their settling is too low and their diffusion is orders of magnitude slower than that of gas molecules.

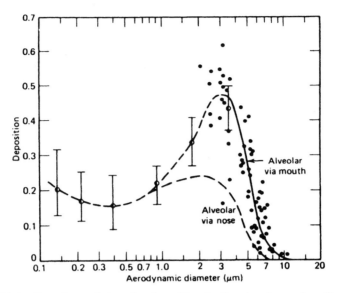

FIGURE 11.4 Experimental data for deposition in the alveolar region. Deposition is expressed as a fraction of mouthpiece inhalation versus aerodynamic diameter (geometric diameter used below 0.5 μm). Reprinted with permission from Lippmann (1977).

The deposition of these particles is controlled by their transfer from the inhaled air to the reserve air in the tracheobronchial region, followed by settling from the trapped reserve air in the alveolar region. Consequently, the rate of alveolar deposition for particles 0.1–1 μm is about 10–20% and is approximately independent of the particle size. As shown in Fig. 11.4, the particle size having the greatest deposition in the alveolar region during mouth breathing is about 3 μm, with approximately 50% of these particles being captured in the alveolar region. During nose breathing, the size for maximum alveolar deposition is reduced to about 2.0 μm, with about 10–20% of these particles being retained.

11.3 DEPOSITION MODELS

Many mathematical models for predicting total and regional deposition have been developed. [See Lippmann (1995).] Two advanced and widely used models are those of the International Commission on Radiological Protection (ICRP, 1994) and the National Council on Radiation Protection and Measurement (NCRP, 1997). These models were developed to estimate the dose to organs and tissue resulting from the inhalation of radioactive particles by typical males and females, both adults and children. The models are based on experimental data, theory, and an earlier ICRP model developed in the 1960s. Each estimates regional and total deposition over a wide range of particle sizes, essentially all aerosol particle sizes, and the full range of breathing conditions. While they apply to typical adults and children, there is large intersubject variability for respiratory deposition, and individuals may have very different deposition patterns. Each model is sufficiently complicated that a computer is required to do the calculations. In addition to modeling deposition, they model clearance, give transport rates between regions and the dose to tissues, including the thoracic lymph nodes.

The two models differ in their approach to calculating regional deposition. The ICRP model uses empirical equations based on experimental data and theory to characterize deposition by settling, inertia, and diffusion in five regions of the respiratory system: the nose and mouth, throat and larynx, upper airways (bronchi), lower airways (bronchioles), and alveolar. The NCRP model uses an empirical equation for deposition in the nose and throat; deposition in the remainder of the respiratory system is calculated using a detailed table of airway geometry for each branching generation and the alveolar region. Deposition is calculated for each generation by aerosol mechanics equations similar to those given in Chapters 3, 5, and 7 for settling, inertia, and diffusion. Data and equations are given by Yeh and Schum (1980) and Yeh et al. (1996). Both models give similar predictions for total deposition and that in the head airways, but differ in the proportioning of deposition between the tracheobronchial region and the alveolar region, especially for particles less than 0.1 μm. However, the differences in predicted deposition by these models are small compared to the differences in deposition among normal individuals.

Figure 11.2 shows the total deposition predicted by the ICRP deposition model (International Commission on Radiological Protection, 1994) for adults at three levels of exercise: sitting, light exercise, and heavy exercise. Calculations were done for spheres of standard density at standard conditions. Results for males and females were averaged. Respiratory parameters used in the model's calculations are given in Table 11.3. Figure 11.2 includes the effect of inhalability, or the fraction of ambient particles of a given size that can enter the mouth or nose. (See Section 11.4.)

TABLE 11.3   Respiratory Parameters Used in the ICRP Model

	Functional Reserve Capacity, FRC (L)	Breathing Rate (m^3/hr)	Breathing Frequency (breaths/min)	Tidal Volume (L)
Female				
Sitting	2.68	0.39	14	0.46
Light Exercise	2.68	1.25	21	0.99
Heavy Exercise	2.68	2.7	33	1.36
Male				
Sitting	3.30	0.54	12	0.75
Light Exercise	3.30	1.5	20	1.25
Heavy Exercise	3.30	3	26	1.92

The inset in the figure shows the total deposition as a fraction of what is inhaled. A comparison of the inset and the main figure reveals that the inhalability effect is quite significant for particles larger than a few micrometers. The figure shows that different levels of exercise have relatively little effect on total deposition, except that sitting yields lower deposition than exercising does for particles in the 1–5-μm range. The figure also shows a pronounced minimum at about 0.3 μm, a size that is too small for significant deposition by inertia and settling and too large for significant deposition by diffusion.

Figure 11.3 shows total and regional deposition for adults engaged in light work, according to the ICRP model. For particles larger than 1 μm, total deposition is dominated by deposition in the head airways. Elsewhere, total deposition reflects deposition in two or more regions. Deposition in any region is affected by deposition in the preceding regions. Thus, only particles that are inhaled can be deposited in the head airways, and only particles passing beyond the head airway region can be deposited in the tracheobronchial region. Deposition in the alveolar region is reduced by deposition in the previous two regions. During exhalation flow direction reverses and deposition in each region is decreased due to the deposition in the preceding deeper-lying regions. Deposition in the head airways increases for particles less than 0.01 μm due to diffusion, especially in the nose. This causes deposition in the lung airways and alveolar regions to drop off for the smallest particles.

The following simplified equations were fitted to the ICRP model for monodisperse spheres of standard density at standard conditions. Data for males and females at three exercise levels were averaged. Deposition fractions predicted by these equations agree with the ICRP model within ±0.03 over the size range of 0.001 to 100 μm. In Eqs. 11.1–11.5, d_p is particle size in μm. The deposition fraction for the head airways DF_{HA} is

$$DF_{HA} = IF\left(\frac{1}{1 + \exp(6.84 + 1.183 \ln d_p)} + \frac{1}{1 + \exp(0.924 - 1.885 \ln d_p)}\right) \quad (11.1)$$

where IF is the inhalable fraction (see next section). Inhalable fraction as used by the ICRP model is given by

$$IF = 1 - 0.5\left(1 - \frac{1}{1 + 0.00076 d_p^{2.8}}\right) \quad (11.2)$$

The deposition fraction for the tracheobronchial region DF_{TB} is

$$DF_{TB} = \left(\frac{0.00352}{d_p}\right)[\exp(-0.234(\ln d_p + 3.40)^2) + 63.9 \exp(-0.819(\ln d_p - 1.61)^2)]$$

$$(11.3)$$

The deposition fraction for the alveolar region DF_{AL} is

$$DF_{AL} = \left(\frac{0.0155}{d_p}\right)[\exp(-0.416(\ln d_p + 2.84)^2) + 19.11\exp(-0.482(\ln d_p - 1.362)^2)]$$

(11.4)

Although IF does not appear explicitly in Eqs. 11.3 and 11.4, they were fitted to data that included the effect of inhalability. The total deposition DF is the sum of the regional depositions, or

$$DF = IF\left(0.0587 + \frac{0.911}{1 + \exp(4.77 + 1.485\ln d_p)} + \frac{0.943}{1 + \exp(0.508 - 2.58\ln d_p)}\right)$$

(11.5)

Figures 11.2 and 11.3 and Eqs. 11.1–11.5 are for spheres of standard density, but can be applied to other particles by using the aerodynamic diameter for particles larger than 0.5 μm and the physical diameter or equivalent volume diameter for particles less than 0.5 μm. For spheres, $d_a = d_p(\rho_p/\rho_o)^{1/2}$; for nonspheres, see Eq. 3.27.

The mass of a given particle size deposited in the respiratory system per minute M_{dep} is

$$M_{dep} = \frac{\pi}{6}N\rho_p d_p^3 V_m(DF)$$

(11.6)

where N is the number concentration of particles of diameter d_p and density ρ_p, V_m is the minute volume or volume inhaled in 1 min, and DF is the total deposition fraction for particle size d_p, Eq. 11.5. For regional deposition, DF is replaced by the appropriate regional deposition fraction, Eqs. 11.1, 11.3, or 11.4.

EXAMPLE

What fraction of *inhaled* 5-μm particles will deposit in the head airway region? Asume standard density and an adult engaged in light work.

$$DF_{HA} = 1\left(\frac{1}{1 + \exp(6.84 + 1.183\ln(5.0))} + \frac{1}{1 + \exp(0.924 - 1.885\ln(5.0))}\right)$$

$$= 1(0.00016 + 0.892) = 0.89.$$

11.4 INHALABILITY OF PARTICLES

The entry of particles into the mouth or nose can be thought of as a sampling process—one that includes elements from isokinetic and still-air sampling. Unlike thin-wall sampling probes, the human head is a *blunt sampler* with a complex geometry. Theoretical approaches and experimental results for blunt samplers have been reviewed by Vincent (1989). As air approaches a blunt sampler, the streamlines display two kinds of distortion: A larger scale divergence occurs as the air flows around the blunt object, and a smaller scale convergence occurs near the inlet. This motion is considerably more complicated than that for isokinetic sampling, and there seems to be no condition that ensures perfect sampling for all particle sizes.

The efficiency of entry of particles into the nose or mouth can be characterized by the *aspiration efficiency, inhalability, or inhalable fraction (IF)*, the fraction of particles originally in the volume of air inhaled that enters the nose or mouth. The inhalable fraction is usually less than one, but can be greater than one under some conditions. It depends on particle aerodynamic diameter and the external wind velocity and direction. Inhalability fraction is determined experimentally, typically in a large, low-velocity wind tunnel with a full-size, full-torso mannequin connected to a mechanical breathing machine. A uniform concentration of dust flows past the mannequin. A series of test dusts with narrow size distributions is used. Together, these dusts cover the range of 5 to 100 μm or more in aerodynamic diameter. Air velocities up to 4 m/s are used to simulate indoor environments and up to 10 m/s to simulate outdoor environments. The mannequin can be positioned at any angle to the wind direction or continuously rotated. Inhaled dust is collected by a sampling filter immediately inside the mouth or nose. Reference samples are usually taken by isokinetic samplers located near the mannequin's head. For a given condition and particle size, inhalable fraction is the ratio of the concentration calculated on the basis of the collected mass and sample volume for the mouth (or nose) filter to that calculated for the isokinetic samplers.

Inhalable fraction data are usually presented as "orientation averaged," meaning that all orientations with respect to the wind direction are equally represented. Orientation-averaged mouth inhalability data are summarized in Figure 11.5, which includes the ACGIH inhalable particulate matter (IPM) sampling criterion (ACGIH, 1996). The curve defines the desired sampling performance of an IPM sampler in terms of the fractional collection for particles up to 100 μm. This criterion is the same as that proposed by the International Standards Organization (ISO 7708) and the Comité Européen de Normalisation (CEN EN481). The equation for the inhalable fraction and inhalable fraction sampling criterion $IF(d_a)$ is

$$IF(d_a) = 0.5(1 + \exp(-0.06 \, d_a)) \quad \text{for } U_0 < 4 \text{ m/s} \tag{11.7}$$

where d_a is the aerodynamic diameter in μm. Vincent et al. (1990) give the following expression for inhalability when the ambient air velocity U_0 is greater than 4 m/s.

$$IF(d_a, U_0) = 0.5(1 + \exp(-0.06 d_a)) + 10^{-5} U_0^{2.75} \exp(0.055 \, d_a) \tag{11.8}$$

where d_a is in μm ($d_a < 100$ μm) and U_0 is in m/s. Equation 11.8 reduces to Eq. 11.7 within ±5% for $d_a < 30$ μm and $U_0 < 10$ m/s and for $d_a < 100$ μm and $U_0 < 3$ m/s. There are fewer data for nasal inhalability IF_N which can be approximated by

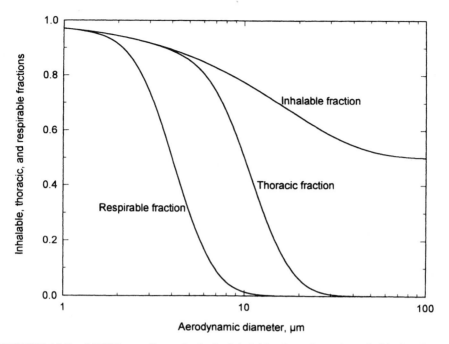

FIGURE 11.5   ACGIH sampling criteria for inhalable, thoracic, and respirable fractions.

$$\text{IF}_N(d_a) = 0.035 + 0.965 \exp(-0.000113d_a^{2.74}) \qquad (11.9)$$

where d_a is the aerodynamic diameter in μm (Hinds et al., 1998).

EXAMPLE

What is the inhalable fraction for 18-μm particles of standard density? Assume that $U_0 < 4$ m/s.
By Eq. 11.7,

$$\text{IF} = 0.5(1 + \exp(-0.06 \times 18)) = 0.67$$

The performance of *inhalable samplers* is usually evaluated in a low-velocity wind tunnel similar to that described for the evaluation of inhalability. The samplers are tested one particle size at a time. A sampler's performance can be expressed in one of two ways: first as the ratio of the concentration measured by the sampler to the true ambient air concentration as measured by the isokinetic samplers or, alternatively, as the ratio of the concentration measured by the sampler to the inhalable concentration. The latter can be determined using a mannequin, or by isokinetic samplers and calculation using the inhalable fraction curve. The first method provides an absolute measure of the ability of the sampler to sample accurately the inhalable fraction of particles of a given size in the ambient environment. The results obtained can be compared directly to the inhalable sampling-criteria curve. The second method provides a relative measure of how well the sampler conforms to the inhalable sampling criterion.

More than a dozen area and personal samplers have been designed to meet the inhalable sampling criterion. An area sampler developed by Mark et al. (1985) conforms well to the inhalable sampling criteria. The device has a cylindrical rotating (2 rpm) sampling head 50 mm in diameter and 60 mm high with a 3 × 15-mm oval inlet on the side. It is battery powered and samples at 0.18 m^3/hr [3 L/min] onto a 37-mm filter mounted in a weighable cassette.

The most commonly used personal inhalable sampler is the Institute for Occupational Medicine (IOM) personal sampler described by Mark and Vincent (1986). The device, shown in Fig. 11.6, has a cylindrical body 37 mm in diameter with a protruding inlet 15 mm in diameter. It is attached to a worker's lapel, so that the inlet is always facing forward. The filter and its lightweight cassette are weighed together so that any particles that deposit in the inlet are included in the sample. The device shows good agreement with the inhalable sampling criterion for $U_0 \leq 1.0$ m/s. A conductive plastic version of this sampler is manufactured by SKC, Inc., of Eighty Four, PA.

Kenny et al. (1997) evaluated the performance of eight designs of personal inhalable samplers. Five performed satisfactorily at 0.5 m/s, two at 1 m/s, and none

FIGURE 11.6 IOM-type personal inhalable aerosol sampler.

TABLE 11.4 Comparison of Performance of IOM Sampler
with that of Inhalable 37-mm in-line Personal Sampler[a]

Aerosol	IOM Concentration / 37-mm Concentration
Mineral dust, flour	2.5
Oil mist, paint spray	2.0
Smelter, foundry	1.5
Smoke, fume, welding	1.0

[a]Adapted from Werner et al. (1996).

at 4 m/s. The commonly used in-line 37-mm plastic filter cassette undersamples particles larger than 30 μm in aerodynamic diameter. The open-face version undersamples particles larger than 30 μm for $U_0 \leq 1.0$ m/s and oversamples all sizes at 4 m/s. Results of side-by-side simultaneous sampling with the IOM personal sampler and 37-mm in-line samplers are summarized in Table 11.4. Inhalable particle samplers are reviewed by Hinds (1999).

11.5 RESPIRABLE AND OTHER SIZE-SELECTIVE SAMPLING

An understanding of regional deposition has spurred the development of health-based, particle size-selective sampling—that is, sampling a subset of the airborne particles on the basis of their aerodynamic size. The subset is chosen to select those particles that can reach a particular region of the respiratory system and potentially deposit there. Other sizes are excluded from the sample. In the field of occupational health, examples are inhalable, thoracic, and respirable sampling, as well as cotton dust sampling. In the field of ambient air quality, examples are PM-10 and PM-2.5.

Respirable sampling was first used in the 1950s to assess occupational exposure to silica dust. For that purpose, we need information on the amount of dust that can deposit at the site of toxic action, the alveolar region of the lungs. Because of the size-selective characteristics of the human respiratory system, particles larger than a certain size are unable to reach the alveolar region and therefore can be considered nonhazardous with respect to alveolar injury. To estimate the potential hazard of silica or other dusts that have their site of toxic action in the alveolar region, we must exclude these nonhazardous particles from our assessment.

Historically, respirable sampling was performed for harmful mineral dusts by using microscopic particle counting. The hazard was assessed by number concentration (dust counting) instead of mass concentration, because the former correlated with the prevalence and extent of respiratory disease observed in miners. The reason mass concentration does not correlate with disease can be seen in Fig. 11.7. More than 60% of the mass of a typical mine aerosol is contributed by particles that are nonrespirable, that is they are too large to reach the alveolar region, whereas about 98% of the *number* of particles can reach the alveolar region. Thus, number concentration more closely reflects the hazard in this situation.

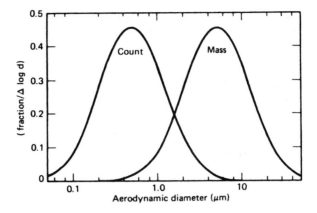

FIGURE 11.7 Typical count and mass distributions for mine dust.

Respirable sampling uses a mechanical device upstream of the sampling filter to aerodynamically remove those particles which are "nonrespirable"—that is, which are unable to reach the alveolar region. The *respirable mass* of an airborne dust is estimated by gravimetric analysis of that portion that passes through the device, or precollector. This approach is called respirable sampling, respirable mass sampling, or size-selective sampling. Unfortunately, the particle size cutoff that separates respirable from nonrespirable particles is not sharp, but extends gradually over the range of 2–10 μm, as shown in Fig 11.5. Respirable sampling is simpler, quicker, and more accurate than counting dust particles.

The *respirable fraction*, RF, is defined by the American Conference of Governmental Industrial Hygienists (ACGIH) particle size-selective sampling criteria as

$$RF = (IF)(1 - F(x)) \tag{11.10}$$

where IF is the inhalable fraction given by Eq. 11.7 and $F(x)$ is the cumulative fraction for a standardized normal variable x.

$$x = 2.466 \ln(d_a) - 3.568 \tag{11.11}$$

in which d_a is aerodynamic diameter in μm.

The quantity x is the number of standard deviations d_a is from the mean of 4.25 μm and $1 - F(x)$ is the fraction of *inhaled* particles that can reach the alveolar region. $F(x)$ can be approximated by

$$F(x) = 0.5(1 - 0.1969x + 0.1152x^2 - 0.0003x^3 + 0.0195x^4)^{-4} \quad \text{for } x \le 0$$

$$F(x) = 1 - 0.5(1 + 0.1969x + 0.1152x^2 + 0.0003x^3 + 0.0195x^4)^{-4} \quad \text{for } x > 0$$
$$\tag{11.12}$$

where x is given by Eq. 11.11. This approximation, together with Eqs 11.10 and 11.11, predicts the respirable fraction to within 1% of the correct value, for all values

TABLE 11.5   Inhalable, Thoracic, and Respirable Fractions[a]

Aerodynamic Diameter (μm)	Inhalable Fraction	Thoracic Fraction	Respirable Fraction
0	1.00	1.00	1.00
1	0.97	0.97	0.97
2	0.94	0.94	0.91
3	0.92	0.92	0.74
4	0.89	0.89	0.50
5	0.87	0.85	0.30
6	0.85	0.81	0.17
8	0.81	0.67	0.05
10	0.77	0.50	0.01
15	0.70	0.19	0.00
20	0.65	0.06	0.00
25	0.61	0.02	0.00
30	0.58	0.01	0.00
35	0.56	0.00	0.00
40	0.55	0.00	0.00
50	0.52	0.00	0.00
60	0.51	0.00	0.00
80	0.50	0.00	0.00
100	0.50	0.00	0.00

[a]ACGIH (1997).

of the respirable fraction greater than 0.005. A simpler approximation permits the direct calculation of RF in terms of IF and d_a,

$$RF = (IF)(1 - \exp(-\exp(2.54 - 0.681\, d_a))) \qquad (11.13)$$

The respirable fraction calculated by Eq. 11.13 differs from that calculated by Eqs. 11.10–11.12 by less than 0.007 for d_a from 0 to 100 μm. Table 11.5 and Figure 11.5 give values of inhalable, thoracic, and respirable fractions, as defined by the ACGIH (ACGIH, 1997). This definition of respirable fraction agrees with the equivalent ISO and CEN protocols. The respirable fraction is the fraction *passing through* or penetrating a respirable precollector, so the collection efficiency of the precollector for a particular particle size $CE_R(d_a)$ is

$$CE_R(d_a) = 1 - RF(d_a) \qquad (11.14)$$

Figure 11.8 compares the respirable fraction criterion with experimentally determined alveolar deposition curves. It should be borne in mind that the criterion attempts to define those particles that *reach* the alveolar region, whereas the deposition curves define those particles that *reach and are deposited in* the alveolar region. A better way to think about the function of a precollector is that it excludes particles from the sample in the same way that the airways of the respiratory system exclude particles from, or prevent particles from *reaching*, the alveolar region.

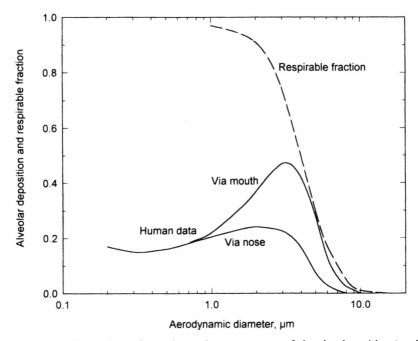

FIGURE 11.8 Comparison of experimental measurements of alveolar deposition (median values) and ACGIH respirable fraction criterion. Human data from Lippmann (1977).

It is recognized that the sampling filter will collect 100% of nonexcluded particles, whereas the alveolar region collects only about 20–40% of these nonexcluded particles, and exposure standards for respirable dust have been set accordingly. The criterion corresponds closely to the mouth-breathing alveolar deposition curve (Fig. 11.8), down to about 4 μm. Respirable sampling would, however, collect more particles greater than 4 μm than are likely to reach the alveolar region of a typical nose-breathing person. The use of respirable mass sampling is relevant only to dusts that have their site of toxic action or are absorbed in the alveolar region of the lung. In the United States, respirable dust standards exist for occupational exposure to cadmium, silica, coal, talc, and other mineral dusts. Nonrespirable particles of heavy metals or pesticides may not be damaging to the alveolar region of the lungs, but can be inhaled and cause serious injury elsewhere in the body.

The most widely used technique for respirable sampling is to use a cyclone for the precollector and a high-efficiency filter for the second-stage collector. Other, less common, approaches use horizontal elutriators, centrifugal inlets, specially designed impactors, open-cell foam, and large-pore capillary-pore-membrane filters as the precollector stage. The characteristics of some respirable precollectors are summarized in Table 11.6 and in Lippmann (1995).

In the United States, the most common precollector is the 10-mm nylon cyclone shown in Fig. 11.9. The dusty air is drawn into the cyclone through an inlet that is tangential to the cylindrical part of the cyclone. The inlet geometry causes the air

TABLE 11.6 Characteristics of Some Precollectors for Respirable Sampling

Name	Type	Sampling Rate (L/min)	Weight (kg)
10-mm nylon cyclone	Cyclone	1.7	0.18[a]
½-in. HASL	Cyclone	9	0.023
1-in. HASL	Cyclone	75	0.11
Uniclone 2 (Aerotec 2)	Cyclone	430	2.0
MRE gravimetric dust sampler	Elutriator	2.5	3.8[b]
Hexlet	Elutriator	50	5.0[b]
P.S. universal impactor	Impactor	6.2	0.1

[a]Includes filter holder and supporting frame.
[b]Includes pump.

to rotate around in the cyclone several times before exiting to the filter at the top center of the cyclone. During the rotation, the larger particles are deposited on walls due to their centrifugal motion. Very large particles and chunks of collected material fall to the "grit pot" at the bottom of the cyclone. Careful experimental evaluations have found that when operated at $0.10 \text{ m}^3/\text{hr}$ [1.7 L/min], this cyclone has a collection efficiency versus aerodynamic diameter that closely matches that of the ACGIH respirable criterion. The cyclone is usually mounted with the filter in a

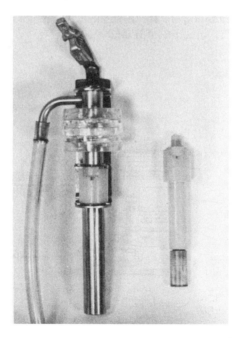

FIGURE 11.9   10-mm nylon cyclone for respirable sampling.

holder and positioned in the breathing zone of the worker whose exposure is being evaluated. The filter is connected by tubing to a battery-powered pump (see Section 10.7) worn on the worker's belt.

The horizontal elutriator shown schematically in Fig. 11.10 is designed for respirable dust sampling. The instrument is bulkier than the nylon cyclone and not suitable for personal sampling. The principle of operation of the horizontal elutriator is covered in Section 3.9. The device must be maintained in a level position during sampling.

Thoracic fraction is another size-selective sampling criterion based on regional deposition. It is defined as the fraction of ambient aerosol particles that will pass beyond the larynx and reach the thorax or chest during inhalation. Sampling in accordance with this criterion is appropriate for materials that are hazardous when deposited in the lung airways or alveolar regions. Thoracic fraction is analogous to respirable fraction, but with a cutsize (size corresponding to a thoracic fraction of 50%) of 10 μm in aerodynamic diameter (See Fig. 11.5.) The thoracic fraction, TF, is defined by the ACGIH criterion as

$$TF = (IF)(1 - F(x')) \tag{11.15}$$

where IF is the inhalable fraction given by Eq. 11.7 and $F(x')$ is the cumulative fraction for the standardized normal variable x',

$$x' = 2.466 \ln(d_a) - 6.053 \tag{11.16}$$

in which d_a is in μm. Equation 11.12 can be used to approximate $F(x')$ by replacing x with x', as given by Eq. 11.16, throughout. A simpler expression permits the direct calculation of TF in terms of IF and d_a.

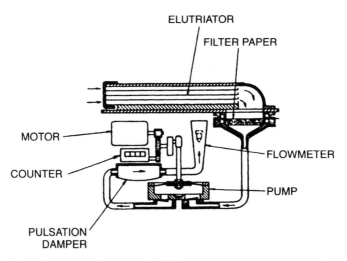

FIGURE 11.10   Schematic diagram of MRE gravimetric dust sampler. Reprinted with permission from *Air Sampling Instruments*, 5th ed., 1978, American Conference of Governmental Industrial Hygienists, 1330 Kemper Meadow Drive, Cincinnati, OH 45240.

$$TF = (IF)(1 - \exp(-\exp(2.55 - 0.249\,d_a)))\qquad(11.17)$$

The thoracic fraction calculated by Eq. 11.17 differs from that calculated by Eqs. 11.15, 11.16, and 11.12 by less than 0.006 for d_a from 0 to 100 μm. The collection efficiency of a thoracic precollector CE_T is

$$CE_T(d_a) = 1 - TF(d_a)\qquad(11.18)$$

Thoracic fraction samplers are reviewed by Baron and John (1999).

Closely related to the thoracic fraction criterion is the PM-10 sampling criteria promulgated by the U.S. Environmental Protection Agency in 1987 as the standard method for ambient particulate sampling. Like thoracic sampling, PM-10 sampling is based on those particles that penetrate to the thorax. The cutoff size is the same, 10 μm in aerodynamic diameter, but there are two important differences. First, PM-10 is a fraction of the total ambient particulate, not a subfraction of the inhalable particulate, as is the thoracic fraction. Second, the cutoff curve that defines PM-10 is considerably sharper than that for the thoracic fraction. The fraction of particles of diameter d_a included in the PM-10 fraction, PF_{10}, can be estimated by

$$
\begin{aligned}
PF_{10} &= 1.0 & &\text{for } d_a < 1.5 \text{ μm}\\
PF_{10} &= 0.9585 - 0.00408 d_a^2 & &\text{for } 1.5 < d_a < 15 \text{ μm}\qquad(11.19)\\
PF_{10} &= 0 & &\text{for } d_a > 15 \text{ μm}
\end{aligned}
$$

where d_a is in μm. PM-10 fractions calculated by Eq. 11.19 differ from those given in the [U.S.] Code of Federal Regulations (40CFR53.43, Revised July 1, 1997) by less than 0.0005. Specifications are also given for wind tunnel testing of new sampler designs and the range of wind velocities (2–24 km/hr [0.55–6.7 m/s]) over which samplers are to meet the PM-10 criteria. PM-10 samplers range from high volume samplers with flow rates of 1.13 m^3/min [1130 L/min] to low-volume dichotomous samplers with flow rates of 1.0 m^3/hr [16.7 L/min]. A PM-10 inlet for a high-volume sampler is shown in Figure 11.11. The dichotomous sampler uses a virtual impactor to cut the PM-10 particle fraction at 2.5 μm for visibility and fine-particle evaluation. Because of concern over the health effects of fine particles in the ambient environment, in 1997 the Environmental Protection Agency adopted new standards for sampling fine particles, PM-2.5. The cutoff curve for PM-2.5 is defined by an impactor cutoff curve similar to that for the virtual impactor used in the dichotomous sampler. The following empirical expression gives the fraction of particles of diameter d_a that are included in the PM-2.5 fraction, $PF_{2.5}$.

$$PF_{2.5} = [1 + \exp(3.233 d_a - 9.495)]^{-3.368}\qquad(11.20)$$

where d_a is particle aerodynamic diameter in μm. PM-2.5 fractions calculated by Eq. 11.20 differ from those given in the [U.S.] Code of Federal Regulations (40CFR53.62, Revised July 18, 1997) by less than 0.001. A personal PM-10 or PM-2.5 sampler utilizing single-stage impaction is available from SKC, Inc. located in Eighty Four, PA. Other devices use cyclones or spiral inlets to achieve the 2.5 μm cutoff.

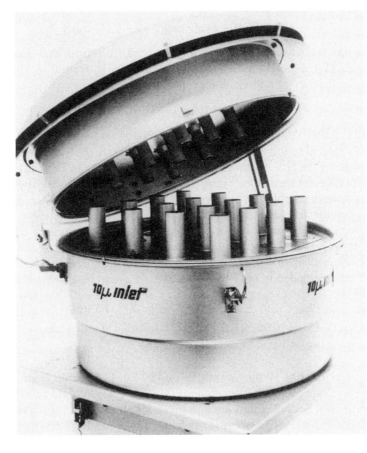

FIGURE 11.11 PM-10 inlet for a high-volume dichotomous sampler. Photo courtesy of Graseby-Andersen, Smyrna, GA.

The U.S. Occupational Safety and Health Administration (OSHA) has recommended a size-selective dust standard for evaluating the health hazards of cotton dust. The standard is based on filter samples taken using a vertical elutriator as a precollector. The elutriator is operated to remove particles larger than 15 μm in aerodynamic diameter from the aerosol stream. An epidemiological correlation has been found between the prevalence of byssinosis among cotton mill workers and the concentration of dust, as measured by the vertical elutriator method. The 15-μm cutoff size was chosen to represent the upper limit for those particles that will deposit in the *alveolar or tracheobronchial region*. Large nonrespirable fibers, called linters, clog the inlets of conventional samplers and can significantly bias mass concentration measurements.

The principle of operation of the vertical elutriator is covered in Section 3.9. The standard cotton dust vertical elutriator, shown in Fig. 11.12, is 15 cm in diameter and 70 cm high. The inlet is at the bottom, and a standard 37-mm filter holder is

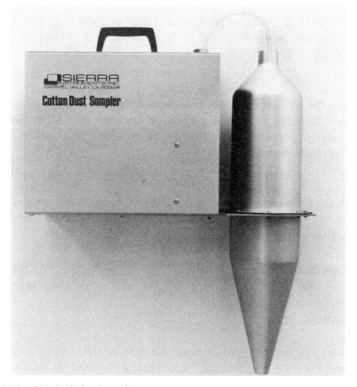

FIGURE 11.12   Vertical elutriator for cotton dust sampling. Courtesy of Graseby Andersen, Inc., Smyrna, GA.

mounted at the top. The recommended flow rate of 7.4 L/min will provide an *average* upward velocity in the widest section of the elutriator of 0.68 cm/s, equal to the terminal settling velocity of a 15-μm sphere of standard density. Apparently, the conical section at the inlet was added to prevent air currents from disrupting the flow in the main section. Unfortunately, the inlet section causes a jet of air to travel along the centerline with sufficient velocity to permit particles of 30 μm in aerodynamic diameter to reach the filter. This nonuniform flow in the elutriator results in a gradual "cutoff" over a wide range in particle size. A further problem may arise because 95-μm particles can enter the inlet, but will be trapped in the elutriator's conical section and act as a floating filter for the air passing through.

PROBLEMS

11.1   According to Table 11.1, what section of the respiratory system has the highest Reynolds number, and what is the Reynolds number at an inhalation flow rate of 3.6 m³/hr [1.0 L/s]?

ANSWER: Trachea, 4200.

11.2 What fraction of 4.0-μm particles will deposit in the head airways? Assume an average adult engaged in light work and a particle density of 1000 kg/m³ [1.0 g/cm³].

ANSWER: 0.82.

11.3 Using the ACGIH respirable fraction criteria, calculate (a) the fraction of *inhaled* particles with an aerodynamic diameter of 4.5 μm that reach the alveolar region and (b) the fraction of *ambient* 4.5-μm particles that reach the alveolar region.

ANSWER: 0.44, 0.39.

11.4 Calculate the inhalable and thoracic fractions for 7-μm spheres of standard density. Assume that the external air velocity is less than 4 m/s.

ANSWER: 0.83, 0.74.

11.5 On the basis of the ICRP model, estimate the fraction of 3.0-μm particles (ρ_p = 4000 kg/m³ [4.0 g/cm³]) that deposits in the alveolar region for an average adult.

ANSWER: 0.045.

11.6 An aerosol is composed of equal numbers of 0.02- and 2.0-μm particles. Using Fig. 11.3, estimate the fraction that will deposit in the alveolar region based on number and based on mass.

ANSWER: 0.31, 0.14.

REFERENCES

ACGIH, *1997 Threshold Limit Values and Biological Exposure Indices*, ACGIH, Cincinnati, (1997).

ACGIH Air Sampling Procedures Committee, *Particle Size-Selective Sampling in the Workplace*, American Conference of Governmental Industrial Hygienists, Cincinnati, OH (1985).

Baron, P. and John, W., "Sampling for Thoracic Aerosols," in *Particle Size-Selective Sampling for Particulate Air Contaminants*, ACGIH, Cincinnati (1999).

European Committee for Standardization (CEN), "Workplace Atmospheres—Size Fraction Definitions for Measurement of Airborne Particles," CEN Standard EN 481, Brussels: European Committee for Standardization, 1993.

Hatch, T. F., and Gross, P., *Pulmonary Deposition and Retention of Inhaled Aerosols*, Academic Press, New York, 1964.

Hinds, W. C., "Inhalable Aerosol Samplers," in *Particle Size-Selective Sampling for Particulate Air Contaminants*, ACGIH, Cincinnati (1999).

Hinds, W. C., Kennedy, N. J., and Tatyan, K., "Inhalability of Large Particles for Mouth and Nose Breathing," *J. Aerosol Sci.*, **29**, S277-S278 (1998).

International Commission on Radiological Protection, "Human Respiratory Tract Model for Radiological Protection," Annals of the ICRP, Publication 66, Elsevier Science, Inc., Tarrytown, NY (1994).

International Standards Organization, "Size Definitions for Particle Sampling: Recommendations of Ad Hoc Working Group Appointed by Committee TC 146 of the International Standards Organization," *Am. Ind. Hyg. Assoc. J.*, **42**, (5) A64–A68 (1981).

Kenny, L. C., Aitkens, R., Chalmers, C., Fabries, J. F., Gonzalez-Fernandez, E., Kronhout, H., Liden, G., Mark, D., Riediger, G., and Prodi, V., "A Collaborative European Study of Personal Inhalable Aerosol Sampler Performance," *Ann. Occup. Hyg.*, **41**, 135–153 (1997).

Lippmann, M., "Regional Deposition of Particles in the Human Respiratory Tract," in Lee, D. H. K., Falk, H. L., Murphy, S. O., and Geiger, S. R. (Eds.), *Handbook of Physiology, Reaction to Environmental Agents*, American Physiological Society, Bethesda, MD, 1977.

Lippmann, M., "Size-Selective Health Hazard Sampling," in *Air Sampling Instruments for Evaluation of Atmospheric Contaminants*, 8th ed., ACGIH, Cincinnati, 1995.

Mark, D., and Vincent, J. H., "A New Personal Sampler for Airborne Total Dust in Workplaces," *Ann. Occup. Hyg.*, **30**, 89–102 (1986).

Mark, D., Vincent, J. H., Gibson, H., and Lynch, G., "A New Static Sampler for Airborne Total Dust in Workplaces," *Am. Ind. Hyg. Assoc. J.*, **46**, 127–43 (1985).

National Council on Radiation Protection and Measurement, *Deposition, Retention and Dosimetry of Inhaled Radioactive Substances*, Report S.C. 57-2, NCRP, Bethesda, MD (1994).

Phalen, R. F., *Inhalation Studies: Foundations and Techniques*, CRC Press, Boca, Raton, FL, 1984.

Vincent, J. H., Mark, D., Miller, B. G., Armbruster, L., and Ogden, J. L., "Aerosol Inhalability at Higher Windspeeds," *J. Aerosol Sci.*, **21**, 577–586 (1990).

Vincent, J. H., *Aerosol Sampling: Science and Practice*, Wiley, Chichester, U.K., 1989.

Walton, W. H., *Inhaled Particles*, vol. 4, Pergamon, Oxford, 1977.

Werner, M.A., Spear, T. M., and Vincent, J. H., "Investigation into the Impact of Introducing Workplace Aerosol Standards Based on the Inhalable Fraction," *Analyst*, **121**, 1207–1214 (1996).

Yeh, H. C. and Schum, G. M., "Models for Human Lung Airways and their Application to Inhaled Particle Deposition," *Bull. Math. Biology*, **42**, 461–480 (1980).

Yeh, H. C. Cuddihy, R. G., Phalen, R. F., and Chang, I-Y., "Comparisons of Calculated Respiratory Tract Deposition of Particles Based on the Proposed NCRP Model and the New ICRP Model," *Aerosol Sci. Tech.*, **25**, 134–140 (1996).

12 Coagulation

Coagulation of aerosols is a process wherein aerosol particles collide with one another, due to a relative motion between them and adhere to form larger particles. The net result is a continuous decrease in number concentration coupled with an increase in particle size. When the relative motion between the particles is Brownian motion the process is called *thermal coagulation*. This type of coagulation is a spontaneous and ever-present phenomenon for aerosols. When the relative motion arises from external forces, such as gravity or electrical forces, or from aerodynamic effects, the process is called *kinematic coagulation*.

Coagulation is the most important interparticle phenomenon for aerosols. The theory of coagulation was originally devised for particles in liquids and was later extended to aerosols. The original name has continued. In the case of solid particles, the process is sometimes called *agglomeration*, and the resulting particle clusters, are known as *agglomerates*. In the production of certain ceramics, pigments, and optical materials, small, solid particles are produced by gas-phase reactions. These particles coagulate under controlled conditions to give the desired particle size. Particle shape is controlled by coalescence (sintering), wherein the particles in an agglomerate partially fuse at high temperature to form solid particles of the desired shape.

The objective of the theory of coagulation is to describe how particle number concentration and size change as a function of time. An exact description is very complicated, particularly for polydisperse aerosols. However, in many cases a simplified theory is adequate for understanding the characteristics of coagulation and estimating changes in number concentration and particle size.

12.1 SIMPLE MONODISPERSE COAGULATION

The simplest kind of coagulation is the thermal coagulation of monodisperse spherical particles with $d_p > 0.1 \mu m$. We assume that particles adhere with every collision, and initially we also make the assumption that particle size changes slowly. This type of coagulation is sometimes referred to as *Smoluchowski coagulation*, after the person who developed the original theory.

The basic approach is to focus attention on a single particle and consider how other particles diffuse to its surface. This is equivalent to the problem of diffusion to surfaces discussed in Section 7.4. The particle flux at the surface of the selected particle—that is, the rate of collisions per unit area per unit time—is given by Fick's first law of diffusion, Eq. 7.1.

$$J = -D\frac{dN}{dx} \tag{12.1}$$

where D is the diffusion coefficient of the particles, N is the number concentration, and dN/dx is the concentration gradient of particles at the collision surface of the selected particle. The collision surface is defined as the imaginary surface around the selected particle in which the centers of other particles are located when they collide with the selected particle. For particles of diameter d_p, the collision surface is a sphere of diameter $2d_p$. Replacing the diffusing particles by their centers simplifies the derivation of the diffusion equations. The rate of collisions between the selected particle and the other particles, dn/dt, is the product of the area A_s of the collision surface and the particle flux,

$$\frac{dn}{dt} = A_s J = -\pi(2d_p)^2 D\frac{dN}{dx} \tag{12.2}$$

The solution giving the concentration gradient at the collision surface of the selected particle (Fuchs, 1964) is, neglecting a transient term,

$$\frac{dN}{dx} = -\frac{2N}{d_p} \quad \text{for } d_p > \lambda_p \tag{12.3}$$

where λ_p is the particle mean free path (Eq. 7.11). Combining Eqs. 12.2 and 12.3 gives the rate of collisions with the selected particle:

$$\frac{dn}{dt} = 8\pi d_p DN \tag{12.4}$$

Since the selected particle is no different from any other particle, the rate for every particle is $8\pi d_p DN$ collisions/s. Because there are N particles in a unit volume, the rate of collisions per unit volume of aerosol dn_c/dt is just N times Eq. 12.4, or

$$\frac{dn_c}{dt} = \frac{N}{2}(8\pi d_p DN) = 4\pi d_p DN^2 \tag{12.5}$$

where the factor 1/2 is included to avoid counting each collision twice.

For each collision, there is a reduction of one in the number of particles in a unit volume, and therefore, the number concentration is reduced by one. The rate of collisions is numerically equal, but of opposite sign, to the rate of change in number concentration, dN/dt.

$$\frac{dN}{dt} = -4\pi d_p DN^2 \tag{12.6}$$

The rate of change of the number concentration can be expressed as

$$\frac{dN}{dt} = -K_0 N^2 \tag{12.7}$$

where K_0 is the coagulation coefficient,

$$K_0 = 4\pi d_p D \tag{12.8}$$

which has units of m^3/s in SI units and cm^3/s in cgs units. Values of K_0 are given in Tables 12.1.

Substituting Eq. 7.7 for the diffusion coefficient in Eq. 12.8, we get

$$K_0 = \frac{4kTC_c}{3\eta} \quad \text{for } d_p > 0.1 \ \mu m \tag{12.9}$$

where particle size cancels out leaving K_0 dependent on the particle size only through the slip correction factor. Thus, K_0 is independent of particle size for large particles with negligible slip correction, but increases as particle size decreases for small particles because of the effect of the slip correction factor. At standard conditions, Eq. 12.9 for the coagulation coefficient can be simplified to

$$K_0 = 3.0 \times 10^{-16} C_c \ m^3/s \quad [3.0 \times 10^{-10} C_c \ cm^3/s] \quad \text{for } d_p > 0.1 \ \mu m \tag{12.10}$$

From Eq. 12.7, the rate of coagulation (the rate of change in number concentration) is proportional to the square of the particle number concentration and to the coagulation coefficient K_0. Usually we want to know the net effect of coagulation over some period of time. We can determine the number concentration as a function of time by integrating Eq. 12.7, assuming K to be a constant:

TABLE 12.1   Coagulation Coefficients at Standard Conditions[a]

Diameter (μm)	Correction Factor, β	K_0 (Eq. 12.9) (m^3/s)	K (Corrected, Eq. 12.13) (m^3/s)
0.004	0.037	168×10^{-16}	6.2×10^{-16}
0.01	0.14	68×10^{-16}	9.5×10^{-16}
0.04	0.58	19×10^{-16}	10.7×10^{-16}
0.1	0.82	8.7×10^{-16}	7.2×10^{-16}
0.4	0.95	4.2×10^{-16}	4.0×10^{-16}
1.0	0.97	3.4×10^{-16}	3.4×10^{-16}
4	0.99	3.1×10^{-16}	3.1×10^{-16}
10	0.99	3.0×10^{-16}	3.0×10^{-16}

[a]For coagulation coefficient in cm^3/s multiply table values by 10^6.

$$\int_{N_0}^{N(t)} \frac{dN}{N^2} = \int_0^t - K_0 dt$$

$$\frac{1}{N(t)} - \frac{1}{N_0} = K_0 t \tag{12.11}$$

where N_0 is the original number concentration at time zero and $N(t)$ is the number concentration at time t. Solving for $N(t)$ gives

$$N(t) = \frac{N_0}{1 + N_0 K_0 t} \tag{12.12}$$

Equations 12.11 and 12.12 are different forms of the same equation. Both are useful for describing simple monodisperse coagulation—coagulation with a constant value of K_0.

Because the rate of coagulation is proportional to N^2, it is rapid at high concentrations but slows as coagulation reduces the concentration of particles. This relationship is shown in Fig. 12.1a, a plot of experimental measurements of the number concentration of a monodisperse aerosol as a function of time. Equation 12.11 is the equation of a straight line for $1/N(t)$ versus t. When the data shown in Fig. 12.1a are replotted as $1/N$ versus t, a straight line is obtained, as shown in Fig. 12.1b. The quantity $1/N$ has units of cm^3 and represents the average gas volume per particle. The slope of the line in Fig. 12.1b is the coagulation coefficient K_0.

Simple monodisperse coagulation is restricted to particles larger than about 0.1 μm because Eq. 12.3 does not correctly describe the concentration gradient within one particle mean free path of the particle's surface. This error increases as particle size decreases and becomes significant for particles less than 0.4 μm in diameter at standard conditions. A factor β to correct for this effect is given by Fuchs (1964). The corrected coagulation coefficient K is

$$K = K_0 \beta \tag{12.13}$$

The corrected coagulation coefficient can be used for particles of any size. Values of β and K are given in Table 12.1. A more complete list of corrected coagulation coefficients is given in Table A11.

Aerosol particle size increases as a direct consequence of the decrease in number concentration due to coagulation. Assuming no losses, the particulate mass of a confined aerosol will remain constant, so mass in a unit volume (mass concentration, C_m) will also remain constant during coagulation. For liquid particles,

$$C_m = N_0 \frac{\pi}{6} \rho_p d_0^3 = N(t) \frac{\pi}{6} \rho_p (d(t))^3 \tag{12.14}$$

$$\frac{d(t)}{d_0} = \left[\frac{N_0}{N(t)} \right]^{1/3} \tag{12.15}$$

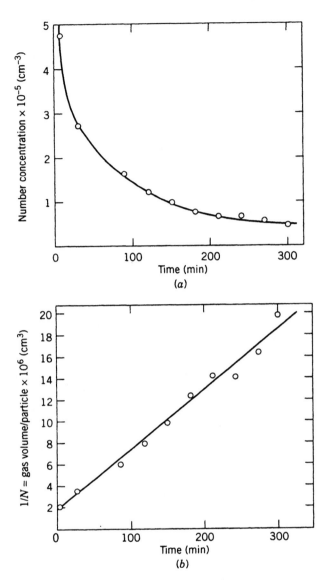

FIGURE 12.1 Coagulation of monodisperse ammonium-chloride aerosol particles, with the same data plotted two ways. (*a*) Number concentration versus time. (*b*) Average gas volume per particle ($1/N$) versus time. After Mercer (1978).

where d_0 is the initial particle size and $d(t)$ *is* the particle size at time t. Particle size increases as the inverse cube root of the number concentration, which changes with time according to Eq. 12.12. An eightfold reduction in number concentration is required to double the particle size by coagulation. For simple monodisperse coagulation, the particle size increases with time according to

$$d(t) = d_0(1 + N_0Kt)^{1/3} \qquad (12.16)$$

Equations 12.15 and 12.16 are approximately correct for solid particles, unless the particles form open structures such as those shown in Fig. 1.3b. These porous agglomerates will grow faster than predicted by Eq. 12.15. Although derived for monodisperse aerosols, Eqs. 12.11–12.15 can, and often are, applied to polydisperse aerosols. Using K calculated for the count median diameter, the equations predict, within 30%, the correct change in number concentration when there is less than an eightfold change in concentration, the count median diameter (CMD) is greater than $0.1\,\mu m$, and the geometric standard deviation (GSD) is less than 2.5.

EXAMPLE

What change in number concentration and particle size will occur during a 10-min period for a 0.8-μm monodisperse aerosol having an initial number concentration of $10^8/cm^3$? Assume simple monodisperse coagulation under standard conditions.

$K_0 = 3.0 \times 10^{-16}(1.2) = 3.6 \times 10^{-16}$ m^3/s $\quad [3.0 \times 10^{-10}(1.2) = 3.6 \times 10^{-10}$ cm^3/s]

$t = 10 \times 60 = 600$ s $\quad N_0 = 10^{14}/m^3$ $[10^8/cm^3]$

$$N(t) = \frac{N_0}{1 + N_0K_0t} = \frac{10^{14}}{1 + 10^{14} \times 3.6 \times 10^{-16} \times 600} = 4.4 \times 10^{12}/m^3$$

$$\left[= \frac{10^8}{1 + 10^8 \times 3.6 \times 10^{-10} \times 600} = 4.4 \times 10^6/cm^3 \right]$$

Thus, the number concentration decreases by a factor of 23, from 10^{14} to $4.4 \times 10^{12}/m^3$. The change in particle size is

$$\frac{d(t)}{d_0} = \left(\frac{N_0}{N(t)} \right)^{1/3} = \left(\frac{10^{14}}{4.4 \times 10^{12}} \right)^{1/3} = 2.8 \quad \left[\left(\frac{10^8}{4.4 \times 10^6} \right)^{1/3} = 2.8 \right]$$

Hence, the particle size increases by a factor of 2.8, from 0.8 to 2.26 μm. This result could have been obtained directly using Eq. 12.16:

$$d(t) = d_0(1 + N_0K_0t)^{1/3} = 0.8(1 + 10^{14} \times 3.6 \times 10^{-16} \times 600)^{1/3} = 2.26 \ \mu m$$

$$= [0.8(1 + 10^8 \times 3.6 \times 10^{-10} \times 600)^{1/3} = 2.26 \ \mu m]$$

Figures 12.2 and 12.3 show number concentration versus time for various initial concentrations, as predicted by Eq. 12.12 for a coagulation coefficient of

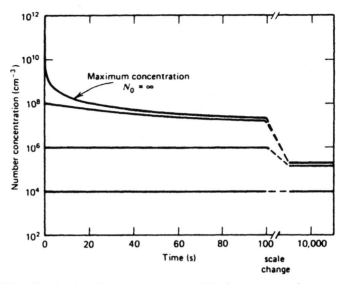

FIGURE 12.2   Simple monodisperse coagulation. Number concentration versus time for various initial concentrations, linear time scale.

5×10^{-16} m³/s [5×10^{-10} cm³/s]. Table 12.2 gives the time required for the number concentration to decrease to one-half of its initial value and the time required for the particle size to double for various values of initial concentration. Both the figures and the table show very rapid coagulation at high concentrations and slow coagulation at low concentrations. It is apparent from the figures that, after a given time has elapsed, the concentration cannot exceed a certain value regardless of its

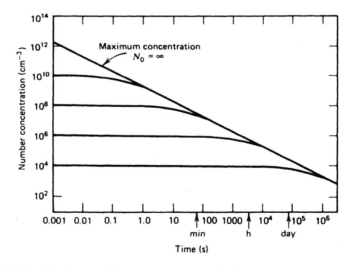

FIGURE 12.3   Simple monodisperse coagulation. Number concentration versus time for various initial concentrations, logarithmic time scale.

TABLE 12.2   Time for Number Concentration to Halve and Particle Size to Double by Simple Monodisperse Coagulation[a]

Initial Concentration, N_0 (number/m³[number/cm³])	Time to Reach $0.5 N_0$	Time for Particle Size to Double $(N = 0.125 N_0)$
$10^{20}[10^{14}]$	20 μs	140 μs
$10^{18}[10^{12}]$	2 ms	14 ms
$10^{16}[10^{10}]$	0.2 s	1.4 s
$10^{14}[10^{8}]$	20 s	140 s
$10^{12}[10^{6}]$	33 min	4 h
$10^{10}[10^{4}]$	55 h	16 days
$10^{8}[10^{2}]$	231 days	4 yr

[a]$K = 5 \times 10^{-16}$ m³/s [5×10^{-10} cm³/s].

initial concentration. The curve marked "maximum concentration" was calculated using Eq. 12.11 for $N_0 = \infty$ and represents the maximum concentration that can exist in an aerosol of the indicated age. For example, after 30 min, N cannot exceed about 10^{12}/m³[10^{6}/cm³], no matter what the initial concentration. Similarly, a global transport of aerosol taking one day will not have a concentration greater than 3×10^{10}/m³[3×10^{4}/cm³].

The two figures and the table define concentration regions where coagulation can be neglected. For example, coagulation will have a negligible effect on particle size during a 5-min measuring period if the concentration is less than 10^{12}/m³[10^{6}/cm³]. If the observation period is two days, the concentration must be less than 10^{9}/m³[10^{3}/cm³] to neglect the effect of coagulation. As a rule of thumb, coagulation is neglected in laboratory experiments and occupational hygiene work if the concentration is less than 10^{12}/m³[10^{6}/cm³]. Equation 12.6 and Fig. 12.2 suggest that the way to stop coagulation and "freeze" the particle size distribution is to rapidly dilute an aerosol to a concentration at which coagulation is negligible.

The preceding discussion considered only the coagulation of spheres. The coagulation of nonspherical particles is enhanced because of their greater surface area compared with that of spheres of the same volume and because of a greater likelihood of collision due to Brownian rotation. The effect is small for particles that are only slightly irregular. Zebel (1966) found by theoretical calculation that ellipsoids with an axial ratio of 10:1 have a coagulation coefficient 35% greater than that of spheres of the same volume.

Attractive electrical forces between particles increase coagulation by enhancing the thermal coagulation mechanism just described. Although aerosol particles with opposite signs have an enhanced likelihood of collision, this is offset by reduced collisions between particles of the same sign. The net result is a negligible change in the coagulation for aerosols with a Boltzmann equilibrium charge distribution. (See Section 15.7.) Nevertheless, there can be a significant increase in coagulation for aerosols with a strong bipolar charge distribution. Charged particles are more likely than uncharged particles to form chainlike agglomerates. Aerosol particles with a unipolar charge distribution are repelled from one another, which slows the rate of coagulation. The mutual repulsion also causes electrostatic dispersion, or the spreading of the aerosol cloud (with deposition on walls if the aerosol is contained), which in turn results in a reduction in the number concentration with a concomitant decrease in the coagulation rate.

12.2 POLYDISPERSE COAGULATION

In Section 12.1, we derived equations describing coagulation for a monodispersed aerosol. We now consider the more complicated situation of a polydisperse aerosol, in which a range of particle sizes are present. Because the rate of coagulation depends on the range of sizes present, the mathematics becomes much more complicated and no explicit solution exists.

Consider first the relatively simple case of an aerosol composed of two particle sizes, d_1 and d_2, both greater than 0.1 μm. Collisions between particles of the same size are characterized by the coagulation equations of the preceding section, with a coagulation coefficient given by Eq. 12.9. Coagulation between particles of different size is characterized by Eqs. 12.11 and 12.12, but the coagulation coefficient, $K_{1,2}$, is given by an expansion of Eq. 12.8 that includes all product combinations of d and D:

$$K_{1,2} = \pi(d_1 D_1 + d_1 D_2 + d_2 D_1 + d_2 D_2) \tag{12.17}$$

If d_2 is large and d_1 is small, then the diffusion coefficient D_1 will be comparatively large and D_2 small. Thus, the product $d_2 D_1$ will be much larger than either of their equivalent monodisperse products, $d_1 D_1$ or $d_2 D_2$, or the product $d_1 D_2$. Therefore, $K_{1,2}$ will be greater than the coagulation coefficient for either d_1 or d_2 alone, and coagulation will proceed faster between particles of different size than between particles of the same size. In a physical sense, the two components of the coagulation process are the absorbing surface and the diffusing particle. Big particles have a large surface, but slow diffusion, whereas small particles have rapid diffusion, but a small absorbing surface. These offsetting factors are the reason that the coagulation coefficient is such a weak function of particle size for monodisperse aerosols. The combination of the large absorbing surface of the big particle and the small particle's rapid diffusion to that surface means that coagulation can take place more rapidly between dissimilar-sized particles than between same-sized particles. The greater the difference in particle size, the greater the effect. Table 12.3 gives coagulation coefficients for various particle size combinations. Coagulation between 0.01- and 1.0-μm particles is 500 times more rapid than for 1.0-μm particles alone and 180 times faster than for 0.01-μm particles alone.

When a 1.0-μm particle (droplet) collides with a 0.1-μm particle, a new particle is formed that has a mass (or volume) 0.1% greater than the original 1.0-μm particle.

$$\frac{\text{new mass}}{\text{original mass}} = \frac{1.0^3 + 0.1^3}{1.0^3} = 1.001$$

TABLE 12.3 Coagulation Coefficients for Coagulation between Aerosol Particles of Different Sizes[a]

d_1 (μm)	Values of $K_{1,2}$[b]			
	$d_2 = 0.01$ μm	$d_2 = 0.1$ μm	$d_2 = 1.0$ μm	$d_2 = 10$ μm
0.01	9.6	122	1700	17000
0.1	122	7.2	24	220
1.0	1700	24	3.4	10.3
10	17000	220	10.3	3.0

[a]Calculated by Eq. 12.17 with Fuchs correction.
[b]For $K_{1,2}$ in m³/s multiply values by 10^{-16} [for cm³/s multiply values by 10^{-10}].

This means that the diameter of the new particle is 0.03% greater than that of the original 1.0-μm particle.

$$\left(\frac{1.001}{1}\right)^{1/3} = 1.0003$$

The size of the new particle is essentially the same as that of the original 1.0-μm particle, but the small particle has changed greatly; in fact, it has disappeared. Thus, large particles act as sinks for the rapid collection of small particles. This is often described as the large particles "mopping up" or "eating up" the small particles.

The situation becomes more complicated when we consider a polydisperse aerosol. Here, we must account for the coagulation of a given particle with particles of every other size. Furthermore, the size distribution and the proportion of particles of a particular size are changing with time. If we approximate the continuous size distribution with a discrete one having k size intervals, we can define an average coagulation coefficient for a polydisperse aerosol, $\overline{K}$, namely,

$$\overline{K} = \sum_{i=1}^{k} \sum_{j=1}^{k} K_{ij} f_i f_j \tag{12.18}$$

where K_{ij} is defined by Eq. 12.17 and f_i and f_j are the fractions of the total number of particles in the ith and jth intervals, respectively. The value of $\overline{K}$ obtained by Eq. 2.18 applies to a given aerosol at a given time. Since the size distribution is changing with time, a new value of $\overline{K}$ must be calculated for every step in the coagulation process. The value of $\overline{K}$ given by Eq. 12.18 can be used in Eq. 12.12 to determine the change in number concentration with time over a period for which the value of $\overline{K}$ does not change appreciably. Lee (1983), Lee and Chen (1984), and Lee et al. (1997) have developed analytical solutions for the coagulation of lognormally distributed aerosols. Lee and Chen (1984) give the following explicit equation for $\overline{K}$ for lognormal distributions with CMD > λ.

$$\overline{K} = \frac{2kT}{3\eta}\left(1 + \exp(\ln^2 \sigma_g) + \frac{2.49\lambda}{\text{CMD}}[\exp(0.5 \ln^2\sigma_g) + \exp(2.5 \ln^2\sigma_g)]\right) \tag{12.19}$$

where λ, CMD, and σ_g are the mean free path, count median diameter, and geometric standard deviation, respectively.

Table 12.4 gives values of $\overline{K}$ for lognormal distributions having various values of CMD and GSD. For a GSD of 1.0, the value is the coagulation coefficient for monodisperse aerosols. This value can be compared with the coagulation coefficient for polydisperse aerosols having the indicated values of GSD. Thus, the differences in coagulation coefficients along a row reflect the effect of polydispersity. Equation 12.19 and the entries in the table apply only to a particular distribution at a particular time. When used in Eq. 12.12, the coagulation coefficients given in the table, will correctly predict the change in number concentration for aerosols having the size distribution shown and will permit the calculation of the change in $d_{\overline{m}}$. As time passes, CMD increases, and the location of the relevant value of $\overline{K}$ moves down the table. Other diameters, such as $d_{\overline{m}}$, increase in direction proportion to CMD if GSD is constant. Table 12.4 (and Fig. 12.5, as will be discussed), permits one to estimate suitable coagulation coefficients for polydisperse aerosols of known properties and allows one to evaluate the error that results from assuming that $\overline{K}$ is a constant. Calculations can be made as a series of steps or by using an average $\overline{K}$ based on starting and ending $\overline{K}$ values.

Regardless of the shape of a size distribution or the change in that shape due to coagulation, the average diameter that can be predicted following a known change in number concentration is the diameter of average mass, $d_{\overline{m}}$. Recall from Section 4.2 that $d_{\overline{m}}$ provides the link between number and mass concentration. If there are no losses in a system, the mass concentration remains constant and is given by

$$C_m = \frac{\pi}{6} \rho_p (d_{\overline{m}})_1^3 N_1 = \frac{\pi}{6} \rho_p (d_{\overline{m}})_2^3 N_2 \qquad (12.20)$$

where N is the number concentration and the subscripts 1 and 2 refer to conditions at times 1 and 2. If ρ_p does not change (it might change for solid particles, but not for liquid particles), then

$$(d_{\overline{m}})_2 = (d_{\overline{m}})_1 \left(\frac{N_1}{N_2} \right)^{1/3} \qquad (12.21)$$

Equation 12.21 holds for any size distribution, no matter how much it has changed, provided that N_1 and N_2 are accurately known, ρ_p is constant, and there are no losses.

If a distribution is sufficiently wide (GSD > 1.5), coagulation will make it narrower with time. This narrowing results from the enhanced coagulation of the smallest particles with the largest and leads to a substantial reduction in the number of small particles and a slight increase in the number of larger particles. The net result is an increase in the diameter of average mass and a narrowing of the size distribution, as shown in Fig. 12.4. On the other hand, the coagulation of monodisperse aerosols will gradually lead to a polydisperse aerosol, as some particles get larger and these experience enhanced coagulation. Theoretical analysis (Friedlander, 1977; Lee, 1983) of the competing mechanisms of narrowing and broadening sug-

TABLE 12.4 Average Coagulation Coefficient $\bar{K}$[b]

Count Median Diameter (μm)	Coagulation Coefficient $\bar{K} \times 10^{16}$ (m³/s) [$\bar{K} \times 10^{10}$ (cm³/s)]					
	GSD = 1.0	GSD = 1.3	GSD = 1.5	GSD = 1.8	GSD = 2	GSD = 2.5
0.002	4.4	4.9	5.9	8.3	11.0	26.2
0.005	6.9	7.8	9.2	13.0	17.2	40.0
0.01	9.6	10.8	12.6	17.5	22.9	50.9
0.02	12.1	13.4	15.4	20.8	26.5	55.1
0.05	10.7	11.7	13.4	17.7	22.2	43.8
0.1	7.5	8.2	9.3	12.2	15.2	29.4
0.2	5.2	5.6	6.3	8.0	9.8	18.2
0.5	3.8	4.0	4.4	5.3	6.1	10.1
1	3.4	3.5	3.8	4.4	5.0	7.4
2	3.2	3.3	3.5	4.0	4.4	6.1
5	3.0	3.2	3.3	3.7	4.1	5.4

[a]Calculated for standard conditions by Eq. 12.18 with β correction.
[b]For $\bar{K}$ in m³/s multiply values by 10^{-16} [for cm³/s multiply values by 10^{-10}].

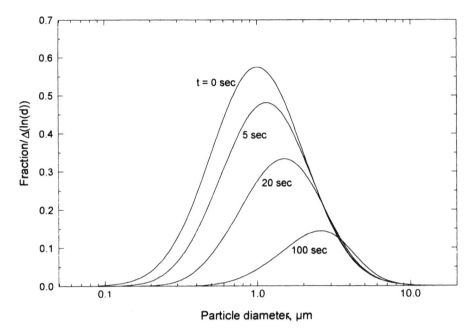

FIGURE 12.4 Effect of coagulation on particle size distribution. Numerical calculation for $N_0 = 10^{14}/m^3$ [$10^8/cm^3$], initial CMD = 1.0 μm, and initial GSD = 2.0.

gests that a stable size distribution will form after a long time, regardless of the initial size distribution. The resulting size distribution is called the *self-preserving size distribution* and is approximately lognormal with a GSD of about 1.32 to 1.36. Figure 12.5, based on an analysis by Lee (1983) for the continuum region, shows the change in GSD that occurs during the coagulation of an aerosol having a lognormal size distribution. Wide distributions become narrower and narrow distributions become wider as they proceed asymptotically toward the self-preserving size distribution with a GSD of 1.32. As a practical matter, long times are required for aerosols to reach self-preserving size distributions, unless the aerosol has a small initial particle size, very high initial number concentrations (about $10^{15}/m^3$ [$10^9/cm^3$] or greater), and relatively narrow initial size distributions (geometric standard deviations less than about 2.0). Another approach to the formation of a stable size distribution is based on the removal of small particles by coagulation and the removal of large particles by settling. This concept is discussed further in Chapter 14.

12.3 KINEMATIC COAGULATION

Kinematic, or orthokinetic, coagulation is coagulation that occurs as a result of relative motion between particles caused by mechanisms other than Brownian motion. Particles of different size settle at different rates under the influence of gravity and

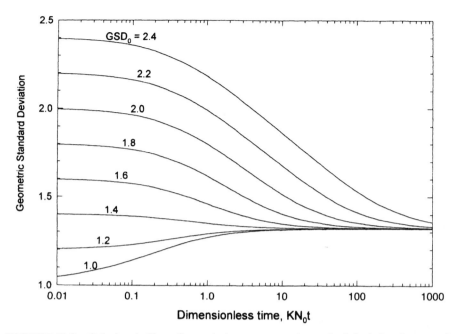

FIGURE 12.5   Calculated effect of coagulation on geometric standard deviation for aerosols with lognormal size distributions. Calculations based on Lee (1983) for the continuum region.

thereby create relative motion between them, leading to collisions and coagulation. An example is the scavenging of atmospheric aerosol particles by raindrops. Consider an aerosol composed of particles of two sizes—small particles that have negligible settling velocity and large particles. As the large particles settle through a cloud of smaller particles, they will collide with some of the smaller particles by the mechanisms of inertia and interception. We define a capture efficiency E, analogous to the single-fiber efficiency used to characterize fibrous filtration, as the ratio of the actual frequency of collisions to the frequency that would occur if the small particles were fixed and not pushed aside by the flow around the large particle. The rate of capture of small particles by a large particle is

$$n_c = \frac{\pi}{4} d_d^2 V_{TS} N E \qquad (12.22)$$

where d_d is the diameter of the large particle, V_{TS} is its settling velocity (or the relative velocity between the large and small particles), and N is the number concentration of small particles.

Capture efficiency is low, except for particles a few micrometers or larger being captured by still larger particles. The theoretical determination of capture efficiencies is a complicated aerosol problem requiring numerical analysis. An empirical expression for the capture efficiency of particles of diameter d_p by droplets of diameter d_d moving at a relative velocity V is, neglecting interception,

$$E = \left(\frac{\text{Stk}}{\text{Stk} + 0.12} \right)^2 \quad \text{for Stk} \geq 0.1 \quad (12.23)$$

where

$$\text{Stk} = \frac{\rho_p d_p^2 C_c V}{18 \eta d_d}$$

The capture efficiency decreases to zero when both particles are the same size, because their relative velocity is zero.

Similar kinematic coagulation occurs when there is relative motion between particles due to centrifugal or electrical fields. Coagulation in these situations can be much more effective than differential settling, because the relative velocities between the particles can be much greater. The droplets emitted by spray nozzles have velocities much greater than their settling velocities and thus can have higher capture efficiencies, particularly for smaller particles. Inside a wet scrubber three coagulation mechanisms may be operating, direct kinematic coagulation by spray droplets, gravitational coagulation by the settling of large spray droplets, and thermal coagulation between the dust particles and the droplets.

Wherever there is airflow, there are flow velocity gradients that give rise to *gradient* or *shear coagulation*. Particles flowing in a velocity gradient, even if they are the same size, will travel at different velocities because of their positions on different (but close) streamlines. As shown in Fig. 12.6, this relative motion leads to collisions as particles on a faster streamline overtake those on nearby slower stream-

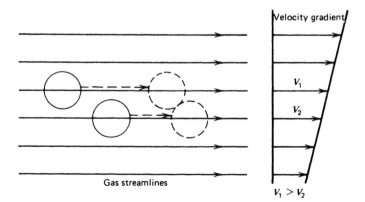

FIGURE 12.6 Diagram showing mechanism for gradient coagulation.

lines. For laminar flow in a tube, the ratio of the rate of gradient to thermal coagulation at any point is

$$\frac{d_p^2 \Gamma}{6\pi D}$$ (12.24)

where Γ is the velocity gradient (dU/dy) at the point (Zebel, 1966). Equation 12.24 predicts that gradient coagulation will exceed thermal coagulation for particles larger than 2.5 μm in a flow of 0.06 m³/hr [1.0 L/min] through a tube 10 mm in diameter.

Turbulent flow produces eddies that cause relative motion between particles. This relative motion can come from velocity gradients in a manner similar to that just described or by the inertial projection of particles across the eddies. The latter becomes particularly important as the scale of the fluctuations (the size of the eddies) approaches the particle stopping distance. For monodisperse aerosols in turbulent flow, Fuchs (1964) gives the ratio of kinematic coagulation to thermal coagulation as,

$$\frac{b d_p^2}{64\pi D} \left(\frac{\rho_g \varepsilon}{\eta} \right)^{1/2}$$ (12.25)

where b is a constant of order 10 and ε is the rate of energy dissipation per unit mass of gas. For turbulent flow in a duct of diameter d_t and average velocity $\overline{U}$, $\varepsilon = 2f\overline{U}^3/d_t$, where f is the friction factor for the duct flow, similar to C_D. For a typical velocity of 10 m/s [1000 cm/s] in a ventilation duct having a diameter of 0.2 m [20 cm], Eq. 12.25 predicts that turbulent coagulation will be more important than thermal coagulation for particles larger than about 1 μm. Generally, turbulent coagulation is negligible for 0.1-μm particles and very important for particles larger than 10 μm. Stirred coagulation is similar to turbulent flow coagulation, but the gradients are established by mechanical stirring rather than by the airflow.

High-energy sound waves will cause particles of different size to oscillate with different amplitudes; small particles will follow the air vibrations, while large particles remain virtually motionless. The coagulation resulting from this relative motion is called *acoustic* or *ultrasonic coagulation*. The term *ultrasonic* is used here to refer to the high intensity (120–160 dB) of the sound used. Frequencies in the audible range, 1–10 kHz, are normally used. There are three forms of acoustic coagulation; orthokinetic coagulation caused by the direct passage of sound waves, coagulation due to standing waves established in a resonant enclosure, and coagulation resulting from acoustically generated turbulence, usually occurring only at very high intensities (> 150 dB). Acoustic coagulation has been used in the laboratory as a pretreatment to increase particle size by coagulation in order to facilitate air cleaning, but such devices have found only limited application in commercial air cleaning.

PROBLEMS

12.1 What is the maximum number concentration for an aerosol 5 h old? Assume a coagulation coefficient of 3×10^{-16} m^3/s [3×10^{-10} cm^3/s].
ANSWER: 1.9×10^{11}/m^3 [1.9×10^5/cm^3].

12.2 The initial number concentration of a magnesium-oxide fume is 10^7/cm^3, and the particles are 0.2 μm in diameter. Determine the time required for the concentration to decrease to 10^6/cm^3. Assume simple monodisperse coagulation at 20°C with a constant K of 5×10^{-16} m^3/s [5×10^{-10} cm^3/s]. What is the average particle diameter at the end of this period?
ANSWER: 30 min, 0.43 μm.

12.3 An investigator reports observing an aerosol for 10 min and then measuring the number concentration as 10^7/cm^3. Is this reasonable? If yes, calculate the original concentration; if not, explain why not. Assume simple monodisperse coagulation with a coagulation coefficient of 3.0×10^{-16} m^3/s [3.0×10^{-10} cm^3/s].
ANSWER: No.

12.4 For an initially monodisperse aerosol of 1.0-μm liquid droplets ($\rho_p = 2000$ kg/m^3) and a number concentration of 10^6/cm^3, determine the percentage of particles that will have twice the initial particle mass after 10 min. Assume simple monodisperse coagulation. Neglect higher order coagulation.
ANSWER: 17%.

12.5 For an initial concentration of 10^{14}/m^3 [10^8/cm^3] and CMD of 1.0 μm, how long does it take for the CMD to double? Assume simple monodisperse coagulation with a constant GSD of 2.0.
ANSWER: 149 s.

12.6 What value of coagulation coefficient is given by the data shown in Fig. 12.1? What order of magnitude of particle size can be inferred from this value?
ANSWER: 9.6×10^{-16} m^3/s [9.6×10^{-10} cm^3/s], 0.05 μm.

12.7 By what factor does the average particle size of tobacco smoke increase due to coagulation during the 2 s it takes for the smoke to travel from the cigarette to your lungs? Assume that the concentration leaving the cigarette is 3×10^9/cm^3 and that $K = 7.0 \times 10^{-10}$ cm^3/s and is constant.
ANSWER: 1.7.

12.8 Calculate the number concentration of a polydisperse aerosol after 200 seconds if the initial concentration is 10^{14}/m^3 and the initial CMD and GSD are

0.2 μm and 2.0, respectively. Assume that the GSD is constant. (a) Assume that the coagulation coefficient is constant and equal to $\overline{K}$ for the initial condition. (b) Recalculate using the average $\overline{K}$ for the starting and ending conditions found in part (a).

ANSWER: $4.9 \times 10^{12}/\text{m}^3$ [$4.9 \times 10^6/\text{cm}^3$], $5.9 \times 10^{12}/\text{m}^3$ [$5.9 \times 10^6/\text{cm}^3$].

12.9 Determine the relative importance of thermal versus gradient (shear) coagulation for the flow of an aerosol ($d = 1.0$ μm) in an 8-in. duct at 2000 ft/min in the region 10 μm from the duct wall. The velocity at a distance y from the duct wall is given by

$$U(y) = \frac{yf\overline{U}^2\rho_g}{2\eta} \quad \text{for } y \ll d_t$$

where f is 0.017 for this particular case.

ANSWER: gradient coagulation = $110 \times$ thermal coagulation.

12.10 For the duct and velocity in problem 12.9, estimate the ratio of kinematic coagulation due to turbulent flow to thermal coagulation for particles with an aerodynamic diameter of 5 μm.

ANSWER: 870.

REFERENCES

Davies, C. N., "Coagulation of Aerosols by Brownian Motion," *J. Aerosol Sci.*, **10**, 151–161 (1979).

Friedlander, S. K., *Smoke, Dust and Haze*, Wiley, New York, 1977.

Fuchs, N. A., *The Mechanics of Aerosols*, Pergamon, Oxford, 1964.

Lee, K. W., and Chen, H., "Coagulation Rate of Polydisperse Particles," *Aerosol Sci. Tech.*, **3**, 327–334 (1984).

Lee, K. W., Lee. Y. J., and Hon, D. S., The Log-Normal Size Distribution Theory for Brownian Coagulation in the Low Knudsen Number Regime, *J. Colloid Interface Sci*, **188**, 486–492 (1997).

Lee, K. W., Changes of Particle Size Distribution during Brownian Coagulation, *J. Colloid Interface Sci.*, **92**, 315–325 (1983).

Mercer, T. T., "Brownian Coagulation: Experimental Methods and Results," in Shaw, D. T. (Ed.), *Fundamentals of Aerosol Science*, Wiley, New York, 1978.

Pruppacher, H. R., and Klett, J. D., *Microphysics of Clouds and Precipitation*, 2d ed., Kluwer, Dordrecht, the Netherlands, 1997.

Seinfeld, J. H., and Pandis, S. N., *Atmospheric Chemistry and Physics*, Wiley, New York, 1998.

Zebel, G., "Coagulation of Aerosols," in Davies, C. N. (Ed.), *Aerosol Science*, Academic Press, New York, 1966.

13 Condensation and Evaporation

The formation and growth of aerosol particles by condensation is the principal method of aerosol production in nature and the most important mass-transfer process between the gas phase and the particulate phase. This process usually requires a supersaturated vapor and is initiated by the presence of small particles (nuclei) or ions that serve as sites for particle formation. The supersaturation for the formation of atmospheric clouds is produced by cooling a saturated vapor by mixing or adiabatic expansion. The supersaturation for the formation of photochemical smog is produced by gas-phase chemical reactions that yield products of low vapor pressure. The other principal method of aerosol production, the disintegration and dispersion of bulk liquids or solids, is covered in Chapter 21.

The reverse of growth by condensation is the closely related process of evaporation. This process is important for spray-drying applications and is involved in the production of nuclei, such as sea salt nuclei, which are originally formed as droplets and evaporate to form nuclei that serve as sites for subsequent condensation.

Some parts of the condensation process—particularly, formation and early growth—are complicated and incompletely understood. Consequently, some of what is presented in this chapter is qualitative.

13.1 DEFINITIONS

The *partial pressure*, p is the pressure that a gas (or vapor) in a mixture of gases would exert if it were to occupy, all by itself, the entire volume that is occupied by the mixture. The sum of the partial pressures of the components equals the total pressure of the mixture (Dalton's law). Consider a mixture of ideal gases A and B in a container at atmospheric pressure. Gas A has a partial pressure of 1/3 atm and gas B 2/3 atm. If only gas A was in the container, the pressure in the container would be 1/3 atm. By the ideal gas law, gas A would have a volume equal to one-third of the container at atmospheric pressure. Similarly, gas B would occupy two-thirds of the container at atmospheric pressure. This means, by Eq. 2.14, that one-third of the molecules are molecules of A and two-thirds are molecules of B. Thus, partial pressure is a way of defining the volume fraction of gases or vapors present in a mixture.

$$\text{volume fraction of gas } A = \frac{p_A}{p_T} \tag{13.1}$$

where p_A is the partial pressure of component A and $p_T = \Sigma p_i$ is the total pressure of the system

The *saturation vapor pressure*, also called the vapor pressure, is the pressure required to maintain a vapor in mass equilibrium with the condensed vapor (liquid or solid) at a specified temperature. When the partial pressure of a vapor equals its saturation vapor pressure, evaporation from the surface of a liquid just equals condensation on that surface, and there is mass equilibrium at the surface. The pressure in any sealed container that contains only a liquid and its vapor is the saturation vapor pressure of that material at the temperature of the container. A sealed container that contains air and liquid water in equilibrium will have a partial pressure of water vapor equal to the saturation vapor pressure of water at the temperature of the container.

In the description of aerosol condensation, the *saturation vapor pressure* for a plane liquid surface is given the symbol p_s. An empirical expression for the saturation vapor pressure of water is

$$p_s = \exp\left(16.7 - \frac{4060}{T - 37}\right) \text{ kPa} \quad \left[= \exp\left(18.72 - \frac{4062}{T - 37}\right) \text{ mm Hg}\right] \quad (13.2)$$

where T is the absolute temperature in K. Equation 13.2 agrees with published data within 0.5% over the temperature range of 273–373 K [0–100°C]. At 293 K [20°C] water has a saturation vapor pressure of 2.34 kPa [17.5 mm Hg], and at 373 K [100°C] water has a vapor pressure of 101 kPa [760 mm Hg], equal to atmospheric pressure, and will boil at sea level. An ambient temperature change of 12 K [12°C] changes p_s by a factor of two.

As a matter of convenience, much of the discussion that follows is presented in terms of water and water vapor, but the same principles apply to any other liquid–vapor system. The equilibrium water vapor pressure for a plane surface p_s is the partial pressure p of water vapor that defines a condition of *saturation*. If, at a given temperature, the partial pressure of water vapor is less than p_s, then the vapor is *unsaturated*, and if p is greater than p_s, the vapor is *supersaturated*.

The *saturation ratio* S_R is the ratio of the partial pressure of vapor in a system to the saturation vapor pressure for the temperature of the system.

$$S_R = \frac{p}{p_s} \quad (13.3)$$

When a saturation ratio is greater than unity, the gas–vapor mixture is supersaturated; when the ratio is equal to unity, the mixture is saturated; and when the ratio is less than unity, the mixture is unsaturated. The *amount of supersaturation*, or simply, the *supersaturation*, refers to that portion of the saturation ratio greater than 1.0. Thus, a supersaturation of 4% corresponds to a saturation ratio of 1.04. In our normal environment, supersaturation of water vapor rarely exceeds a few percent, because at that level condensation occurs by the mechanisms described in Section 13.5 and prevents any further increase in vapor concentration. The more familiar term for saturation ratio is *relative humidity*. When expressed in percent, relative humidity is equal to 100 times the saturation ratio. Relative humidities of 110% and 50% correspond to saturation ratios of 1.10 and 0.50, respectively.

Supersaturation is usually produced by cooling a saturated vapor. Because the partial pressure of the vapor remains constant and p_s decreases with decreasing temperature, a saturation ratio greater than 1.0 can be achieved. One procedure for producing a uniform supersaturation throughout a vapor, without having to establish temperature gradients, is through cooling by *adiabatic expansion*—that is, an expansion that does not allow any heat input from the surroundings. In the laboratory, this is accomplished by the rapid expansion of a gas in an insulated chamber or by the expansion of a gas as it exits a nozzle. Clouds are formed in the atmosphere as large masses of humid air rise by buoyancy and expand adiabatically with the decrease in pressure due to altitude. In this case, the mass of air is so large that there is no practical transfer of heat from the surrounding air into the bulk of the buoyant air mass. Adiabatic expansion is also responsible for the thin wisp of "smoke" that is sometimes observed in the neck of a wine bottle when its cork is withdrawn rapidly. The cork acts as a piston, expanding adiabatically the saturated vapor above the wine. A similar process occurs during forced exhalation by whales to produce a visible spout of condensed moisture.

For adiabatic expansion, the relationship between the absolute temperature and the total volume of a gas or vapor before and after expansion is given by the thermodynamic equation

$$\frac{T_2}{T_1} = \left(\frac{v_1}{v_2}\right)^{\kappa - 1} \tag{13.4}$$

where κ is the specific heat ratio (1.40 for air) and the subscripts 1 and 2 refer to the conditions before and after the expansion. The ratio of the final to the initial volume, v_2/v_1, is called the *expansion ratio*. For a known initial temperature and expansion ratio, the resulting saturation ratio (Eq. 13.3) can be determined by calcu-

TABLE 13.1   Expansion Ratio, Final Temperature, and Saturation Ratio for Adiabatic Expansion of Saturated Air Initially at 293 K [20°C][a]

Expansion Ratio v_2/v_1	T_2 (K)	Saturation Ratio $(p_s)_1/(p_s)_2$
1.0	293	1.0
1.1	282	2.0
1.2	272	4.0[b]
1.3	264	8.4[b]
1.4	256	17.0[b]
1.5	249	33.3[b]

[a]Assumes that no condensation or crystal formation occurs. Subscripts 1 and 2 refer to conditions before and after expansion, respectively.
[b]$p_s = \exp(22 - 6145/T)$ for $T_2 < 273$ K.

lating the saturation vapor pressure, using Eq. 13.2, at the initial temperature to get p and at the final temperature, as given by Eq. 13.4, to get p_s. Table 13.1 gives the saturation ratios resulting from adiabatic expansion of saturated air initially at 293 K [20°C].

EXAMPLE

Air saturated with water vapor at 20°C is expanded adiabatically by 18%. What are the new temperature and saturation ratio?

$$p = (p_s)_{20°C} = 2.34 \text{ kPa} \quad [17.5 \text{ mm Hg}]$$

$$\text{expansion ratio} = \frac{v_2}{v_1} = 1.18$$

$$T_2 = T_1 \left(\frac{v_1}{v_2}\right)^{\kappa-1} = 293\left(\frac{1}{1.18}\right)^{1.40-1} = 274.2 \text{ K} \quad [= 1.2° \text{ C}]$$

$$p_s = \exp\left(16.7 - \frac{4060}{T_2 - 37}\right) = 0.66 \text{ kPa} \quad \left[\exp\left(18.72 - \frac{4062}{T_2 - 37}\right) = 4.95 \text{ mm Hg}\right]$$

$$S_R = \frac{p}{p_s} = \frac{2.34}{0.66} = 3.5 \quad \left[\frac{17.5}{4.95} = 3.5\right]$$

13.2 KELVIN EFFECT

The saturation vapor pressure is defined as the equilibrium partial pressure for a plane (flat) liquid surface at a given temperature. If the liquid surface is sharply curved, such as the surface of a small droplet, the partial pressure required to maintain mass equilibrium is greater than that for a flat surface. The curvature of the surface modifies slightly the attractive forces between surface molecules, so that the smaller the droplet, the easier it is for molecules to leave the droplet surface. To prevent this evaporation—that is, to maintain mass equilibrium—the partial pressure of vapor surrounding the droplet must be greater than p_s.

For *pure* liquids, the relationship between the saturation ratio required for equilibrium (no growth or evaporation), called the *Kelvin ratio* K_R, and the droplet size is given by the Kelvin or Thomson–Gibbs equation,

$$S_R \text{ required for mass equilibrium for size } d^* = K_R = \frac{p_d}{p_s} = \exp\left(\frac{4\gamma M}{\rho R T d^*}\right) \quad (13.5)$$

where γ, M, and ρ are, respectively, the surface tension, molecular weight, and density of the droplet liquid. $d*$ is the Kelvin diameter, the diameter of the droplet that will neither grow nor evaporate when the partial pressure of vapor at the droplet surface is p_d. Equation 13.5 defines a specific saturation ratio required to maintain mass equilibrium for a given particle size $d*$. These specific saturation ratios are called Kelvin ratios K_R where $K_R = f(d*)$. For every size of droplet, there is one saturation ratio that will exactly maintain that size of the particle, namely, the Kelvin ratio; too great a saturation ratio, and the particle grows; too small, and it evaporates. Conversely, for a given Kelvin ratio only those particles having a diameter $d*$ are stable; smaller ones evaporate, and larger ones grow.

The right-hand side of Eq. 13.5 is always positive, so the Kelvin ratio is always greater than unity, and supersaturation is required to prevent droplets of pure liquid from evaporating. This effect is called the *Kelvin effect* and, for water droplets, is significant only for particles less than 0.1 μm. It follows from Eq. 13.5 that $p_d = K_R p_s$ and $p_d \geq p_s$. The relationship between the Kelvin ratio and $d*$ is shown in Fig. 13.1 for two materials, water and di (2-ethylhexyl) phthalate (DOP), that span the range of normally encountered liquids. The line for each material represents the boundary between growth (condensation) and evaporation. The region above each line is a growth, region and that below the line an evaporation region. Although the lines define the required saturation ratio for mass equilibrium for a given size of particle, they do not define the conditions for stability of a droplet . For a fixed saturation ratio, droplets of pure material just above the line will grow, theoretically without limit, and those just below the line will evaporate completely.

EXAMPLE

What Kelvin ratio is necessary to maintain the size of 0.05-μm pure water droplets at 293 K [20°C]? What is p_d for this situation?

$$K_R = \exp\left(\frac{4\gamma M}{\rho R T d*}\right) = \exp\left(\frac{4 \times 0.0727 \times (18/1000)}{1000 \times 8.31 \times 293 \times 0.05 \times 10^{-6}}\right) = 1.044$$

$$\left[K_R = \exp\left(\frac{4 \times 72.7 \times 18}{1 \times 8.31 \times 10^7 \times 293 \times 0.05 \times 10^{-4}}\right) = 1.044\right]$$

$$p_d = K_R p_s = 1.044(2.34) = 2.44 \text{ kPa} \quad [1.044(17.5) = 18.3 \text{ mm Hg}]$$

One implication of the Kelvin equation, shown in Fig. 13.1, is that at saturation $(S_R = 1.0)$, small droplets of *pure* liquid are not stable, but will evaporate and eventually disappear. A slightly supersaturated environment is required to prevent these droplets from evaporating. In the preceding example, the 0.05-μm-diameter water droplets require a Kelvin ratio of greater than 1.044, equivalent to a relative humidity of 104.4%, to prevent evaporation. Note that the Kelvin equation applies only to

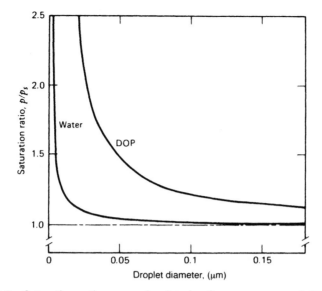

FIGURE 13.1   Saturation ratio versus droplet size for pure water and di(2-ethyhexyl) phthalate (DOP) at 293 K [20°C]. The region above each curve is a growth region and that below, an evaporation region.

pure materials; in the usual ambient situation, covered in Section 13.5, impurities and electric charge modify the situation greatly.

13.3   HOMOGENEOUS NUCLEATION

Homogeneous nucleation is the formation of particles from a supersaturated vapor without the assistance of condensation nuclei or ions. The process is also called *self-nucleation*. This type of particle formation is rare for water vapor in the atmosphere, but can occur for other gases and vapors, and can be produced readily in the laboratory to study the process of formation and growth. We cover homogenous nucleation here because it serves as a framework for understanding the more common heterogeneous nucleation processes discussed in Section 13.5.

   An examination of Eq. 13.5 suggests that, for a given saturation ratio greater than unity, a particle must initially reach a diameter d^* to grow and become a stable droplet. A theoretical saturation ratio of 220 would be required for growth to commence on an individual water molecule ($d^* = 0.0004$ μm). Experimental measurements of pure materials have shown that particle formation and growth occur at saturation ratios far less than this—at less than 10 for pure materials and less than 1.02 for nucleated condensation. Clearly, the homogeneous nucleation process does not start with individual molecules.

   Even in unsaturated vapor, the attractive forces between molecules, such as van der Waals forces, lead to the formation of molecular clusters. The clusters are

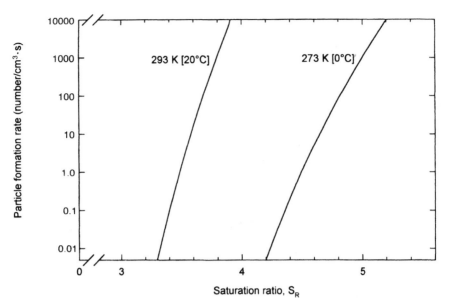

FIGURE 13.2   Particle formation rate by homogeneous nucleation versus saturation ratio at 293 and 273 K [20 and 0°C].

formed continuously, but are unstable and continuously disintegrate. When the vapor is supersaturated, the number concentration of clusters increases to the point where they collide with one another frequently. This process is similar to coagulation, except that the "agglomerates" disintegrate soon after being formed. The greater the supersaturation, the greater the number concentration of clusters and the more frequent is the formation of transient "agglomerates" having a size that exceeds d^*. Once such an "agglomerate" exceeds d^*, even momentarily, it becomes stable and grows by condensation to form a large particle. For a given vapor and temperature, the supersaturation required for this to happen occurs at a well-defined point called the *critical saturation ratio*.

Pure water vapor at 293 K [20°C] will spontaneously form particles by homogeneous nucleation when the saturation ratio exceeds 3.5. This value corresponds to a Kelvin diameter of 0.0017 μm, the diameter of a molecular cluster containing 90 water molecules. At 273 K [0°C], a saturation ratio of 4.5 is required for homogeneous nucleation of water vapor. As shown in Fig. 13.2, homogeneous nucleation is nearly an all-or-nothing phenomenon. It produces a high number concentration of submicrometer particles with a narrow, but not monodisperse, size distribution.

In photochemical smog formation, certain gas phase reactions are promoted by ultraviolet light and form low-vapor-pressure reaction products. Because of their low vapor pressure, these products exist at high supersaturations and can form particles by homogeneous nucleation. When an increase in aerosol mass concentration occurs in the atmosphere by this mechanism, it is called *gas-to-particle conversion*.

13.4 GROWTH BY CONDENSATION

Once a stable droplet is formed—that is, once the droplet diameter exceeds $d*$ for a particular saturation ratio—the droplet has passed a threshold and will grow by condensation. The rate of growth depends on the saturation ratio, particle size, and particle size relative to the gas mean free path. When a particle first starts to grow, its size will likely be less than the mean free path. For this condition, the rate of particle growth is governed by the rate of random molecular collisions between the particle and the vapor molecules. The rate of collisions is given by the kinetic theory of gases, Eq. 2.15. Combining this equation with Eq. 2.22 for the mean molecular velocity and expressing n, the molecular concentration, in terms of the ambient partial pressure of vapor p_∞ by the ideal gas law ($n = p_\infty/kT$) gives

$$z = \frac{p_\infty}{\sqrt{2\pi mkT}} \qquad (13.6)$$

where m is the mass of a vapor molecule. This equation gives the rate of arrival of vapor molecules per unit area of droplet surface. To characterize growth, we need the *net* rate of arrival after deducting for molecules that leave the surface due to evaporation and for those arriving molecules that do not stick. The rate of loss is given by Eq. 13.6 with p_∞ replaced by p_d, the partial pressure at the droplet surface as given by Eq. 13.5. The fraction of arriving molecules that stick is called the *condensation coefficient* α_c, for which a value of 0.04 is often used although there is uncertainty as to the correct value. See Barrett and Clement (1988) and Seinfeld and Pandis (1998). The *net* rate of arrival of molecules to the entire droplet surface n_z is $(z_{in} - z_{out}) A_s$, where A_s is the surface area of the droplet.

$$n_z = \frac{\pi d_p^2 \alpha_c (p_\infty - p_d)}{\sqrt{2\pi mkT}} \qquad (13.7)$$

The rate of change of particle volume is

$$\frac{dv}{dt} = n_z v_m \qquad (13.8)$$

where v_m is the volume of a molecule, which is given by

$$v_m = \frac{M}{\rho_p N_a} \qquad (13.9)$$

where M is the molecular weight of the liquid, ρ_p is the density of the liquid, and N_a is Avogadro's number. Taking the derivative of $v = (\pi/6)d_p^3$ and combining the result with Eqs. 13.7–13.9 gives the rate of particle growth:

$$\frac{d(d_p)}{dt} = \frac{2M\alpha_c(p_\infty - p_d)}{\rho_p N_a \sqrt{2\pi mkT}}, \quad \text{for } d_p < \lambda \qquad (13.10)$$

where p_∞ is the partial pressure of vapor in the gas surrounding the droplet, but away from its immediate surface, and p_d is the partial pressure of vapor at the droplet surface, as given by the Kelvin equation, Eq. 13.5.

For particles larger than the gas mean free path, growth depends, not on the rate of random molecular collisions, but on the rate of diffusion of molecules to the droplet surface. This situation is completely analogous to the coagulation of aerosol particles covered in Section 12.1. The rate of collisions of vapor molecules with a droplet of diameter d_p by diffusion can be obtained from Eq. 12.4. However, since Eq. 12.4 was derived for particles of equal size and diffusion coefficient, diffusing to each other, we must adjust it to account for the fact that a molecule has a negligible diameter compared with a droplet and that the droplet has a negligible diffusion coefficient compared with that of a vapor molecule, D_v. Making these adjustments, Eq. 12.4 becomes

$$n_z = 2\pi d_p D_v N \tag{13.11}$$

where N must be replaced with the net concentration of vapor molecules, expressed, as before, in terms of the partial pressure p_∞ and temperature T_∞ away from the droplet surface and the partial pressure p_d and temperature T_d at the droplet surface.

$$n_z = \frac{2\pi d_p D_v}{k}\left(\frac{p_\infty}{T_\infty} - \frac{p_d}{T_d}\right) \tag{13.12}$$

where k is Boltzmann's constant. Following the procedure used to obtain Eq. 13.10, we see that the rate of particle growth for an isolated droplet of pure liquid becomes

$$\frac{d(d_p)}{dt} = \frac{4D_v M}{R\rho_p d_p}\left(\frac{p_\infty}{T_\infty} - \frac{p_d}{T_d}\right)\phi \quad \text{for } d_p > \lambda \tag{13.13}$$

where R is the gas constant and ϕ is the Fuchs correction factor, explained later in this section.

Under conditions of slow growth, with $1.0 < S_R < 1.05$, the droplet temperature will be approximately the same as the ambient temperature ($T_d \cong T_\infty$), and p_d can be calculated by Eq. 13.2 using the ambient temperature. The Kelvin effect can usually be neglected when $d_p > \lambda$. For more rapid growth, such as that which occurs in a condensation nuclei counter (see Section 13.6), droplet heating takes place due to the release of latent heat of vaporization during the condensation process. An equilibrium droplet temperature is established by balancing the heat acquired by the droplet due to condensation with that lost by conduction to the cooler surrounding air. The resulting steady-state temperature elevation $T_d - T_\infty$, is independent of the droplet size and is given by

$$T_d - T_\infty = \frac{D_v M H}{R k_v}\left(\frac{p_d}{T_d} - \frac{p_\infty}{T_\infty}\right) \tag{13.14}$$

where H is the latent heat of evaporation and k_v is the thermal conductivity of the gas. For T_∞ between 0 and 50°C, the quantities D_v, H, and k_v can be evaluated at T_∞ with less than a 5% error in the calculated temperature elevation. Because of the dependence of p_d on T_d, Eq. 13.14 cannot be solved explicitly. Instead, the following empirical equation permits direct calculations of $T_d - T_\infty$ for S_R of 0–5 and an ambient temperature of 273–313 K [0–40°C].

$$T_d - T_\infty = \frac{(6.65 + 0.345T_\infty + 0.0031T_\infty^2)(S_R - 1)}{1 + (0.082 + 0.00782T_\infty)S_R} \qquad (13.15)$$

where T_∞ is in °C. $T_d - T_\infty$, as calculated by Eq.13.15, agrees with numerical calculations using Eq. 13.14 within 0.2 K [0.2°C] for S_R of 0.3–3.6 and T_∞ of 273–303 [0–30°C] and within 1 K [1°C] for S_R of 0–5 and T_∞ of 273–313 K [0–30°C].

Figure 13.3 shows $T_d - T_\infty$ for water droplets at different ambient conditions. The particle temperature is elevated relative to its surroundings during growth ($S_R > 1.0$) and depressed during evaporation ($S_R < 1.0$). During growth, the effect of the elevated temperature is to increase the partial pressure of vapor at the droplet surface and slow the rate of condensation and growth.

Equation 13.13 is based on the diffusion of molecules to the surface of the droplet. Both the diffusion equation and the concept of a gradient break down in the re-

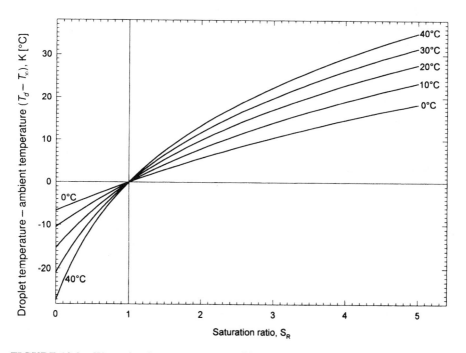

FIGURE 13.3 Water droplet temperature–ambient temperature ($T_d - T_\infty$) versus saturation ratio for ambient temperatures of 273–313 K [0 – 40°C].

gion within one mean free path of the droplet surface. In this region, transport is controlled by kinetic processes and different equations apply. The Fuchs correction factor ϕ (also called the Knudsen correction) corrects Eq. 13.13 for this effect (Fuchs, 1959). The factor is given by Davies (1978) as

$$\phi = \frac{2\lambda + d_p}{d_p + 5.33(\lambda^2/d_p) + 3.42\lambda} \tag{13.16}$$

This effect results in a slowing of the rate of growth of small particles. It is significant for particles less than 1 μm ($\phi = 0.91$) and very important for particles less than 0.1 μm ($\phi = 0.42$).

13.5 NUCLEATED CONDENSATION

Nucleated condensation, or heterogeneous nucleation, is a process of particle formation and growth that is promoted by the presence of condensation nuclei or ions. Whereas homogeneous nucleation usually requires saturation ratios of 2–10, nucleated condensation can occur at supersaturations of only a few percent. For soluble nuclei, condensation can occur in unsaturated conditions, and droplets that are stable in size can be formed. Nucleated condensation is the primary mechanism for the formation of clouds in the atmosphere. We consider three cases: insoluble nuclei, ions, and soluble nuclei.

Insoluble nuclei can provide a passive site for the condensation of a supersaturated vapor. At a given level of supersaturation, an insoluble nucleus with a wettable surface will have an adsorbed layer of vapor molecules on its surface. If its diameter is greater than d^*, the nucleus will behave like a droplet of that size and grow by condensation, as described in the previous section. The actual situation is more complicated because a particle's ability to nucleate condensation depends on many factors, including its size, shape, chemical composition, surface structure, and surface charge. Under favorable conditions, the size of the nucleus required for growth at a given supersaturation is given by d^* in the Kelvin equation, Eq. 13.5. For example, in a 5% supersaturated atmosphere at 20°C, nuclei larger than 0.05 μm in diameter will grow to form large droplets.

Closely related to droplet formation is the ability of certain finely dispersed materials, such as silver iodide crystals, to serve as ice nuclei. These materials, because of their crystal structure, cause ice particles to form and grow in a supercooled cloud in a manner analogous to droplet nucleation. This topic is covered in books on cloud physics [see for example Pruppacher and Klett (1997)] and is not covered here.

Our normal atmosphere contains about 1000 ions/cm^3. These ions are continuously created by the action of cosmic radiation and radioactive gases that emanate from the ground. Ions are singly charged clusters of about 30 air molecules that differ from the clusters described in Section 13.3 in that they are stable. The presence of a charged molecular cluster slightly distorts the curve of the Kelvin ratio versus d^* (Fig. 13.1), facilitating the formation and growth of droplets at supersaturations greater than about 2.0. Since our normal atmosphere contains numerous condensation nuclei that can cause particle formation and growth at much lower levels of supersaturation, air ions usually have little effect on the formation of particles larger than 0.01 μm in the atmosphere.

The ability of air ions to promote particle formation and growth at high saturation ratios has been used in Wilson cloud chamber experiments to observe the tracks associated with radioactive decay. A level of supersaturation below the threshold for significant spontaneous growth (homogeneous nucleation), but sufficient to permit ions to promote particle formation and growth, is established in a chamber. When a radioactive disintegration produces an alpha "particle," it travels a few centimeters through the air, leaving behind a wake of thousands of air ions. These ions grow rapidly to micrometer-sized particles, and the path of the alpha particle can be seen with suitable lighting.

The most important particle formation mechanism is condensation on *soluble nuclei*. The discussion that follows holds for *any* soluble nuclei, but only the most common soluble nucleating material, sodium chloride, is considered. Sodium-chloride nuclei are formed in great numbers by the action of waves and bubbles in the oceans and exist throughout the global atmosphere.

As is well known, salt dissolved in water raises the boiling point of the water. The salt does this by lowering the equilibrium vapor pressure above the water surface, which, in the case of droplets, allows growth by condensation to occur at a lower saturation ratio than for pure water. The affinity of the dissolved salt for water also allows the formation of stable droplets in a saturated or unsaturated environment.

Because soluble nuclei permit particle growth at supersaturations of only a few percent, these nuclei are the primary vehicle for the formation of water droplets in the atmosphere. Supersaturations will not normally exceed a few percent, because particle formation and growth on the ever-present soluble nuclei will consume the excess vapor (relieve the supersaturation) and prevent other mechanisms (or smaller nuclei) from becoming active at greater supersaturations. For a droplet containing a soluble nucleus, two competing effects control the relationship between the Kelvin ratio and particle size. First, the salt concentration in a droplet will increase with decreasing droplet size because the mass of salt remains constant and only the water evaporates. Thus, a given mass of dissolved salt serves to reduce the vapor pressure at the droplet surface to a greater degree as the particle size decreases. In competition with this trend is the Kelvin effect, which causes an *increase* in vapor pressure at the droplet surface as the particle size decreases. The relationship between the Kelvin ratio and the particle size for droplets containing dissolved materials is

$$K_R = \frac{p_d}{p_s} = \left(1 + \frac{6imM_w}{M_s\rho\pi d_p^3}\right)^{-1} \exp\left(\frac{4\gamma M_w}{\rho RTd_p}\right) \tag{13.17}$$

where m is the mass of the dissolved salt having a molecular weight M_s, M_w is the molecular weight of the solvent, usually water, ρ is the density of the solvent, and

i is the number of ions each molecule of salt forms when it dissolves (2 for NaCl). Equation 13.17 applies when the saturation ratio is near 1.0. The first factor in brackets is due to the effect of the dissolved salt, and the exponential term is the Kelvin equation.

These competing effects are best understood by their effect on the graph of saturation ratio versus particle size shown in Figs. 13.4 and 13.5. With no dissolved salt, Eq. 13.17 is equivalent to the Kelvin equation, and the relationship between the particle size and the Kelvin ratio is as shown in Fig. 13.4 by the line labeled "pure water." Each of the other lines, called Kohler curves, corresponds to droplets containing a fixed mass of dissolved sodium chloride. This mass, the mass of the original salt nucleus, remains constant during growth or evaporation of the droplet. As in Fig. 13.1, the region above a line in Fig. 13.4 is a growth region and that below the line an evaporation region. As shown in Fig. 13.4 the presence of the dissolved salt dramatically alters the shape of the curve separating evaporation from growth.

Figure 13.5 shows the growth or evaporation that occurs at a constant saturation ratio in the particular case of a sodium-chloride nucleus of mass 10^{-16} g. This mass of sodium chloride corresponds to a sphere 0.045 μm in diameter. For a supersaturation above 0.36% (RH = 100.36%), the nucleus will form a droplet and grow as long as the supersaturation is maintained. This relationship is shown by the arrow labeled A in Fig. 13.5. Such droplet growth will occur for all sodium-chloride nuclei of mass equal to or greater than 10^{-16} g when the saturation ratio is greater than

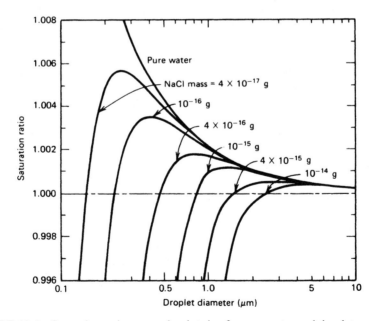

FIGURE 13.4 Saturation ratio versus droplet size for pure water and droplets containing the indicated mass of sodium chloride at 293 K [20°C]. The region above each curve is a growth region and that below, an evaporation region.

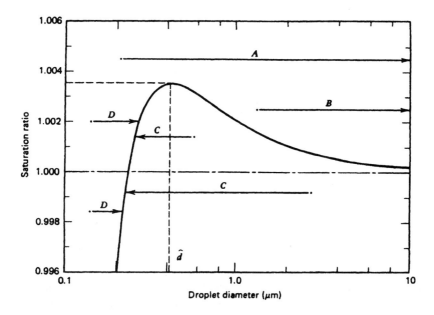

FIGURE 13.5   Saturation ratio versus droplet size for droplets containing 10^{-16} g of sodium chloride at 293 K [20°C]. Growth and evaporation at a constant saturation ratio are shown by arrows.

that corresponding to the maximum of the curve at $\hat{d}$. Similar growth will occur for a droplet size and saturation ratio shown by arrow B. Droplets below the curve will evaporate and eventually reach the curve at a size less than $\hat{d}$, as shown by the arrows labeled C. Droplets at a condition that puts them above the curve and to the left of $\hat{d}$ will grow until they, too, reach the curve, as shown by the arrows labeled D. In all cases for the arrows labeled C and D, the droplets will reach a stable size at the curve to the left of $\hat{d}$. This holds true even if the saturation ratio is less than unity (i.e., if we have unsaturated conditions). Thus, our 0.045-μm salt nucleus will end up as a 0.23-μm droplet at saturated conditions. This is a true stability condition, and the droplet will neither grow nor evaporate unless the saturation ratio (relative humidity) changes. If that happens, the particle will change to a new equilibrium size.

Figures 13.6 and 13.7 are plots of relative humidity versus particle size, similar to Figs. 13.4 and 13.5, but covering a much wider range of saturation ratios. Figure 13.6 shows the size of a droplet containing a sodium chloride nucleus of the indicated mass versus relative humidity. Figure 13.7 shows, for a sodium chloride nucleus of 10^{-14} g, the transition from a salt crystal (0.21 μm) to a droplet (0.38 μm) at 75% relative humidity when the humidity is increasing and the recrystallization at about 40% relative humidity when humidity is decreasing. This delayed recrystallization is called *hysteresis*. It is different for different salts, and the characteristic signature (Fig. 13.7), can be used to identify the chemical species present. Summertime haze is a result of such growth brought about by high summer humidi-

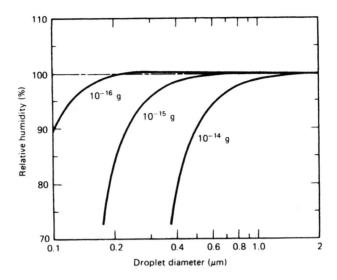

FIGURE 13.6 Relative humidity versus droplet size for droplets containing the indicated mass of sodium chloride at 293 K [20°C]. Figure 13.4 is an enlarged portion of this figure.

ties that cause typical salt nuclei to reach droplet sizes of about 0.3 μm in diameter, a size that is very effective at scattering light and, therefore, can cause these visibility effects. (See Chapter 16.)

13.6 CONDENSATION NUCLEI COUNTERS

The ability of minute particles to grow to micrometer-sized droplets in a supersaturated environment can be exploited to measure their number concentration. Instruments that do this are called *condensation nuclei counters* (CNCs) or *condensation particle counters* (CPCs). All of these instruments saturate an aerosol by water or alcohol vapor and then cool it by adiabatic expansion or flow through a cold tube to create the level of supersaturation required for growth. The nuclei that are present are all exposed to the same level of supersaturation for the same length of time and grow to about 10 μm in diameter, regardless of their initial size. Since each nucleus grows to a droplet, the number concentration of droplets and nuclei remain the same. The number concentration of droplets is usually measured by suitably calibrated light-transmission measurements (see Section 16.2) or by single-particle optical counters (see Section 16.5).

To a first approximation, the size of the smallest particles that will grow is equal to d^* in the Kelvin equation, Eq. 13.5, for the level of supersaturation achieved. Instruments used to study atmospheric clouds achieve a supersaturation of only a few percent, because this level is the maximum that occurs during cloud formation. The nuclei measured with such instruments are called cloud condensation nuclei (CCN). More common kinds of instruments (CN counters) achieve supersaturations of

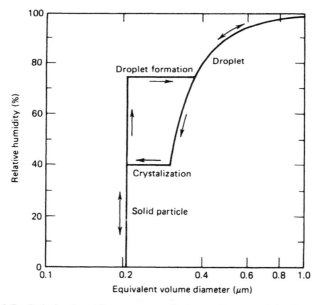

FIGURE 13.7 Relative humidity versus equivalent volume particle diameter showing the transition from a 10^{-14}-g sodium-chloride particle to a droplet with increasing humidity and recrystallization with decreasing humidity. Transitions are approximate.

200-400%, which corresponds to a theoretical lower limit of detectability of about 0.002 µm in diameter. Depending on the type of instrument, particle number concentrations from 100 to 10^{13}/m^3 [0.0001 to 10^7/cm^3] can be measured. Higher concentrations can be measured by diluting the aerosol with nuclei-free air before measurement. Some manually operated instruments require about 2 min for each measurement; other instruments operate automatically to provide real-time information on number concentration. These instruments can be calibrated by allowing the 10-µm droplets formed to settle onto a coated slide and counting the number of spots with a microscope. There is uncertainty as to the actual size of the smallest detectable nuclei; it depends on the instrument and on the chemical nature of the nuclei. Figure 13.8 shows measured counting efficiency versus particle size for a continuous-flow CPC using *n*-butyl alcohol as the working fluid.

The reference instrument against which others are calibrated is the Pollak counter, shown schematically in Fig. 13.9. The device consists of a vertical tube 60 cm high with a water-saturated ceramic lining, a light source at the top, and a photodetector at the bottom. The aerosol sample is drawn into the tube, and the tube is sealed and pressurized with nuclei-free air. After a delay to allow the air to become saturated, a valve is opened to permit adiabatic expansion down to atmospheric pressure. The change in light transmission before and after growth is related to the particle number concentration by a suitable calibration curve. The concentration range for this instrument is 10^8–5×10^{11}/m^3 [100–500,000/cm^3].

Automated versions of the Pollak counter operate on the same principle as the manual version, but at a faster rate of one to five operating cycles per second. The

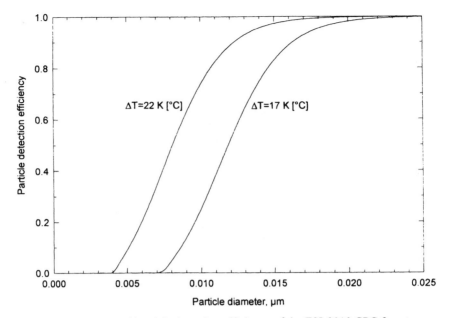

FIGURE 13.8 Measured particle detection efficiency of the TSI-3010 CPC for a tempera-ture difference of 22 and 17 K [22 and 17°C] between saturator and condenser. Data from Wiedensohlet et al. (1997).

flow rates range from 20 cm^3/s to 100 cm^3/s, and concentration ranges are from 3×10^8 to 10^{13}/m^3 [300 to 10^7/cm^3].

A continuous-flow CPC is shown in Fig. 13.10. A supersaturation of n-butyl alcohol is achieved by saturating the aerosol stream as it flows through an alcohol-saturated, felt-lined tube and subsequently cooling it to 283 K [10°C] in a chilled tube. Droplets exiting the tube can be counted individually for low concen-trations by monitoring the scattered light pulses with an optical particle counter or by monitoring the change in transmitted light, as described earlier, for high con-centrations. Different versions of this instrument are available with flow rates of 30 to 1000 cm^3/s and concentration limits of 100–10^{13}/m^3 [10^{-4}–10^7/cm^3]. The use of alcohol condensation favors the detection of very small oil and combustion particles that are hydrophobic. The Portacount (TSI, Inc., St. Paul, MN), a handheld, battery-powered portable instrument, operates on the same principle, but uses isopropyl alcohol as the condensing vapor. It can detect particles as small as 0.02 μm and has a concentration range of 0.1 to 10^5 particles/cm^3. Continuous-flow condensation nuclei counters are used with diffusion batteries (Section 7.5) to measure the num-ber size distribution of aerosols in the range of 0.002–0.2 μm.

13.7 EVAPORATION

In evaporation, which is the reverse of growth by condensation, more molecules leave the particle surface than arrive. With evaporation, there is no counterpart to

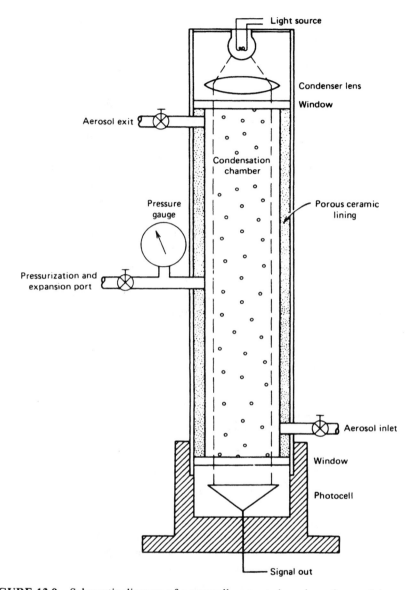

FIGURE 13.9 Schematic diagram of a manually operated condensation nuclei counter.

the threshold of formation that must be reached for condensation to commence, and droplets of pure liquid will evaporate completely. The objective here is to determine the *rate of evaporation* (the rate of change of particle size with time), $d(d_p)/dt$, and the time required for a droplet to evaporate completely, the *droplet lifetime* or *drying time*.

For particles much larger than the mean free path, the rate of evaporation is controlled by the rate at which vapor can diffuse away from the droplet. This rate is given by Eq. 13.13, the equation that governs the rate of growth by condensation,

$$\frac{d(d_p)}{dt} = \frac{4D_vM}{R\rho_p d_p}\left(\frac{p_\infty}{T_\infty} - \frac{p_d}{T_d}\right)\phi \quad \text{for } d_p > \lambda \tag{13.18}$$

where ϕ, the Fuchs correction factor, is needed for $d_p < 1.0$ μm. When the partial pressure of vapor well away from the droplet surface, p_∞, is less than the partial pressure of vapor at the droplet surface, p_d, the right-hand side of Eq. 13.18 will be negative, and the particle size will be decreasing with time; that is, net evaporation will occur. As with growth by condensation we must take into account the effect evaporation has on the droplet temperature T_d. Here, the droplet is cooled by the heat required for evaporation. This cooling lowers the partial pressure of vapor at the droplet surface, p_d, and the rate of evaporation, $d(d_p)/dt$. The resulting equilibrium temperature depression, $T_d - T_\infty$, is given by Eqs. 13.14 and 13.15 and shown in Figure 13.3. The partial pressure at the droplet surface, p_d, is given by Eq. 13.2, evaluated at T_d. The Fuchs effect also slows the rate of evaporation for $d_p < 1.0$ μm. For large particles with negligible Fuchs correction, Eq.13.18 can be integrated to get the time t required for a droplet of a given size d_1 to evaporate completely— that is, the lifetime, or drying time, of the droplet:

$$\int_{d_1}^{0} d_p d(d_p) = \int_0^t \frac{4D_vM}{R\rho_p}\left(\frac{p_\infty}{T_\infty} - \frac{p_d}{T_d}\right) dt$$

$$d_1^2 = \frac{8D_vMt}{R\rho_p}\left(\frac{p_d}{T_d} - \frac{p_\infty}{T_\infty}\right)$$

$$\tag{13.19}$$

$$t = \frac{R\rho_p d_p^2}{8D_vM\left(\dfrac{p_d}{T_d} - \dfrac{p_\infty}{T_\infty}\right)} \quad \text{for } d_p > 1.0 \text{ μm}$$

Equation 13.19 is accurate for calculating the drying times of large droplets, from 10 to 50 μm, because, as shown in Figure 13.11, such droplets are larger than 1.0 μm during most of their drying time. Under these conditions, lifetimes are proportional to the square of the initial particle size.

For particles smaller than about 0.1 μm, the partial pressure of vapor at the droplet surface, p_d, is given by the Kelvin equation evaluated at T_d; otherwise p_d is equal to p_s, as given by Eq. 13.2 evaluated at T_d.

EXAMPLE

How long does it take a 20-μm droplet of pure water to evaporate completely in dry air at 293 K [20°C]?

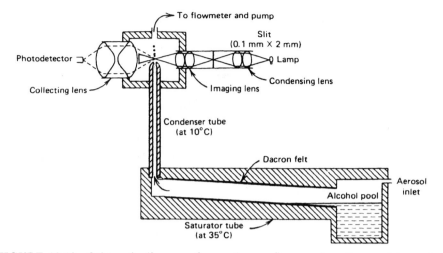

FIGURE 13.10 Schematic diagram of a continuous flow condensation nuclei counter. Reprinted with permission from Agarwal, J. K., and Sem, G. J. (1980).

Use Eq. 13.19. T_d is given by Eq. 13.15 or Figure 13.3, p_d is evaluated at T_d, and $p_\infty = 0$.

$$T_d - 20 = \frac{(6.65 + 0.345(20) + 0.0031(20^2))(0 - 1)}{1 + (0.082 + 0.00782(20))0}$$

$$T_d = 20 - 14.8 = 5.2°C = 278.2 \text{ K}$$

$$p_d = \exp\left(16.7 - \frac{4060}{278.2 - 37}\right) = 0.876 \text{ kPa} = 876 \text{ Pa}$$

$$t = \frac{8.31 \times 1000 \times (20 \times 10^{-6})^2}{8 \times 2.4 \times 10^{-5} \times 0.018(876/278.2)} = 0.31 \text{ s}$$

In cgs units,

$$\left[p_d = \exp\left(18.72 + \frac{4062}{278.2 - 37}\right) = 6.55 \text{ mm Hg} \right]$$

$$\left[t = \frac{62,400 \times 1 \times (20 \times 10^{-4})^2}{8 \times 0.24 \times 18(6.55/278.2)} = 0.31 \text{ s} \right]$$

For particles larger than 50 μm, an additional correction must be applied to compensate for the effect of settling velocity on the evaporation rate. This "wind" velocity modifies the mass diffusion and thermal conduction at the particle surface.

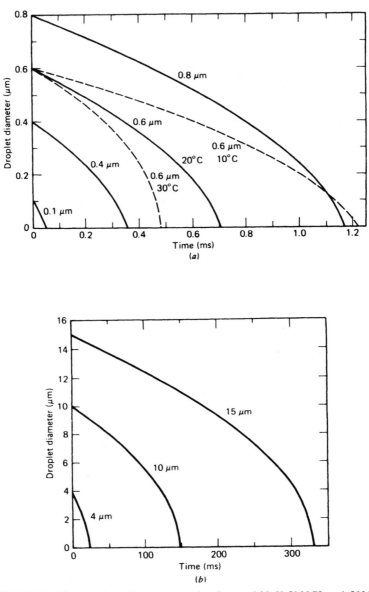

FIGURE 13.11 Evaporation of pure water droplets at 293 K [20°C] and 50% relative humidity. (*a*) Droplet diameters of 0.1–0.8 μm. Dashed lines show the effect of the ambient temperature on evaporation. (*b*) Droplet diameters of 4–15 μm.

For water droplets less than 50 μm in aerodynamic diameter, the evaporation rate is increased less than 10% by the "wind" velocity effect. For 100-μm droplets, the evaporation rate is 31% greater than is predicted by Eq. 13.18. Values for this "wind" factor for water droplets up to 2 mm are given by Davies (1978).

Equation 13.18 gives a rate of particle shrinkage that is directly proportional to the difference between the ambient partial pressure of vapor, p_∞ (proportional to the ambient relative humidity for the case of water vapor in air), and the partial pressure of vapor at the droplet surface, p_d. The effect of ambient relative humidity is shown in Fig. 13.12, a plot of water droplet lifetimes at an ambient temperature of 293 K [20°C] versus particle size. The figure shows a small difference in droplet lifetimes between 0 and 50% relative humidity and much slower drying times at saturated conditions. Although relative humidity can affect droplet lifetimes by a factor of 100–1000, ambient temperature has a much smaller effect, as shown by the dashed lines for 10 and 30°C in Fig. 13.11a.

The droplet lifetimes shown in Fig. 13.12 were calculated by numerical integration of Eq. 13.18 from the original size to zero size. The results shown in the figure include corrections for the Fuchs effect, the Kelvin effect, and droplet temperature depression. The significance of these corrections for water droplet lifetimes for particle of various sizes is shown in Table 13.2. The Fuchs effect and droplet temperature depression act to slow evaporation, while the Kelvin effect speeds up the evaporation of small particles. The correction for temperature depression is significant for water for all particle sizes, but is less significant for less volatile liquids. The Fuchs and Kelvin effects become important for droplets smaller than 1.0 and 0.1 μm, respectively. Ignoring the Kelvin effect (that is, using p_s rather than p_d in Eq. 13.18) leads to the incorrect prediction that droplets will not evaporate in a saturated environment.

To a first approximation, droplet evaporation rates depend on the material properties group, $D_v M p_s / \rho_p$, a quantity that varies widely for different liquids. Table 13.3 gives droplet lifetimes for four liquids at standard conditions and shows the extremely wide range of lifetimes.

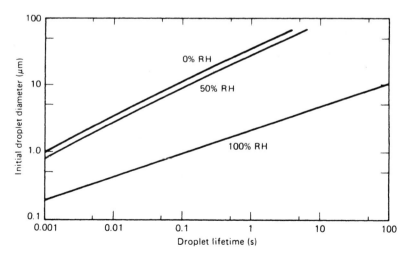

FIGURE 13.12 Water droplet lifetimes as function of droplet size for 0, 50, and 100% relative humidity at 293 K [20°C].

TABLE 13.2　Effect of Fuchs Effect, Kelvin Effect, and Temperature Depression Corrections on Calculated Lifetimes of Water Droplets at 293 K [20°C] and 50-Percent Relative Humidity[a,b]

Droplet Diameter (μm)	Droplet Lifetime (s)				
	Including All Corrections	Omitting Fuchs Effect Correction	Omitting Kelvin Effect Correction	Omitting Temperature Depression Correction	No Corrections $p_d = p_s$ $T_d = T_\infty$ $\phi = 1$
0.01	1.6×10^{-6}	6.0×10^{-8}	5.0×10^{-6}	9.1×10^{-7}	6.0×10^{-8}
0.04	1.4×10^{-5}	1.7×10^{-6}	2.1×10^{-5}	6.6×10^{-6}	9.6×10^{-7}
0.1	4.7×10^{-5}	1.3×10^{-5}	5.8×10^{-5}	2.1×10^{-5}	6.0×10^{-6}
0.4	3.6×10^{-4}	2.2×10^{-4}	3.8×10^{-4}	1.5×10^{-4}	9.6×10^{-5}
1.0	1.7×10^{-3}	1.4×10^{-3}	1.8×10^{-3}	7.4×10^{-4}	6.0×10^{-4}
4.0	0.024	0.023	0.024	0.010	9.6×10^{-3}
10	0.15	0.14	0.15	0.062	0.060
40	2.3	2.3	2.3	0.97	0.96

[a] Calculated with corrections as indicated by numerical integration of Eq. 13.18.

[b] Error in calculated lifetime exceeds 20% above the line in each column.

TABLE 13.3 Droplet Lifetimes for Ethyl Alcohol, Water, Mercury, and Di (2-ethyl-hexyl) Phthlate (DOP) in Vapor-Free Air at 293 K [20°C][a]

Initial Droplet Diameter (μm)	Droplet Lifetime (s)			
	Ethyl Alcohol	Water	Mercury	DOP
0.01	4×10^{-7}	2×10^{-6}	0.005	1.8
0.1	9×10^{-6}	3×10^{-5}	0.3	740
1	3×10^{-4}	0.001	14	3×10^{4}
10	0.03	0.08	1200	2×10^{6}
40	0.4	1.3	2×10^{4}	4×10^{7}

[a]Calculated by Eq. 13.18 with appropriate corrections.

Since most droplets are formed by nucleated condensation, they will dry, not to zero diameter, but to the diameter of their original nucleus. As the droplet size gets close to the nucleus size the drying rate for droplets with soluble nuclei will be slower than that for pure water, because the solution concentration increases and causes a reduction in p_d, which in turn reduces $|p_\infty - p_d|$. Droplets in the atmosphere may adsorb contaminants which form surface films that can greatly reduce evaporation rates.

PROBLEMS

13.1 What is the equilibrium droplet size at a relative humidity of 100% for a NaCl cubic crystal 0.077 μm on a side?

ANSWER: 0.8 μm.

13.2 At what air temperature will exhaled breath condense by self-nucleation? Assume that a saturation ratio of 4.3 is required.

ANSWER: 285.6 K [12.6°C].

13.3 By what factor do 0.1-μm diameter NaCl aerosol particles increase in size when the relative humidity is increased from 0 to 100%? The density of NaCl is 2200 kg/m^3 [2.2 g/cm^3].

ANSWER: 8.5.

13.4 Calculate the maximum expansion ratio for adiabatic expansion of saturated water vapor without homogeneous nucleation. Assume that the initial temperature is 20°C and that a saturation ratio of 4.3 is required for homogeneous nucleation.

ANSWER: 1.21.

13.5 What is the minimum relative humidity necessary to keep a 0.2-μm-diameter water droplet from evaporating? $T = 274$ [1°C].
ANSWER: 101.2%.

13.6 What is the growth rate of a 2-μm droplet of pure water in an environment with 2% supersaturation? $T = 293$ K [20°C]. Neglect the Fuchs effect and droplet heating.
ANSWER: 17 μm/s.

13.7 Determine the drying time for a 7-μm-diameter water droplet at 0% relative humidity and 293 K [20°C]. Calculate an answer with and without correction for droplet temperature depression.
ANSWER: 0.041 s, 0.015 s.

13.8 To generate a test aerosol, a suspension of polystyrene spheres in water is aerosolized as 20-μm droplets and dried by mixing with a large volume of air at 60% RH. How much residence time is required after mixing to ensure complete drying of the droplets? $T = 293$ K [20°C].
ANSWER: 0.98 s.

13.9 What change in aerosol mass concentration occurs in a condensation nuclei counter when 0.02-μm salt nuclei at a concentration of $10^4/cm^3$ grow to 10-μm water droplets? If the volume of the condensation chamber is 240 cm^3, how much water is needed for this process?
ANSWER: 6×10^7-fold increase, 1.3 mg.

13.10 Air is cooled by expansion in a nozzle to 293 K [20°C] and would produce a saturation ratio of 1.3 in the absence of condensation. If there are 10^4 nuclei/cm^3 present in the air downstream of the nozzle, how large will each nucleus grow? There is no additional input of water. [*Hint*: The supersaturation represents the amount of water vapor available for particle growth.]
ANSWER: 10 μm.

REFERENCES

Agarwal, J. K. and Sem, G. J., *J. Aerosol Sci.*, **11**, 343 (1980).

Barrett, J. C., and Clement, C. F., "Growth Rates for Liquid Drops," *J. Aerosol Sci.*, **9**, 223–242 (1988).

Davies, C. N., "Evaporation of Airborne Droplets," in Shaw, D. T. (Ed.), *Fundamentals of Aerosol Science*, Wiley, New York, 1978.

Ferron, G. A., and Soderholm, S. C., "Estimation of the Times for Evaporation of Pure Water Droplets and for Stabilization of Salt Solution Particles," *J. Aerosol Sci*, **3**, 415–429 (1990).

Fuchs, N. A., *Evaporation and Droplet Growth in Gaseous Media*, Pergamon, Oxford, 1959.

Green, H. L., and Lane, W. R., *Particulate Clouds*, Spon, London, 1964.

Hidy, G. M., and Brock, J. R., *The Dynamics of Aerocolloidal Systems*, Pergamon, Oxford, 1971.

Lundgren, D. A., Harris, F. S., Marlow, W. H., Lippmann, M., Clark, W. E., and Durham, M. D., *Aerosol Measurement*, University Presses of Florida, Gainesville, FL, 1979.

Mason, B. J., *The Physics of Clouds*, 2d ed., Clarendon Press, Oxford, 1971.

Pruppacher, H. R., and Klett, J. D., *Microphysics of Clouds and Precipitation*, 2d ed., Kluwer, Dordrecht, The Netherlands, 1997.

Sienfeld, J. H., and Pandis, S. N., *Atmospheric Chemistry and Physics*, Wiley, New York, 1998.

Wiedensohlet, A., Orsini, D., Covert, D. S., Coffmann, D., Cantrell, W., Havlicek, M., Brechtel, F. J., Russell, L. M., Weber, R. J., Gras, J., Hudson, J. G., and Litchy, M., "Intercomparison Study of Size-dependent Counting Efficiency of 26 Condensation Particle Counters, *Aerosol Sci. Tech.* *27*, 224–242 (1997).

14 Atmospheric Aerosols

The atmospheric aerosol—that is, the particles that are normally found in our atmosphere—is a complex and dynamic mixture of solid and liquid particles from natural and anthropogenic sources. We consider first the natural background aerosol—the aerosol that would be present in the absence of human activity. At the other extreme the urban aerosol is dominated by anthropogenic sources. In both cases, primary particles are continuously emitted into, and secondary particles are formed in, the atmosphere. Both kinds of particles may undergo growth, evaporation, or chemical reactions and are subject to various removal mechanisms. The size distributions one finds in the atmosphere reflect the complex interaction of all these processes. We omit from this discussion the largest and most important atmospheric aerosol: natural clouds of water droplets.

14.1 NATURAL BACKGROUND AEROSOL

Sources of natural background and anthropogenic (related to human activity) aerosols are shown in Table 14.1. The wide range of the estimates given for natural aerosol sources reflects the uncertainty of the assumptions upon which these estimates are based. Soil dust includes windblown soil particles from arid regions and the natural weathering of rock. A comparison of sources can be misleading, because their contribution to the atmospheric aerosol varies greatly due to differences in transport distances and removal mechanisms. The majority of the particulate mass from soil dust, volcano aerosol emissions, and anthropogenic direct emissions is likely to be large particles that fall out near the source. The other sources shown—fires and secondary particles—produce fine particles that remain suspended for many days, long enough to travel global distances. Most of the natural photochemical smog is formed by the action of sunlight on isoprene and monoterpene vapor emissions from trees.

A significant fraction of the particulate mass in the atmosphere, both natural and anthropogenic is formed in the atmosphere from gaseous emissions. Although there are large uncertainties in the emission estimates given in Table 14.1, on a global scale natural sources exceed anthropogenic sources by a wide margin. The anthropogenic sources contribute from less than 10 to 50% of the global particulate emissions. The natural sources are well distributed around the globe, and most of their mass contribution comes from vast area sources. The anthropogenic sources, on the other hand, are smaller in amount, but are concentrated in a small portion of the globe:

TABLE 14.1   Sources and Estimates of Global Emissions of Atmospheric Aerosols[a]

Source	Amount, Tg/yr [10^6 metric tons/yr]	
	Range	Best Estimate
Natural		
Soil dust	1000–3000	1500
Sea salt	1000–10000	1300
Botanical debris	26–80	50
Volcanic dust	4–10000	30
Forest fires	3–150	20
Gas-to-particle conversion[b]	100–260	180
Photochemical[c]	40–200	60
Total for natural sources	2200–24000	3100
Anthropogenic		
Direct emissions	50–160	120
Gas-to-particle conversion[d]	260–460	330
Photochemical[e]	5–25	10
Total for anthropogenic sources	320–640	460

[a]Data primarily from Andreae (1995) and SMIC (1971).
[b]Includes sulfate from SO_2 and H_2S, ammonium salts from NH_3, and nitrate from NO_x.
[c]Primarily photochemical particle formation from isoprene and monoterpenes vapors from trees.
[d]Includes sulfate from SO_2 and nitrate from NO_x.
[e]Primarily photochemical particle formation from anthropogenic volatile organic compounds.

the industrialized regions of the world. In these areas, the human contribution exceeds that from natural sources.

Once emitted into or formed in the atmosphere, particles can grow by vapor condensation or by coagulation with other particles. Particles are removed after growing to micrometer or larger sizes by settling or impaction on surfaces such as tree leaves. Small particles may diffuse to surfaces or serve as nucleation sites for raindrops (rainout). Larger particles may be swept out by falling rain or snow (washout). Small particles have lifetimes of days in the troposphere; large particles ($d_a > 20$ μm) are removed in a matter of hours.

Although most of the atmospheric particulate mass is confined to the troposphere (region below an altitude of 11 km), the *stratospheric aerosol* can have significant effects on climate. This subject has been reviewed by Pueschel (1996). The primary source of particulate in the stratosphere (altitude from 11 to 50 km) is the formation of sulfuric acid droplets by gas-to-particle conversion of SO_2 injected into the stratosphere by major volcanic eruptions. These droplets are formed by homogeneous nucleation involving photochemical reactions of SO_2 and water vapor. They spread widely over the hemisphere (north or south) in which they originated.

The ozone layer in the lower stratosphere absorbs solar radiation and heats up relative to the surrounding air. This heating creates a temperature inversion and a region of atmospheric stability. The sulfuric acid droplets formed in the stratosphere accumulate in this stable layer at 18–20 km. This aerosol layer is called the *Junge layer*, after its discoverer.

Major volcanos can increase the stratospheric particulate concentration by two orders of magnitude. The Mount Pinatubo eruption in 1991 injected 14–20 Tg of SO_2 into the stratosphere, causing the aerosol concentration to increase from 2–5 $\mu g/m^3$ to 20–100 $\mu g/m^3$. Before this eruption the particle size distribution had a CMD of 0.14 μm and a GSD of 1.6; after, it had a CMD of 0.66 μm and a GSD of 1.5. The only significant anthropogenic source of particles in the stratosphere is soot from high-altitude aircraft, but this represents less than 1 percent of the total stratospheric aerosol.

Because of the stability and low moisture content of the stratosphere, clouds do not form in it. The absence of cloud formation prevents particle removal by rainout. Also, since particle concentrations are only a few per cubic centimeter, growth by coagulation with subsequent removal by settling is negligible. Similarly, growth by condensation is slow because of low vapor concentration. Consequently, particle lifetimes in the stratosphere are 1–2 years, compared with 1–2 weeks in the lower troposphere. Climatic and global effects of the stratospheric aerosol are described in Section 14.3.

The natural background aerosol in the troposphere depends on direct emissions and gas-to-particle formation from natural sources, as shown in Table 14.1. On a mass basis, direct emissions from deserts, the oceans, and vegetation are the largest sources of particles. Eighty percent of direct emissions go to the lowest kilometer of the troposphere. Consequently, the *tropospheric background aerosol* varies significantly with altitude. Above an altitude of a few kilometers, particle concentrations are influenced little by direct emissions from the earth's surface, and the size distribution in this region resembles that of the background aerosol. A further distinction can be made, between continental and marine background aerosols. Whitby (1978) summarizes available data on the size distribution of average ground-level *background* aerosol in terms of three modes, nuclei, accumulation, and coarse particle. The number concentrations for the three modes are 6400/cm^3, 2300/cm^3, and 3/cm^3 respectively. Each mode can be represented as a lognormal distribution. The size distributions for the modes are as follows: nuclei—CMD = 0.015 μm, GSD = 1.7; accumulation—CMD = 0.076 μm, GSD = 2.0; and coarse particle—CMD = 1.02 μm, GSD = 2.16.

Measurements by Nyeki et al. (1997) of the background aerosol at an elevation of 8.5 km found the accumulation mode to have an average CMD of 0.12 μm and a GSD of 1.57. Concentrations ranged from $1.5 \times 10^7/m^3$ [15/cm^3] in winter to $6.6 \times 10^7/m^3$ [66/cm^3] in summer. Figure 14.1, based on Jaenicke (1986), summarizes available data on background aerosol size distributions. Defining the term "clean air" is difficult because of the substantial contribution of natural background aerosol. A useful, although arbitrary, definition is "air with a total particle content, measured by a condensation nuclei counter, of less than 700/cm^3." Natural background aerosol at remote locations and at altitudes greater than 2 km usually fit this definition of clean air.

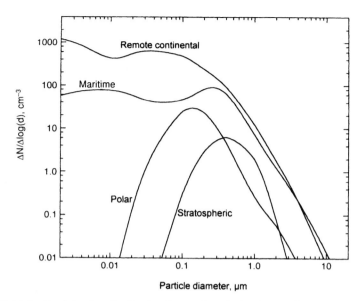

FIGURE 14.1   Particle size distribution by number for background atmospheric aerosols. Data from Jaenicke (1986).

Ions are formed continuously in the atmosphere by the action of cosmic radiation and radioactive gases emanating from soil. Over land, about 10 ion pairs (a positive and a negative ion) are formed in every cubic centimeter every second. These ions attach to aerosol particles and recombine with each other at a rate that yields an equilibrium concentration of small ions (charged molecular clusters) of 100–5000/cm^3 with an average value of 8×10^8/m^3 [800/cm^3].

14.2   URBAN AEROSOL

The urban aerosol—that found in the lowest kilometer of the atmosphere over large cities—is dominated by anthropogenic sources. Aerosol mass concentrations in urban areas range from a few tens of micrograms per cubic meter to 1 mg/m^3 during air pollution episodes in heavily polluted cities in developing countries. The horizontal distribution of aerosol concentration in urban areas varies greatly, depending on the proximity to natural and anthropogenic sources. Atmospheric stability and the thickness of the mixing layer (usually 0–2 km) also affect local concentrations. Table 14.2 gives historical particulate concentrations for different types of land areas.

The particle size distribution of urban aerosol is complex, because it is a mixture of aerosols from various sources, each with a different size distribution, and is modified by size-dependent processes of growth, evaporation, and removal. The most common way to present particle size distribution data for the urban aerosol is in terms of the three modes, the nuclei, accumulation, and coarse particle modes.

**TABLE 14.2 Aerosol Concentrations for
Various Types of Areas of the United States[a]**

Location	Concentration (μg/m^3)
Background[b]	20
Rural areas[b]	40
Urban areas[c]	
Population <10^5	86
10^5–10^6	104
>10^6	154

[a]Data from Corn (1976). Concentrations are total suspended particulate (TSP).
[b]1966–1967.
[c]1957–1963, geometric means.

Each mode has different sources, size range, formation mechanisms, and chemical compositions. Each can be described by a lognormal distribution, see Table 14.3. Figure 14.2a shows the average urban aerosol size distribution on a log–log graph. The combined distribution, represented by the solid line, masks the contributions of the individual modes, shown by dashed lines and does not show the nature of the surface or mass distributions.

The more or less straight-line portion of the distribution between 0.1 and 10 μm can be represented by an inverse power law distribution (the Junge distribution). For the data shown in Figure 14.2a,

$$\frac{\Delta N}{\Delta \log(d_p)} = 24.0 d^{-3.08} \qquad (14.1)$$

for d in μm. Equation 14.1 is empirical; there is no fundamental or theoretical reason that the atmospheric aerosol should have this distribution function.

Figure 14.2b presents the urban aerosol size distribution as cumulative number and volume distributions on a log-probability graph. (See Section 4.5.) The distributions show significant departures from a lognormal distribution, and the contributions of the individual modes are hidden. Figures 14.2c and 14.2d respectively show particle number and volume per unit log interval on an arithmetic scale ver-

TABLE 14.3 Modal parameters for average urban aerosol.[a]

Mode	CMD (μm)	GSD	C_N (cm^{-3})	C_{vol} (μm^3/cm^3)
Nuclei	0.014	1.80	106,000	0.63
Accumulation	0.054	2.16	32,000	38.4
Coarse Particle	0.86	2.21	5.4	30.8

[a]Data from Whitby (1978).

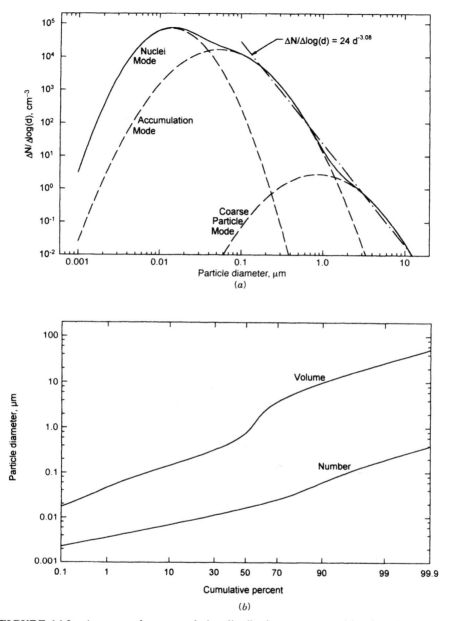

FIGURE 14.2 Average urban aerosol size distributions represented by three lognormal distributions. The modal parameters are given in Table 14.3. (a) Log–log plot of number distribution. (b) Log-probability plot of number and volume distributions. (c) Linear–log plot of number distribution. (d) Linear–log plot of volume distribution. Data from Whitby (1978).

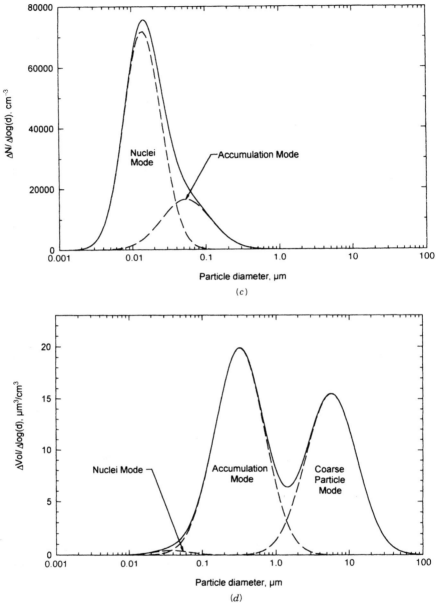

FIGURE 14.2 Continued.

sus particle size on a log scale. These show clearly the modal nature of the urban aerosol and the contribution of each mode to the distribution. Most information about atmospheric aerosol size distributions is now presented in the form of Fig. 14.2c or Fig. 14.2d. These figures show that most of the particles are in the nuclei mode and most of their volume (and mass) is split between the accumulation and coarse-particle modes. The volume (or mass) distribution shown in Fig. 14.2d is typical of most urban aerosols containing photochemical smog. Such distributions are usually bimodal, with a saddle point in the 1–3-μm-diameter range. Particle surface area and light scattering are associated primarily with particles in the accumulation mode.

The *nuclei mode* consists primarily of combustion particles emitted directly into the atmosphere and particles formed in the atmosphere by gas-to-particle conversion. This mode is not always present, but is usually found near highways and other sources of combustion. Because of their high number concentration, especially near their source, these small particles coagulate rapidly with each other and with particles in the accumulation mode. Consequently, nuclei particles have relatively short lifetimes in the atmosphere and end up in the accumulation mode. Nuclei particles may serve as sites for the formation of cloud droplets and may subsequently be removed from the atmosphere as rain droplets (rainout).

The *accumulation mode* includes combustion particles, smog particles, and nuclei-mode particles that have coagulated with accumulation-mode particles. The smog particles are formed in the atmosphere by photochemical reactions of volatile organics and oxides of nitrogen in the presence of strong sunlight. There is significant overlap between the nuclei and accumulation modes. As the name suggests, particles accumulate in this mode because removal mechanisms are weak. Particles can be removed by rainout or washout, but they coagulate too slowly to reach the coarse-particle mode. The nuclei and accumulation modes together constitute "fine" particles. The size range of the accumulation mode includes the wavelengths of visible light, and these particles account for most of the visibility effects of atmospheric aerosols. (See Section 16.4.)

Under conditions of high humidity, such as in a cloud or fog, the accumulation mode may itself have two submodes: a condensation mode with MMAD of 0.2–0.3 μm and a droplet mode with MMAD of 0.5–0.8 μm. The droplets are formed by the growth of hygroscopic condensation-mode particles. This process may be facilitated by chemical reactions in the droplet.

The *coarse-particle mode* consists of windblown dust, large salt particles from sea spray, and mechanically generated anthropogenic particles such as those from agriculture and surface mining. Because of their large size, the coarse particles readily settle out or impact on surfaces, so their lifetime in the atmosphere is only a few hours or days. The dividing line between coarse and fine particles is the saddle point between 1 and 3 μm. The fine-particle mode contains from one-third to two-thirds of the total mass, with the remainder in the coarse particle mode. The ratio varies according to the region of the country. Fine and coarse-particles have different chemical compositions, sources, and lifetimes in the atmosphere. There is comparatively little mass exchange between the two modes. In some locations, there is an inverse correlation between mass concentration in the two modes. Low wind velocity reduces the concentration of windblown soil particles, but favors photochemical particle formation, and high wind velocity does the opposite.

The size distribution of the urban aerosol, shown in Figs. 14.2a–14.2d, is the steady-state result of the different sources of particles and the processes of formation (gas-to-particle conversion and photochemical reactions), growth (coagulation, condensation and evaporation) and removal (settling, deposition, rainout, and washout) that they are subject to. Figure 14.3 shows these processes schematically.

The chemical compositions of fine and coarse particles differ greatly. Because there is little mass transfer between the fine and coarse particles, they exist together in the atmosphere as two chemically distinct aerosols. As a group, the fine particles are acidic and contain most of the sulfates, ammonium compounds, hydrocarbons, elemental carbon (soot), toxic metals, and water in the atmosphere. The coarse particles are basic and contain most of the crustal materials and their oxides, such as silicon, iron, calcium, and aluminum, as well as large sea salt particles and vegetation debris. Table 14.4 gives data on the chemical composition of fine and coarse particles. Data for the table were obtained using the dichotomous virtual impactor that separates fine and coarse particles during sampling. (See Section 5.7.)

14.3 GLOBAL EFFECTS

The atmospheric aerosol, described in Section 14.1, has a significant influence on two important global atmospheric processes: global warming and ozone depletion. The stratospheric aerosol concentration can increase by up to two orders of magnitude following major volcanic eruptions. These particles have half-lives in the stratosphere of about a year. This stratospheric haze directly affects the earth's radiation balance by scattering incoming solar radiations back into space. (See Section 16.3.) This represents a change in the earth's albedo (reflectivity). Stratospheric particles have little effect on long-wave terrestrial radiation, so the net result is a cooling of the troposphere and the earth's surface. Pueschel (1996) estimates that the eruption of Mount Pinatubo in June 1991 caused a change in solar radiation reaching the earth's surface and troposphere of -2.7 W/m^2 [-2700 erg/s $\cdot$ cm^2] by September 1991. (This was partially offset by an increase in tropospheric clouds, due to the increase in tropospheric nuclei from the eruption, that reduced terrestrial radiation to space.) By comparison, the effect of the CO_2 buildup in the atmosphere since the Industrial Revolution has caused a net change of $+1.25$ W/m^2 [$+1250$ erg/s $\cdot$ cm^2]. Thus, the stratospheric aerosol can have an effect on global surface temperature that is the same order of magnitude, but in the opposite direction, as greenhouse gases and must be included in any analysis of global warming.

The tropospheric aerosol can also affect climate. Tropospheric particles scatter incoming solar radiation back into space by two mechanisms. The first is direct scattering by the tropospheric aerosol. The second mechanism is due to an increase in cloud reflectively because of a higher number concentrations of cloud droplets, resulting from an increased concentration of tropospheric condensation nuclei. Both

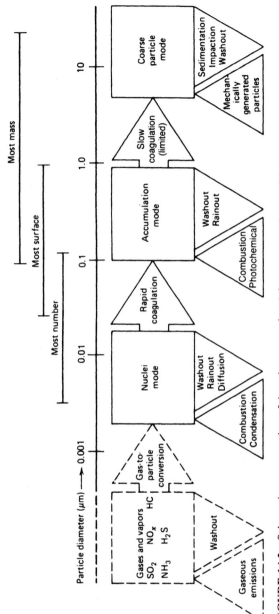

FIGURE 14.3 Schematic representation of the urban aerosol and the processes that modify it.

TABLE 14.4   Average composition of fine and coarse particles in $\mu g/m^3$ at an urban and a rural site.[a]

	Urban		Rural	
	Fine	Coarse	Fine	Coarse
Total Mass	42	27	24	5.6
SO_4^-	17	1.1	12.0	—
NO_3^-	0.25	1.8	0.30	—
NH_4^+	4.3	<0.19	2.3	—
H^+	0.067	<0.01	0.114	—
C	7.6	3.3	3.3	1.3
Al	0.095	1.4	0.020	0.20
Si	0.20	3.8	0.038	0.58
S	—	—	3.7	0.20
Ca	0.15	3.1	0.016	0.32
Fe	0.17	0.73	0.028	0.12
Pb	0.48	0.13	0.097	0.014

[a]Data from Finlayson-Pitts and Pitts (1986).

mechanisms modify the earth's radiation balance so as to cool the earth's surface. The combined effect due to anthropogenic aerosols is called the "whitehouse" effect, as an analogue to the greenhouse effect. The uncertainty of the magnitude of the whitehouse effect is greater than that for the greenhouse effect. Schwartz (1996) estimates the cooling due to the whitehouse effect to be 20–100% of the heating due to the greenhouse effect. The particles responsible for the whitehouse effect have a lifetime in the troposphere of about a week, whereas the gases responsible for the greenhouse effect have atmospheric lifetimes of decades.

Stratospheric particles play a key role in ozone depletion. In the cold temperature of the polar stratosphere during winter, nitric acid and water vapor condense to form polar stratospheric clouds. The surfaces of these cloud particles serve as sites for the catalytic conversion of stratospheric chlorine compounds, such as anthropogenic chlorofluorocarbons (CFCs), to molecular chlorine (Cl_2) and hypochlorous acid (HOCl). In the polar spring, the sun photodissociates these compounds to atomic chlorine (Cl), which reacts with ozone (O_3) to form oxygen (O_2) and chlorine monoxide (ClO). The latter is photolyzed back to atomic chlorine, and the cycle keeps repeating with the continued destruction of ozone. Volcanic eruptions enhance this process by increasing the stratospheric aerosol, which migrates to the poles and provides additional surface for the catalytic activation of chlorine. See Seinfeld and Pandis (1998).

PROBLEMS

14.1 For the modal data given in Table 14.3, what is the particle number per unit mass for each mode? Assume standard particle density.

ANSWER: 1.4×10^{20}/kg, 8.7×10^{17}/kg, 1.8×10^{14}/kg [1.4×10^{17}/g, 8.7×10^{14}/g, 1.8×10^{11}/g].

14.2 Estimate how long it would take for the diameter of average mass of the nuclei mode to equal the CMD of the accumulation mode if the nuclei mode particles were allowed to coagulate by themselves. Use modal data given in Table 14.3. Assume that the GSD is constant at its initial value and K is constant at 30×10^{-16} m^3/s [30×10^{-10} cm^3/s] (an interpolated average value from Table 12.4).

ANSWER: 3.3×10^4 s (9.1 hrs).

REFERENCES

Andreae, M. O., "Climatic Effects of Changing Atmospheric Aerosol Levels," in *Future Climates of the World: A Modelling Perspective*, A. Henderson-Sellers (Ed.), Elsevier, Amsterdam, 1995.

Corn, M., "Aerosols and the Primary Air Pollutants—Nonviable Particles, Their Occurrence, Properties, and Effects," in Stern, A. C. (Ed.), *Air Pollution*, Academic Press, New York, 1976.

Finlayson-Pitts, B. J., and Pitts, J. N., "*Atmospheric Chemistry: Fundamentals and Experimental Techniques*, Wiley, New York, 1986.

First, M. W., "Aerosols in Nature," *Arch. Intern. Med.*, **131**, 24–32 (1973).

Hidy, G., Mueller, P., Grosjean, D., Appel, B., and Wesolowski, J., *The Character and Origins of Smog Aerosols*, Wiley, New York, 1980.

Jaenicke, R., "Physical Characterization of Aerosols," in Lee, S. D., Schneider, T., Grant, L. D., and Verkerk, P. J., (Eds.), *Aerosols: Research, Risk Assessment, and Control Strategies*, Lewis Publishers, Chelsea, MI, 1986.

Kiehl, J. T. and Rodhe, H., "Modeling Geographical and Seasonal Forcing due to Aerosols," in Charlson, R. J. and Heintzenberg, J., (Eds.), *Aerosol Forcing of Climate*, Wiley, New York, 1995.

National Research Council, *Airborne Particles*, University Park Press, Baltimore, 1979.

Nyeki, S., Li, F., Rosser, D., Colbeck, I., and Baltensperger, U., "The background aerosol size distribution at a high alpine site: An analysis of the seasonal cycle," *J. Aerosol Sci.*, **28**, 5211–5212 (1997).

Pueschel, R. F., "Stratospheric aerosols: Formation, properties, effect," *J. Aerosol Sci.*, **27**, 383–402 (1996).

Schwartz, S. E., "The Whitehouse Effect—shortwave Radiative Forcing of Climate by Anthropogenic Aerosols: an Overview," *J. Aerosol Sci.*, **27**, 359–382 (1996).

SMIC, *Inadvertent Climate Modification, Report of the Study of Man's Impact on Climate*, MIT Press, Cambridge, MA, 1971.

Sienfeld, J. H. and Pandis, S. N., *Atmospheric Chemistry and Physics*, Wiley, New York, 1998.

Whitby, K. J., "The Physical Characteristics of Sulfur Aerosol," *Atmos. Env.*, **12**, 135–159 (1978).

15 Electrical Properties

In aerosol mechanics, the most important electrostatic effect is the force exerted on a charged particle in an electrostatic field. Most aerosol particles carry some electric charge, and some may be highly charged. For highly charged particles, the electrostatic force can be thousands of times greater than the force of gravity. The motion induced by electrostatic forces forms the basis for important types of air-cleaning equipment and aerosol sampling and measuring instruments.

15.1 UNITS

A charged particle is acted upon by an electrostatic force near charged surfaces or other charged particles. The force acts remotely through air or vacuum and does not require the flow of current. The charge on a particle can be negative or positive, depending on whether the particle has an excess or deficiency, respectively, of electrons. In this section, we review the basic principles of electrostatics and the system of units used to express electrostatic quantities.

The fundamental equation of electrostatics is *Coulomb's law*, which gives the electrostatic repulsive force F_E between two point charges of like sign separated by a distance R:

$$F_E = K_E \frac{qq'}{R^2} \qquad (15.1)$$

where q and q' represent the amount of charge at the two points, and K_E is a constant of proportionality that depends on the system of units used.

Equation 15.1 takes on a different form in the SI and cgs systems of units, because of the way the amount of charge is defined in each system.

In the SI system of units, the ampere (A), one of the seven SI base units, is defined as the current required to produce a specified force between two parallel wires 1 m apart. The units of charge and potential difference are derived from the ampere. The unit of charge, the coulomb (C), is defined as the amount of charge transported in 1 s by a current of 1 A. The unit of potential difference, the volt (V), is defined as the potential difference between two points along a wire carrying 1 A and dissipating 1 watt (W) of power between the points. In the SI system of units, the value of K_E in Eq. 15.1 is

$$K_E = \frac{1}{4\pi\epsilon_0} = 9.0 \times 10^9 \quad \text{N} \cdot \text{m}^2/\text{C}^2 \qquad (15.2)$$

where ϵ_0 is the permittivity of a vacuum, 8.85×10^{-12} $C^2/N \cdot m^2$. Combining Eqs. 15.1 and 15.2 gives

$$F_E = 9.0 \times 10^9 \frac{qq'}{R^2} \quad \text{in SI units} \tag{15.3}$$

where F_E is expressed in N, q and q' are in C, and R is in m.

In the cgs electrostatic system of units (esu), the amount of charge is expressed in terms of the statcoulomb (stC), which is defined by Eq. 15.1 as the amount of charge that causes a repulsive force of 1 dyn when two equal charges are separated by 1 cm. The prefix "stat" is used to distinguish cgs units from SI units. Thus, in the cgs system, $K_E = 1$ (dimensionless) and Eq. 15.1 can be simplified to

$$F_E = \frac{qq'}{R^2} \quad \text{in cgs units} \tag{15.4}$$

where F_E is in dyn, q and q' are in stC, and R is in cm. Having defined charge in this way, we must define unit current and unit potential difference in consistent units. Unit current, the statampere (stA), is equal to a flow of charge of 1 stC/s along a wire. Unit potential difference, the statvolt (stV), is defined as the voltage between two points that causes 1 erg of work to be done moving 1 stC of charge between the two points.

The cgs system has the advantage of simplified equations, but the disadvantage of having to use "stat" units. The SI system has the advantage of using common electrical units, but requires including the constant K_E in the equations. Table 15.1 gives conversion factors relating the two systems of units. From the definitions given and Coulomb's law, the following relationships can be derived:

$$
\begin{array}{ll}
A = C/s & stA = stC/s \\
V = N \cdot m/C & stV = dyn \cdot cm/stC \\
N = K_E C^2/m^2 & dyn = stC^2/cm^2 \\
V = K_E C/m & stV = stC/cm
\end{array} \tag{15.5}
$$

where $K_E = 9.0 \times 10^9$ $N \cdot m^2/C^2$, as defined by Eq. 15.2.

TABLE 15.1 Conversion Factors and Constants for SI and cgs Electrostatic Units

Quantity	SI	cgs
Charge	1 C	3.0×10^9 stC
Current	1 A	3.0×10^9 stA
Potential difference	1 V	0.0033 stV
K_E	9.0×10^9 $N \cdot m^2/C^2$	1
Charge on an electron, e	1.60×10^{-19} C	4.80×10^{-10} stC
Electrons per unit charge	$6.3 \times 10^{18}/C$	$2.1 \times 10^9/stC$
Electrical mobility	1 $m^2/V \cdot s$	3.0×10^6 $cm^2/stV \cdot s$
Field strength	1N/C or V/m	3.3×10^{-5} dyn/stC or stV/cm

15.2   ELECTRIC FIELDS

An electric field exists in the space around a charged object and causes a charged particle in this space to be acted upon by a force—the electrostatic force. We express the strength of such a field in terms of the magnitude of the force F_E produced per unit charge on the particle. The field strength, or intensity, E is

$$E = \frac{F_E}{q} \tag{15.6}$$

where q is the charge on the particle. The units of field strength are N/C [dyn/stC]. Field strength is a vector that has the same direction as the force F_E. It is common to express the amount of charge q as n multiples of the smallest unit of charge, the charge on an electron e, 1.6×10^{-19} C [4.8×10^{-10} stC].

$$q = ne \tag{15.7}$$

Thus, the force on a particle with n elementary units of charge in an electric field E is

$$F_E = qE = neE \tag{15.8}$$

Equation 15.8 is the basic equation for the electrostatic force acting upon an aerosol particle. The application of the equation is straightforward if the values of n and E are known. The central problem in the application of electrostatic theory to aerosols is the determination of these two quantities. Both n and E can change with time and position, and the magnitude of one can affect the other. Sections 15.1–15.7 address this problem.

Because Coulomb's law provides a relationship between charge, force, and geometry, it can be used to determine the field strength at any point near a charged surface. An imaginary unit charge is positioned at the desired location, and the total electrostatic force on this unit charge is calculated by the vector sum of the Coulombic force due to each charge on the charged surface. The field strength is then computed from the total force by Eq. 15.6. The utility of this method is limited, because we usually do not know the location and magnitude of all the charges on a surface. In theory, the field strength at any point could be determined by placing a real test charge at the point and measuring the electrostatic force on it, but there are many practical difficulties with such measurements.

An alternative definition of field strength is based on voltage or potential, a relatively easy quantity to measure. The potential difference ΔW between two points can be defined as the work required to move a unit charge between the two points. This work is equal to the force per unit charge (the field strength) times the distance between the points, Δx:

$$\Delta W = \frac{F_E \Delta x}{q} \tag{15.9}$$

When F_E is expressed by Eq. 15.8 and Eq. 15.9 is rearranged, an alternative definition of field strength is obtained:

$$E = \frac{\Delta W}{\Delta x} \tag{15.10}$$

Thus, the field strength in any direction at a point is equal to the potential gradient in that direction at the point. The determination of field strength is primarily a problem of determining the potential gradient near charged surfaces.

The solution for an electric field (that is, an equation that gives the field strength at any point), can be determined for simple geometries. Three such cases are considered here. The field around a *single point charge q* can be determined easily by placing an imaginary test charge q' in the field and determining the force on it by Coulomb's law and the field strength by Eq. 15.6.

$$E = \frac{F_E}{q'} = \frac{K_E q}{R^2} \tag{15.11}$$

The field strength between two oppositely charged, closely spaced *parallel plates* is uniform in the region between the plates (neglecting edge effects) and is given by

$$E = \frac{\Delta W}{x} \tag{15.12}$$

where ΔW is the algebraic difference in voltage between the two plates and x is the separation between the plates. In a uniform field, the electrostatic force on a charged particle is constant everywhere between the plates. For a positively charged particle, the direction of the force is toward the more negatively charged plate; for a negatively charged particle, the force is directed toward the more positively charged plate.

The field strength inside a *cylindrical tube* with a wire along its axis is given by

$$E = \frac{\Delta W}{R \ln(d_t/d_w)} \tag{15.13}$$

where ΔW is the algebraic difference in voltage between the wire and the tube, R is the radial position for which the field is being calculated, and d_t and d_w are the diameters of the tube and the wire, respectively. Equation 15.13 indicates that the field strength goes to infinity as R goes to zero. The finite diameter of the conducting wire precludes a field at $R = 0$. The maximum field will be at the surface of the wire and will increase as the wire diameter decreases. Equation 15.13 can also be used for the field strength in the space between two concentric cylinders. It becomes much more difficult to calculate the field strength for complex geometries or when there is significant space charge (ions or charged particles in the field region).

In the troposphere, near the earth's surface there is an electric field caused by the difference in potential between the earth's surface (negative) and the upper layers of the atmosphere (positive). In normal, clear weather, the average field strength at sea level is 120 V/m [1.2 V/cm], but it can be 10,000 V/m [100 V/cm] beneath thunderclouds and much higher at the site of a lightning discharge.

EXAMPLE

A particle is positioned 4 cm from a point charge of 10^{-14} C. What is the field strength at the particle? What is the electrostatic force on the particle if it has 100 excess elementary charges?

$$E = \frac{K_E q}{R^2} = \frac{9 \times 10^9 \times 10^{-14}}{(0.04)^2} = 0.056 \text{ V/m}$$

$$F = qE = neE = 100 \times 1.6 \times 10^{-19} \times 0.056 = 9.0 \times 10^{-19} \text{ N}$$

In cgs units,

$$[q = 10^{-14} \times 3 \times 10^9 = 3 \times 10^{-5} \text{ stC}$$

$$\left[E = \frac{1 \times 3 \times 10^{-5}}{4^2} = 1.88 \times 10^{-6} \text{ stV/cm} \right]$$

$$[F = 100 \times 4.8 \times 10^{-10} \times 1.88 \times 10^{-6} = 9.0 \times 10^{-14} \text{ dyn}]$$

15.3 ELECTRICAL MOBILITY

When a charged particle is placed in an electric field, it is acted upon by a force F_E (Eq. 15.8). The particle velocity resulting from this force can be determined in a manner similar to that for terminal settling velocity. (See Sections 3.3 and 3.7) For particle motion in the Stokes region, the terminal electrostatic velocity V_{TE} is obtained by equating the electrostatic force to Stokes drag (Eq. 3.18) and solving for velocity:

$$neE = 3\pi\eta V d/C_c \tag{15.14}$$

$$V_{TE} = \frac{neEC_c}{3\pi\eta d} \tag{15.15}$$

Equation 15.15 can be written in terms of the particle mechanical mobility B (Eq. 3.16) as

$$V_{TE} = neEB \tag{15.16}$$

To determine V_{TE} when the particle motion is outside the Stokes region, it is necessary to follow a procedure analogous to that used to calculate V_{TS} for Re > 1. (See Section 3.7.) A value of Re > 1 is more common for electrostatic motion than for settling, because the electrostatic force can be much greater than the force of gravity. Equating Eqs. 15.8 and 3.4, we obtain

$$neE = C_D \frac{\pi}{8} \rho_g d^2 V_{TE}^2 \tag{15.17}$$

and solving for C_D gives

$$C_D = \frac{8neE}{\pi \rho_g d^2 V_{TE}^2} \tag{15.18}$$

Multiplying both sides of Eq. 15.18 by Re^2 gives

$$C_D Re^2 = \frac{8neE\rho_g}{\pi \eta^2} \tag{15.19}$$

$C_D Re^2$ can be calculated without knowing the velocity or Reynolds number at the terminal electrostatic velocity. The terminal electrostatic velocity can then be determined from the value of $C_D Re^2$ by Eq. 3.33 with V_{TS} replaced by V_{TE}, or by the graphical or tabular procedures given in Section 3.7.

The Millikan oil-drop experiment, one of the classical experiments of physics in the early 1900s, used the preceding electrostatic principles to measure the charge of an individual electron. The apparatus used, called a Millikan cell, is shown in Fig. 15.1. A spherical, submicrometer-sized particle (an oil droplet) is introduced between two horizontal parallel plates. The potential difference between the plates can be carefully controlled. The particle is illuminated by a beam of light between the plates and observed by a horizontal microscope positioned perpendicular to the

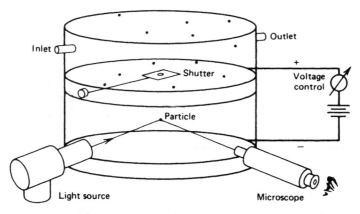

FIGURE 15.1 Apparatus for Millikan oil-drop experiment.

beam . The illuminated particle is seen as a tiny dot of light against a dark background. With no voltage applied to the plates, the diameter of the oil droplet is determined by measuring its settling velocity, as described for the sedimentation cell in Section 3.9. Because of the spray-generation method used, the droplets will ordinarily carry one or more excess electrons. By adjusting the voltage on the upper plate (positive with respect to the lower plate), the upward electrostatic force can be made to balance the force of gravity and the particle held stationary. In this state,

$$neE = \frac{\rho_p \pi d^3 g}{6} \qquad (15.20)$$

From thousands of measurements, Millikan found that the value of the charge on a particle was always a whole-number multiple (e, $2e$, $3e$, etc.) of a fundamental unit of charge: the elementary charge, or the charge of an electron, e.

It is convenient to express the ability of a particle to move in an electric field in terms of the particle's *electrical mobility* Z, the velocity of a particle with a charge ne in an electric field of unit strength. Electrical mobility is given by

$$Z = \frac{V_{TE}}{E} = \frac{neC_c}{3\pi\eta d} \quad \text{for Re} < 1 \qquad (15.21)$$

$$V_{TE} = ZE \quad \text{for Re} < 1 \qquad (15.22)$$

Particle mobility, as given by Eq. 15.21, is based on Stokes drag, and if motion is outside the Stokes region (Re > 1), the true velocity will be greater than that calculated by Eqs. 15.21 and 15.22. Mobility is usually expressed in units of $m^2/V \cdot s$ [$cm^2/stV \cdot s$].

Table 15.2 gives values of mobilities for electrons, ions, and charged aerosol particles with minimum and maximum charge. Ions are charged molecular clusters with an excess or deficiency of one or more electrons, called negative or positive ions, respectively. Air ions are usually singly charged and are formed from air molecules by flames, radiation, or corona discharge. (See Section 15.5.) The ability of highly charged particles to reach high velocities in a moderate electric field is apparent from the table.

Electrical and mechanical mobility are related by

$$Z = qB = neB \qquad (15.23)$$

Acceleration and deceleration of particles in an electric field are characterized by the particle relaxation time τ and follow the relationships given in Section 5.2.

EXAMPLE

A 0.6-μm particle acquires 40 excess electrons. What is its electrical mobility? If the particle is placed between two parallel plates 1 cm apart, with one maintained at + 1000 V and the other at −1000 V, what is its terminal electrostatic velocity?

TABLE 15.2   Electrical Mobility of Electrons, Ions, and Aerosol Particles at Standard Conditions

Particle Diameter (μm)	Electrical Mobility ($m^2/V \cdot s$)[a]	
	Singly Charged	Maximum Charge[b]
Electron	6.7×10^{-2}	—
Negative air ion	1.6×10^{-4}	—
Positive air ion	1.4×10^{-4}	—
0.01	2.1×10^{-6}	7.3×10^{-4}
0.1	2.7×10^{-8}	9.3×10^{-4}
1.0	1.1×10^{-9}	$(2.5 \times 10^{-3})^c$
10	9.7×10^{-11}	$(6.7 \times 10^{-3})^c$
100	9.3×10^{-12}	$(1.1 \times 10^{-2})^c$

[a]For mobility in $cm^2/stV \cdot s$, multiply the value shown by 3×10^6.
[b]Based on the ion limit. (See Section 15.6.)
[c]Velocity (m/s) in a unit electric field, but because $Re > 1.0$, Eq. 15.22 does not hold.

The particle's mechanical mobility can be obtained from Table A11.

$$Z = neB = 40 \times 1.6 \times 10^{-19} \times 1.23 \times 10^{10} = 7.9 \times 10^{-8} \text{ m}^2/V \cdot s$$

$$[40 \times 4.8 \times 10^{-10} \times 1.23 \times 10^7 = 0.24 \text{ cm}^2/stV \cdot s]$$

$$E = \frac{\Delta W}{x} = \frac{+1000 - (-1000)}{0.01} = 200,000 \text{ V/m} \quad \left[\frac{2000 \times 0.0033}{1} = 6.6 \text{ stV/cm} \right]$$

$$V_{TE} = ZE = 7.9 \times 10^{-8} \times 200,000 = 0.016 \text{ m/s} \quad [0.24 \times 6.6 = 1.6 \text{ cm/s}]$$

$$Re = 66,000 \times 0.016 \times 0.6 \times 10^{-6} = 6.3 \times 10^{-4} \quad [6.6 \times 1.6 \times 0.6 \times 10^{-4} = 6.3 \times 10^{-4}]$$

Note: $V_{TE}/V_{TS} = 0.016/1.37 \times 10^{-5} = 1170$

15.4   CHARGING MECHANISMS

The principal mechanisms by which aerosol particles acquire charge are flame charging, static electrification, diffusion charging, and field charging. The last two require the production of unipolar ions, usually by corona discharge (Section 15.5), and are used to produce highly charged aerosols. It is often necessary to estimate the charge on a particle based on the conditions under which it acquires the charge.

Flame charging occurs when particles are formed in or pass through a flame. At the high temperature of the flame, direct ionization of gas molecules creates high concentrations of positive and negative ions and thermionic emissions of electrons or ions from particles. The net charge acquired by the particles depends on the material and is usually symmetric with respect to polarity (equal numbers of positively and negatively charged particles).

Static electrification causes particles to become charged by mechanical action as they are separated from the bulk material or other surfaces. This charging mechanism can produce highly charged particles under the right circumstances, but is not reliable for aerosol charging. It can be extremely important, for example, when dealing with explosive dusts. Particles are usually charged by static electrification during their formation, resuspension, or high-velocity transport. The three primary mechanisms of static electrification that can charge aerosol particles during their generation are electrolytic charging, spray electrification, and contact charging.

Electrolytic charging results when liquids with a high-dielectric constant are separated from solid surfaces. During atomization these liquids strip off charge from the surfaces of the atomizer and produce slightly to moderately charged droplets as the liquids separate from the surfaces. Pure water is a high dielectric liquid that can become charged during atomization. *Spray electrification* results from the disruption of charged liquid surfaces. Due to surface effects, some liquids have a charged surface layer, and when the surface is disrupted during the formation of droplets by atomization or bubbling, charged droplets are produced. *Contact charging*, or triboelectrification, occurs during the separation of dry, nonmetallic particles from solid surfaces. When a particle contacts a surface, charge is transferred between the particle and the surface such that the particle acquires a net positive or negative charge when it separates from the surface. The particle's polarity and the amount of charge on the particle depend on the materials involved and their relative positions in the triboelectric series. Friction increases the amount of charge acquired. Rubbing leather shoes on a woolen rug is a common example of this mechanism. Because it requires dry surfaces, contact charging becomes ineffective at relative humidities greater than about 65%. Most methods of resuspending dry powders involve some friction between the powder and the apparatus and, consequently, produce charged particles.

When an ion collides with a particle, it sticks, and the particle acquires its charge. Particles mixed with unipolar ions (all of the same sign) become charged by random collisions between the ions and the particles. This process is called *diffusion charging* because the collisions result from the Brownian motion of the ions and particles. This mechanism does not require an external electrical field and, to a first approximation, does not depend on the particle material. As the charge accumulates, it produces a field that tends to repel additional ions, reducing the charging rate. The ions, being in equilibrium with the gas molecules, have a Boltzmann distribution of velocities. As the charge on the particle increases, fewer and fewer ions have sufficient velocity to overcome the repulsive force, and the charging rate slowly approaches zero. It never reaches zero, however, because the Boltzmann distribution of velocities has no upper limit.

An approximate expression for the number of charges $n(t)$ acquired by a particle of diameter d_p by diffusion charging during a time t is

$$n(t) = \frac{d_p kT}{2K_E e^2} \ln\left[1 + \frac{\pi K_E d_p \bar{c}_i e^2 N_i t}{2kT}\right] \qquad (15.24)$$

where $\bar{c}_i$ is the mean thermal speed of the ions ($\bar{c}_i = 240$ m/s [2.4×10^4 cm/s] at standard conditions) and N_i is the concentration of ions. At standard conditions, for $N_i t > 10^{12}$ s/m³ [10^6 s/cm³], Eq. 15.24 is accurate to within a factor of two for particles from 0.07 to 1.5 μm and, for $N_i t > 10^{13}$ s/m³ [10^7 s/cm³], to within a factor of two for particles from 0.05 to 40 μm. A more accurate, but less convenient, expression requiring numerical integration is given by Lawless (1996). Even in the presence of an electrostatic field, diffusion charging is the predominant mechanism for charging particles less than 0.2 μm in diameter. A process similar to diffusion charging, but utilizing bipolar ions, is used to discharge highly charged aerosols, see Section 15.7.

Field charging is charging by unipolar ions in the presence of a strong electric field. This mechanism is depicted in Figs. 15.2a–15.2c, in which the negatively charged plate is at the left and negative ions are present. The rapid motion of ions in an electric field results in frequent collisions between the ions and the particles. When an uncharged spherical particle is placed in a uniform electric field, it distorts the field, as shown in Fig. 15.2a. The field lines shown represent the trajectories of ions. The extent of the distortion of the field lines depends on the relative permittivity (dielectric constant) ε of the particle material and the charge on the particle. For an uncharged particle, the greater the value of ε, the greater the number of field lines that converge on the particle.

Ions in this electric field travel along the field lines and collide with the particle where the field lines intersect the particle. All ions on intersecting lines to the left of the particle will collide with the particle and transfer their charge to it. As the particle becomes charged, it will tend to repel the like-charged incoming ions. This situation is shown in Fig. 15.2b, in which the particle is partially charged. The presence of the charge on the particle reduces the field strength and the number of field lines converging on the particle from the left and increases these two quantities on the right. Because of these changes, the rate of ions reaching the particle decreases as the particle becomes charged.

Ultimately, the charge builds up to the point where no incoming field lines converge on the particle (Fig. 15.2c) and no ions can reach the particle. At this maximum-charge condition, the particle is said to be at *saturation charge*.

When diffusion charging can be neglected, the number of charges, n, acquired by a particle during a time t in an electric field E with an ion number concentration N_i is

$$n(t) = \left(\frac{3\varepsilon}{\varepsilon + 2}\right)\left(\frac{Ed^2}{4K_E e}\right)\left(\frac{\pi K_E e Z_i N_i t}{1 + \pi K_E e Z_i N_i t}\right) \qquad (15.25)$$

where ε is the relative permittivity of the particle and Z_i is the mobility of the ions, approximately 0.00015 m²/V · s [450 cm²/stV · s]. In this equation, the first two fac-

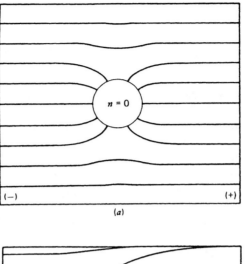

(a)

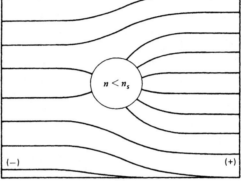

(b)

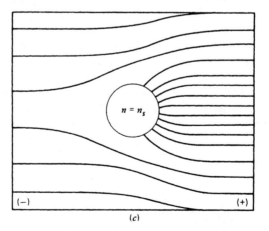

(c)

FIGURE 15.2 Electric field lines for a conducting particle in a uniform field (negative ions and negative plate at left). (a) An uncharged particle. (b) A partially charged particle. (c) A particle at saturation charge.

tors represent the saturation charge n_s reached after sufficient time at a given charging condition.

$$n_s = \left(\frac{3\varepsilon}{\varepsilon + 2} \right) \left(\frac{Ed^2}{4K_Ee} \right) \tag{15.26}$$

The first factor on the right in Eqs. 15.25 and 15.26 depends only on the material of the particle and ranges from 1.0 for $\varepsilon = 1$ to 3.0 for $\varepsilon = \infty$. The relative permittivity, or dielectric constant ε, reflects the strength of the electrostatic field produced in different materials by a fixed potential relative to that produced in a vacuum under the same conditions. For most materials, $1 < \varepsilon < 10$; ε is 1.0 for a vacuum, 1.00059 for air, 5.1 for DOP, 4.3 for quartz, 80 for pure water, and is infinite for conducting particles. The second factor in Eq. 15.25 indicates that the saturation charge is proportional to the surface area of the particle and to the electrostatic field strength. The final factor is a time-dependent term that reaches a value of 1 when $\pi K_E e Z_i N_i t \gg 1.0$. The rate of charging does not depend on the particle size or field strength, but only on the ion concentration. When particles are intentionally charged by field charging, the ion concentration is usually $10^{13}/m^3$ $[10^7/cm^3]$ or greater, so charging will be 95% complete in 3 s or less.

The charge that is acquired is proportional to d_p^2 in field charging and to d_p in diffusion charging, so field charging is the dominant mechanism for particles larger than 1.0 μm, and diffusion charging is the dominant mechanism for particles less than 0.1 μm, even in the presence of an electric field. Between these sizes, both mechanisms are operating and the situation is much more complicated. Equations 15.24 and 15.25, while explicit, involve restrictive assumptions and are somewhat oversimplified. More accurate expressions exist, but they are not explicit and usually require computer calculation. Lawless (1996) presents approximations for charging rates that are applicable to diffusion, field, and combined charging. The charging rate obtained must be numerically integrated to get the charge acquired by a particle in a given period. Table 15.3 gives the number of charges acquired in 1 s by particles of various sizes by the diffusion, field, and combined charging mechanisms at an ion concentration of $10^{13}/m^3$ $[10^7/cm^3]$. For the conditions of the table calculations, the field-charging equation, Eq. 15.25, is accurate for particles larger than 5 μm, and Eq. 15.24 is useful from 0.2 to 2 μm. Figure 15.3 shows the number of charges acquired by a particle having $\varepsilon = 5.1$ for various charging conditions.

Figure 15.4 shows particle mobility versus particle size for a typical field-charging condition. Although particle charge decreases with particle size, mechanical mobility increases rapidly with decreasing size; consequently, there is a size for minimum electrical mobility in the submicrometer range. Also, mobility under typical charging conditions is a relatively weak function of particle size in the size range 0.1–1 μm.

EXAMPLE

How many elementary charges are acquired by a 2.0-μm water droplet in 1 s by
(a) diffusion charging in an ion concentration of $10^{13}/m^3$ $[10^7/cm^3]$ and (b) field

TABLE 15.3 Comparison of Calculation Methods for Charging by Field, Diffusion, and Combined Charging at $N_i t = 10^{13}$ s/m³ [10^7 s/cm³]. $\varepsilon = 5.1$.

Particle Diameter (µm)	Number of Elementary Units of Charged Acquired			
	Diffusion Charging		Field Charging E = 500 kV/m [5 kV/cm]	Combined Charging E = 500 kV/m [5 kV/cm]
	Eq. 15.24	Numerical Solution[a]	Eq. 15.25	Numerical Solution[a]
0.01	0.10	0.41	0.02	0.42
0.04	0.79	1.6	0.26	1.9
0.1	2.7	4.1	1.6	5.6
0.4	15.7	16.3	25.9	40
1.0	47	41	162	162
4.0	237	163	2580	2680
10	673	407	16,200	16,540
40	3180	1630	259,000	264,000

[a]Lawless (1996).

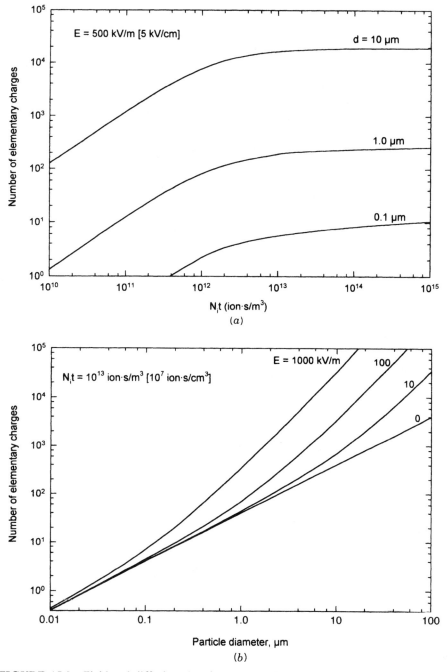

FIGURE 15.3 Field and diffusion charging. (*a*) Number of charges acquired versus $N_i t$ for particle diameters of 0.1, 1, and 10 μm at a field strength of 500 kV/m [5 kV/cm]. (*b*) Number of charges acquired versus particle diameter for field strengths of 0, 100, 1000, and 10,000 V/cm at $N_i t = 10^{13}$ s/m³ [10⁷ s/cm³]. ε = 5.1.

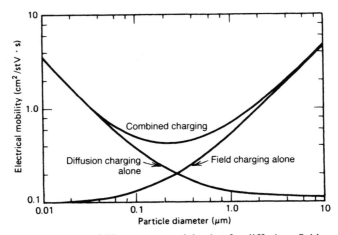

FIGURE 15.4 Electrical mobility versus particle size for diffusion, field, and combined charging at $E = 500$ kV/m [5 kV/cm] and $N_i t = 10^{13}$ s/m³ [10^7 s/cm³].

charging in an ion concentration of 10^{13}/m³ [10^7/cm³] and an electric field of 600 kV/m [6 kV/cm]?

(a) In SI units,

$$n = \frac{2 \times 10^{-6} \times 1.38 \times 10^{-23} \times 293}{2 \times 9 \times 10^9 \times (1.6 \times 10^{-19})^2}$$

$$\times \ln\left(1 + \frac{\pi \times 9 \times 10^9 \times 2 \times 10^{-6} \times 240 \times (1.6 \times 10^{-19})^2 \times 10^{13} \times 1}{2 \times 1.38 \times 10^{-23} \times 293}\right)$$

$$= 17.6 \ \ln(1 + 430) \quad \text{m} \cdot \text{N} \cdot \text{m} \cdot \text{K}^{-1} \cdot \text{K}/(\text{N} \cdot \text{m}^2 \cdot \text{C}^{-2} \cdot \text{C}^2) = 106$$

In cgs units,

$$n = \frac{2 \times 10^{-4} \times 1.38 \times 10^{-16} \times 293}{2 \times 1 \times (4.8 \times 10^{-10})^2}$$

$$\times \ln\left(1 + \frac{\pi \times 1 \times 2 \times 10^{-4} \times 2.4 \times 10^{-4} \times (4.8 \times 10^{-10})^2 \times 10^7 \times 1}{2 \times 1.38 \times 10^{-16} \times 293}\right)$$

$$= 17.6 \ \ln(1 + 430) \quad \text{cm} \cdot \text{dyn} \cdot \text{cm} \cdot \text{K}^{-1} \cdot \text{K}/\text{stC}^2 = 106$$

(b) In SI units,

$$\frac{3\varepsilon}{\varepsilon + 2} = \frac{3 \times 80}{80 + 2} = 2.93$$

$$\frac{Ed^2}{4K_Ee} = \frac{600,000 \times (2 \times 10^{-6})^2}{4 \times 9 \times 10^9 \times 1.6 \times 10^{-19}} = 417 \; \frac{\text{V} \cdot \text{m}^{-1} \cdot \text{m}^2}{\text{N} \cdot \text{m}^2 \cdot \text{C}^{-2} \cdot \text{C}}$$

$$\pi K_E e Z_i N_i t = \pi \times 9 \times 10^9 \times 1.6 \times 10^{-19} \times 0.00015 \times 10^{13} \times 1 = 6.79 \; \frac{\text{N} \cdot \text{m}^2 \cdot \text{C} \cdot \text{m}^2 \cdot \text{s}}{\text{C}^2 \cdot \text{V} \cdot \text{s} \cdot \text{m}^3}$$

$$n = 2.93 \times 417 \times \left(\frac{6.79}{6.79 + 1}\right) = 1065$$

In cgs units,

$$\frac{3\varepsilon}{\varepsilon + 2} = \frac{3 \times 80}{80 + 2} = 2.93$$

$$\frac{Ed^2}{4K_Ee} = \frac{(6000/300) \times (2 \times 10^{-4})^2}{4 \times 1 \times 4.8 \times 10^{-10}} = 417 \; \text{stV} \cdot \text{cm}^{-1} \cdot \text{cm}^2/\text{stC}$$

$$\pi K_E e Z_i N_i t = \pi \times 1 \times 4.8 \times 10^{-10} \times 450 \times 10^7 \times 1 = 6.79 \; \frac{\text{stC} \cdot \text{cm}^2 \cdot \text{s}}{\text{stV} \cdot \text{s} \cdot \text{cm}^3}$$

$$n = 2.93 \times 417 \times \left(\frac{6.79}{6.79 + 1}\right) = 1065$$

15.5   CORONA DISCHARGE

The field- and diffusion-charging methods described in the previous section require high concentrations of unipolar ions. Because of the mutual repulsion and high mobilities of those ions, their lifetime is brief. Consequently, ions must be continuously produced to use these charging methods. Ions can be created in air by radioactive discharge, ultraviolet radiation, flames, and corona discharge. Only the corona discharge can produce unipolar ions at a high enough concentration to be useful for aerosol charging.

To produce a corona discharge, one must establish a nonuniform electrostatic field, such as that between a needle and a plate or between a concentric wire and a tube, as described in Section 15.2. Air and other gases are normally very good insulators, but in a region of sufficiently high field strength, air undergoes an electrical breakdown and becomes conductive. Depending on the geometry of the field, this breakdown can be an arc or a corona discharge. For a wire-and-tube geometry, the only region with sufficient field strength for breakdown is a thin layer at the surface of the wire. The field strength required for breakdown depends on the wire diameter d_w and is given by the empirical equation (White, 1963),

$$E_b = 3000 + 127d_w^{-1/2} \text{ kV/m} \quad [30 + 12.7d_w^{-1/2} \text{ kV/cm}] \quad (15.27)$$

where d_w is in m [cm]. For a wire diameter of 1 mm, E_b is 7000 kV/m [70 kV/cm].
Figure 15.5 shows the field around a 1-mm wire maintained at 50 kV in a tube with
a diameter of 0.2 m [20 cm]. A region 0.8 mm thick around the wire exceeds 7000
kV/m [70 kV/cm] and is the corona discharge region. The low field strength near
the tube prevents arcing. This would not be the case for the uniform field between
two parallel plates where electrical breakdown would occur in the entire region between the plates and would cause an arc.

In the corona region, electrons are accelerated to a velocity sufficient to knock
an electron from an air molecule upon collision and thereby create a positive ion
and a free electron. Within the corona region, this process takes place in a self-
sustaining avalanche that produces a dense cloud of free electrons and positive ions
around the wire called *corona discharge*. The process is initiated by electrons and
ions created by natural radiation. If the wire is positive with respect to the tube, the
electrons will move rapidly to the wire, and the positive ions will stream away from
the wire to the tube in a unipolar "ion wind." If the wire is negative, the positive
ions will go to it, and the electrons will be repelled toward the tube. As their velocity slows with the decreasing field strength at greater distances from the wire,
the electrons attach to air molecules (if an electronegative gas such as oxygen is
present) and form negative ions, which stream across to the tube. In either case, the
ions migrate from the wire to the tube in high concentrations of 10^{12}–10^{15}/m^3

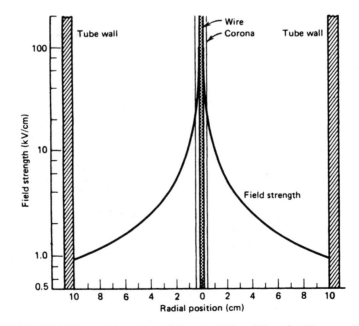

FIGURE 15.5   Field strength in a tube of diameter 0.2 m [20 cm] with a concentric wire
1 mm in diameter at 50 kV. Corona region not to scale.

$[10^6 – 10^9/\text{cm}^3]$ and at high velocities of about 75 m/s [7500 cm/s] for the conditions of Fig. 15.5.

Positive and negative coronas have quite different properties and appearances. With positive corona, the entire region around the wire has a stable, glowing sheath with a characteristic bluish-green color. With negative corona, the corona glow exists in tufts or brushes that appear to be in a dancing motion over the surface of the wire. These tufts may be several millimeters in length. There is sufficient energy in the corona region to produce ozone from oxygen. A negative corona produces about 10 times as much ozone as a positive corona. Indoor and recirculating electrostatic precipitators use positive corona for this reason. Industrial electrostatic precipitators usually use negative corona, because they can be operated at higher voltages and thereby achieve higher efficiencies. Gas temperature, pressure, and composition affect corona generation.

The introduction of aerosol particles into the space between the wire and the tube will result in field charging of the particles to the same polarity as the wire. The field used to create the corona and the ion wind also causes the field charging. If clean air is blown through the tube at high velocity, it will carry the unipolar ions out of the field region, where they may be mixed with aerosol particles for diffusion charging.

15.6 CHARGE LIMITS

There are fundamental limits on the maximum amount of charge that can be acquired by an aerosol particle of a given size. In the presence of strong external electrostatic fields, aerosol particles can exhibit corona discharge. The loss of charge by this mechanism sets an upper limit on the charge for a given particle size that is on the order of the corona onset voltage for a sphere of that size. This type of corona discharge has been observed surrounding raindrops during a thunderstorm and in the laboratory. In the more usual case, where there is no strong external field, a much higher charge can be achieved before the limit of spontaneous charge loss is reached.

For negatively charged solid particles, the maximum charge is reached when the self-generated field at the surface of a particle (see Eq. 15.11) reaches the value required for spontaneous emission of electrons from a surface. When this limit is exceeded, the crowding of electrons on the surface of the particle causes electrons to be ejected from the particle by the force of mutual repulsion. For spherical particles, this limit is

$$n_L = \frac{d_p^2 E_L}{4 K_E e} \tag{15.28}$$

where E_L is the surface field strength required for spontaneous emission of electrons, 9.0×10^8 V/m $[3 \times 10^4$ stV/cm]. This charge limit is proportional to the surface area of the particle, as expected.

A similar limit exists for positively charged particles; the only difference is that a positive ion must be emitted instead of an electron. However, emitting a positive ion is a more difficult process and requires a greater surface field strength. Equation 15.28 can be used with a value of $E_L = 2.1 \times 10^{10}$ V/m [7 × 10^5 stV/cm] to estimate the limit. Values for the positive and negative limits are shown as a function of particle size in Fig. 15.6.

A different type of limit, called the Rayleigh limit, exists for liquid droplets. When the mutual repulsion of electric charges within a droplet exceeds the confining force of surface tension, the droplet shatters into smaller droplets. The limiting charge is given by

$$n_L = \left(\frac{2\pi\gamma d_p^3}{K_E e^2} \right)^{1/2} \tag{15.29}$$

where γ is the surface tension of the droplet liquid. As shown in Fig. 15.6, the Rayleigh limit is the controlling limit for most water droplets, because it will be reached before the surface emission limits are reached. A moderately charged droplet will become highly charged as it evaporates to a smaller size. Eventually, the droplet will reach the Rayleigh limit and disintegrate. When the droplet disintegrates, the fragments are below the Rayleigh limit because the same amount of charge is now distributed over a larger surface.

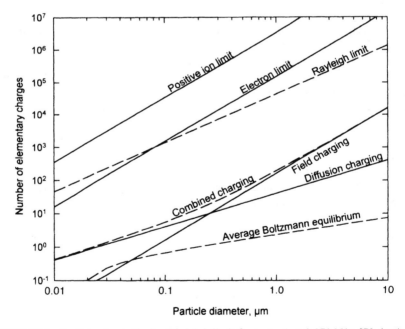

FIGURE 15.6 Particle charge limits. Rayleigh limit for water ($\gamma = 0.073$ N/m [73 dyn/cm]), diffusion charging at $N_i t = 10^{13}$ s/m^3 [10^7 s/cm^3] and field charging at $E = 500$ kV/m [5 kV/cm] and $N_i t = 10^{13}$ s/m^3 [10^7 s/cm^3]. $\varepsilon = 5.1$.

EXAMPLE

What is the maximum charge for a negatively charged 1-μm sphere?

$$n_L = \frac{d_p^2 E_L}{4 K_E e} = \frac{(10^{-6})^2 \times 9 \times 10^8}{4 \times 9 \times 10^9 \times 1.6 \times 10^{-19}} = 156,000$$

$$\left[= \frac{(10^{-4})^2 \times 3 \times 10^4}{4 \times 1 \times 4.8 \times 10^{-10}} = 156,000 \right]$$

15.7   EQUILIBRIUM CHARGE DISTRIBUTION

The minimum charge an aerosol particle can have is zero. This condition is rarely achieved, because of random collisions of aerosol particles with the omnipresent air ions. As explained in Chapter 14, every cubic centimeter of air contains about 10^3 ions with approximately equal numbers of positive and negative ions. Aerosol particles that are initially neutral will acquire charge by collision with ions due to their random thermal motion. Aerosol particles that are initially charged will lose their charge slowly as the charged particles attract oppositely charged ions. These competing processes eventually lead to an equilibrium charge state called the *Boltzmann equilibrium* charge distribution (also called the steady-state, stationary, or bipolar equilibrium charge distribution). The Boltzmann equilibrium charge distribution represents the charge distribution of an aerosol in charge equilibrium with bipolar ions. This minimum amount of charge is very small, with a statistical probability that some particles will have no charge and others will have one or more charges. For equal concentrations of positive and negative ions, a reasonable first approximation for normal air, the fraction of particles f_n of a given size having n positive (or n negative) elementary units of charge is given by

$$f_n = \frac{\exp(K_E n^2 e^2 / d_p kT)}{\sum_{n=-\infty}^{\infty} \exp(K_E n^2 e^2 / d_p kT)} \tag{15.30}$$

For particles larger than 0.05 μm, Eq. 15.30 becomes identical to the equation representing the normal distribution and can be written in an easier-to-use form as

$$f_n = \left(\frac{K_E e^2}{\pi d_p kT} \right)^{1/2} \exp\left(\frac{-K_E n^2 e^2}{d_p kT} \right) \tag{15.31}$$

This equation agrees with Eq. 15.30 to within 7% for particles larger than 0.02 μm and to within 0.04% for particles larger than 0.05 μm. For particles less than about 0.05 μm, both Eq. 15.30 and 15.31 underestimate the fraction of charged particles, and more complicated procedures must be used. [See Hoppel and Frick (1986).]

Table 15.4 gives the percentage of particles of a given size having the indicated charge. The distribution is symmetrical about zero; that is, the fraction of particles with n positive charges equals the fraction with n negative charges. As a practical matter, for particles less than 0.1 μm in diameter, the distribution consists of a fraction that is uncharged and a remaining fraction, which is singly charged with equal fractions charged positively and negatively. The average charge given in the table is the average charge calculated without regard to sign. The average Boltzmann equilibrium charge is compared with other charge conditions in Fig. 15.6. An empirical approximation for the average number of charges $\bar{n}$ is

$$\bar{n} \approx 2.37\sqrt{d_p} \qquad (15.32)$$

where d_p is in μm. Equation 15.32 provides satisfactory accuracy (±5%) for particles larger than 0.2 μm.

EXAMPLE

What fraction of 1-μm particles at Boltzmann equilibrium carry +2 charges?
By Eq. 15.31,

$$f_n = \left(\frac{9 \times 10^9 \times (1.6 \times 10^{-19})^2}{\pi \times 10^{-6} \times 1.38 \times 10^{-23} \times 293} \right)^{1/2} \exp\left(\frac{-9 \times 10^9 \times 2^2 \times (1.6 \times 10^{-19})^2}{10^{-6} \times 1.38 \times 10^{-23} \times 293} \right) = 0.107$$

$$\left[\left(\frac{1 \times (4.8 \times 10^{-10})^2}{\pi \times 10^{-4} \times 1.38 \times 10^{-16} \times 293} \right)^{1/2} \exp\left(\frac{-1 \times 2^2 \times (4.8 \times 10^{-10})^2}{10^{-4} \times 1.38 \times 10^{-16} \times 293} \right) = 0.107 \right]$$

The rate at which an aerosol reaches the Boltzmann equilibrium charge distribution depends on the concentration of bipolar ions and is given by

$$\frac{n(t)}{n_0} = \exp(-4\pi K_E e Z_i N_i t) \qquad (15.33)$$

where $n(t)$ is the number of charges on a particle after it has been exposed to a bipolar ion concentration N_i for a time t if it had n_0 charges at time $t = 0$. The fractional discharge rate does not depend on the particle size or the initial charge. Equation 15.33 is most useful when there is an excess of bipolar ions, so that N_i can be considered constant. The discharge of aerosols to Boltzmann equilibrium depends on the product $N_i t$; that is, the same result is reached for different ion concentrations when the value of $N_i t$ is the same.

Equation 15.33 indicates that rapid discharge of a highly charged aerosol can be achieved by mixing the aerosol with a high concentration of bipolar ions. One method uses an alternating-current corona discharge in a sonic velocity jet of air. The bipolar ions that are produced are swept out of the field region by the sonic

TABLE 15.4 Distribution of Charge on Aerosol Particles at Boltzmann Equilibrium

Particle Diameter (μm)	Average Number of Charges	Percentage of Particles Carrying the Indicated Number of Charges								
		< -3	-3	-2	-1	0	+1	+2	+3	> +3
0.01	0.007				0.3	99.3	0.3			
0.02	0.104				5.2	89.6	5.2			
0.05	0.411			0.6	19.3	60.2	19.3	0.6		
0.1	0.672		0.3	4.4	24.1	42.6	24.1	4.4	0.3	
0.2	1.00	0.3	2.3	9.6	22.6	30.1	22.6	9.6	2.3	0.3
0.5	1.64	4.6	6.8	12.1	17.0	19.0	17.0	12.1	6.8	4.6
1.0	2.34	11.8	8.1	10.7	12.7	13.5	12.7	10.7	8.1	11.8
2.0	3.33	20.1	7.4	8.5	9.3	9.5	9.3	8.5	7.4	20.1
5.0	5.28	29.8	5.4	5.8	6.0	6.0	6.0	5.8	5.4	29.8
10.0	7.47	35.4	4.0	4.2	4.2	4.3	4.2	4.2	4.0	35.4

jet and into a mixing chamber with the charged particles. A more common approach is to use a radioactive source, such as polonium-210 or krypton-85 gas in a thin stainless-steel tube, to ionize the air molecules inside a chamber through which the aerosol to be discharged flows. The volume of the chamber is such that it provides sufficient residence time for the aerosol to be "neutralized," or brought to the Boltzmann equilibrium charge distribution. A value of $N_i t$ of 6×10^{12} ion · s/m^3 [6×10^6 ion · s/cm^3] is required for complete neutralization of highly charged particles. In commercial radioactive aerosol neutralizers, this level of neutralization is achieved in about 2 s. In the atmosphere, where the normal bipolar ion concentration is about 10^9 ions/m^3 [10^3 ions/cm^3], the same effect is achieved in 100 min.

15.8   ELECTROSTATIC PRECIPITATORS

Electrostatic precipitators use electrostatic forces to collect charged particles for aerosol sampling and air cleaning. These forces act only on the particles and can be much greater than the corresponding gravitational or inertial forces. The principles are the same for sampling and air cleaning, but the size scale differs greatly, by a factor of 10^8 in the extreme. The two basic steps are to charge the particles and to subject them to an electric field so that their electrostatic migration velocity causes them to deposit on a collection surface. A third step, that of removing the deposited particles, is necessary for continued operation of air-cleaning precipitators. As a class, these devices are characterized by low pressure drop, high efficiency for small particles, and the ability to handle high dust concentrations.

In an electrostatic precipitator, particles are usually charged by field charging using a corona discharge. Two-stage precipitators have separate sections for charging and precipitation. In a single-stage precipitator the corona wires are in the precipitation section, and the same field that creates the corona also causes the precipitation. Normally, the field and the ion concentration are such that the particles reach their saturation charge in less than a second.

The charged aerosol flows through an electric field oriented perpendicular to the direction of flow and to the collection surface, which is usually the inside of a metal tube or a flat plate. For a laminar-flow precipitator, particle collection is analogous to the gravity-settling chamber described in Section 3.9 and shown in Fig. 3.6. All particles with $V_{TE} > HV_x/L$, where H is the distance from wires to collection plates in a single-stage precipitator and between the plates in a two-stage precipitator, V_x is the flow velocity, and L is the dimension along the flow direction, will be collected with 100% efficiency. For a given charging condition, there is a particle size with minimum mobility (see Fig. 15.4) a laminar-flow precipitator designed to completely capture this size particle will be 100% efficient for all particle sizes.

In air-cleaning electrostatic precipitators, the flow is turbulent and the collection situation is analogous to stirred settling, as described in Section 3.8. Consider the wire-and-tube geometry shown in cross section in Fig. 15.7. In a period dt that is brief compared with that required for turbulent mixing, all particles within a distance $V_{TE}dt$ of the tube wall will be removed. For simplicity, we assume that par-

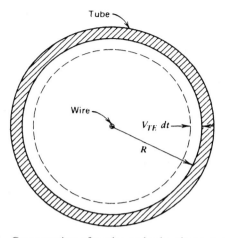

FIGURE 15.7   Cross section of a wire-and-tube electrostatic precipitator.

ticles stick if they touch the wall and are not reentrained. The fraction of particles removed during this period, dN/N, *is* the negative ratio of the area of the annulus, $2\pi R V_{TE} dt$, to the total cross-sectional area of the tube, πR^2, and is given by

$$\frac{dN}{N} = -\frac{2\pi R V_{TE} dt}{\pi R^2} = -\frac{2V_{TE} dt}{R} \tag{15.34}$$

This equation can be integrated to obtain the number concentration after a time t.

$$\int_{N_0}^{N(t)} \frac{dN}{N} = \int_0^t \frac{-2V_{TE}}{R} dt \tag{15.35}$$

$$\frac{N(t)}{N_0} = \exp\left(\frac{-2V_{TE} t}{R}\right) \tag{15.36}$$

For a tube of length L and a volumetric flow rate Q, the residence time in the tube, $\pi R^2 L/Q$, can be substituted into Eq. 15.36 to get the penetration—the fraction of the inlet concentration that exits the tube:

$$\mathbf{P} = \frac{N_{out}}{N_0} = \exp\left(\frac{-2\pi V_{TE} RL}{Q}\right) \tag{15.37}$$

It is customary to express Eq. 15.37 in terms of collection efficiency and the area of the collection surface, $A_c = 2\pi RL$. Doing so gives the Deutsch–Anderson equation,

$$\mathbf{E} = 1 - \exp\left(\frac{-V_{TE} A_c}{Q}\right) \tag{15.38}$$

This is the basic scaling equation for electrostatic precipitators. It assumes that the particles are uniformly distributed across every cross section and become fully charged as soon as they enter the precipitator. It predicts the change in efficiency that will result from changes in precipitator size, flow rate, and V_{TE}. Equations 15.36–15.38 suggest that any change in the precipitator size or flow rate that increases the aerosol residence time in the precipitator will increase its efficiency. Similarly, increasing V_{TE} by increasing the charge on the particles or the field strength will increase the efficiency of the precipitator. Efficiencies for laminar- and turbulent-flow electrostatic precipitators are shown in Fig. 15.8. The 100% efficiency point for laminar-flow precipitators corresponds to the $1 - 1/e$ point for a turbulent-flow precipitator. Although Eq. 15.38 predicts that any efficiency (up to 100%) can be achieved, the efficiency of high-efficiency air-cleaning units is not controlled by that equation, but by secondary effects such as a nonuniform flow distribution, reentrainment, and excess resistivity in the layer of deposited dust.

The point-to-plane type of electrostatic precipitator is used for sampling aerosols for electron microscope studies and for piezoelectric measurement of mass. As shown in Fig. 15.9, the corona and precipitating fields are formed between a corona needle (the point) and a flat collection surface (the plane). Depending on the design, the separation between the point and the plane is from 3 to 40 mm, flow rates are from 0.5 to 5 L/min, and voltages are from 2 to 15 kV. When used for sampling for electron microscopy, the particles are deposited directly onto a carbon-coated electron microscope grid 3 mm in diameter. From simple geometric consideration, this small target results in a low collection efficiency—as low as a few percent. The absolute collection efficiency is not too important, because samples are usually taken to determine the particle size distribution and morphology. More important is the possibility of a particle size bias in the collection efficiency. In the point-to-plane precipitator, charging and precipitation must occur during a fraction of a second, so differences in mobility under these charging conditions can lead to

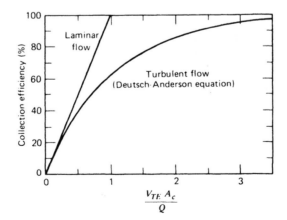

FIGURE 15.8   Efficiency curves for laminar- and turbulent-flow electrostatic precipitators.

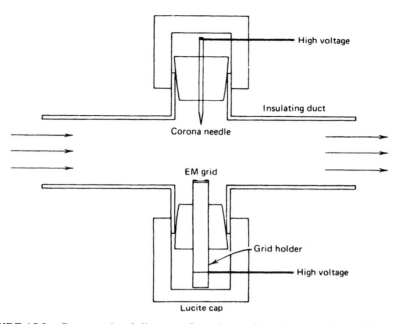

FIGURE 15.9 Cross-sectional diagram of a point-to-plane electrostatic precipitator used to obtain aerosol samples for electron microscopy.

a sampling bias that favors the collection of larger particles (Cheng et al., 1981). The use of a low sampling flow rate and a moderate ion current reduces this bias.

15.9 ELECTRICAL MEASUREMENT OF AEROSOLS

A simple electrical mobility analyzer is shown in Fig. 15.10. Aerosol particles are introduced along the centerline between two oppositely charged parallel plates. Filter samples (or other aerosol measurements) are taken downstream of the device while a specific voltage is maintained on the plates. For a given voltage, all particles with mobility greater than a certain amount will migrate to the plates and be trapped. Those with lower mobility will get through, to be collected on the downstream fil-

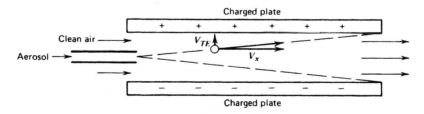

FIGURE 15.10 Diagram of a simple electrical mobility analyzer.

ter. Uncharged particles pass through unaffected. By comparing the mass (or other quantity) sampled when there is no voltage on the plates with that for different voltages, one can determine the fraction of particles that are uncharged, the fraction with mobility in a certain range, and the distribution of electrical mobilities.

If the particles have been charged by diffusion charging under well-defined and carefully controlled conditions, the charge on a particle of a given size is known and a unique electrical mobility is associated with every particle size. Under these conditions the distribution of particle size can be determined by measuring the distribution of electrical mobility. This is the principle of the *electrical aerosol analyzer* (EAA®) manufactured by TSI, Inc., of St. Paul, Minnesota, and shown schematically in Fig. 15.11. Aerosol is drawn into the instrument at 5 L/min and passes through a unipolar-ion diffusion charger. The time in the charger is fixed by the flow rate, and the ion concentration is held constant at $4.6 \times 10^{13}/m^3$ [$4.6 \times 10^7/cm^3$] by a feedback circuit measuring the ion current. A fixed value of $N_i t$ of 10^{13} ion $\cdot$ s/m^3 [10^7 ion $\cdot$ s/cm^3] ensures that particles of a given size receive a predictable number of charges, as shown in Fig. 15.3b. The charged aerosol flows to the mobility analyzer as a thin annular stream at the periphery of a laminar flow of clean air between two concentric cylinders. All particles with mobility less than the cutoff mobility, determined by the voltages on the tubes, pass out of the analyzer section and are collected in a high-efficiency conductive filter.

An electrometer continuously monitors the current produced by the capture of the charged particles in the filter. Since there is a monotonic relationship between mobility and particle size, the difference in current measured at two analyzer voltage settings is related to the number of particles in the size (mobility) range defined by the cutoff sizes of the two voltage settings. The instrument operates with preset voltages to provide the number of particles in each of 10 size ranges from 0.003 to 1.0 μm. In automatic operation, the instrument steps through the 10 size ranges in 2 min. The aerosol must have a stable number concentration and size distribution during this period, and the particles must be solid or nonvolatile liquids. Below 0.02 μm, a substantial, but predictable, fraction of the particles is uncharged. This is compensated for in the calibration of the instrument. The curve of mobility versus size becomes quite flat for particles larger than 0.3 μm (there are only two size ranges greater than this size), and the size resolution deteriorates. The concentration range depends on particle size and goes from 10^7 to $10^{11}/m^3$ [10 to10^5/cm^3] for 0.5-μm particles to 2×10^9 to $5 \times 10^{13}/m^3$ [2000 to 5×10^7/cm^3] for 0.0l-μm particles.

A related instrument is the differential mobility analyzer (DMA®), shown in Figure 15.12. The input aerosol first passes through an impactor to remove particles larger than 10 μm that might cause data inversion problems. The aerosol is then neutralized to the Boltzmann equilibrium charge distribution before entering the electrostatic classifier section. The geometry and flow in the classifier section are similar to the geometry and flow in the analyzer section of the EAA. A laminar flow of clean air is surrounded by a thin annular layer of aerosol as the two fluids travel axially between the central rod and the coaxial tube. The tube is grounded and the voltage on the central rod controlled between 20 and 10,000 V. Near the bottom of the central rod is a gap that allows a small airflow to leave the classifier and exit

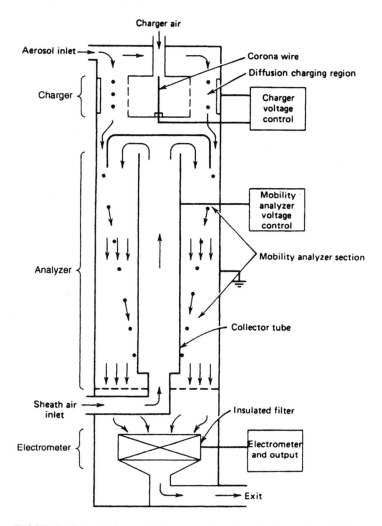

FIGURE 15.11 Schematic diagram of an electrical aerosol analyzer.

through a central tube. Only particles with a narrow range of mobilities can enter the gap. Particles with greater mobility migrate to the central rod before reaching the gap, while those with lower mobility go beyond the gap and are filtered out. The exiting aerosol is nearly all singly charged and nearly monodisperse. Its size is controlled by adjusting the voltage on the central rod.

The DMA was originally intended to be used as a monodisperse aerosol generator to produce submicrometer-sized aerosols for testing and calibration. However, it is used more commonly to measure particle size distribution with high resolution in the submicrometer size range. By monitoring the exit aerosol stream with a CNC (see Section 13.6), the number concentration in a narrow range of mobilities, and thus the particle size, can be determined. The entire submicrometer particle size dis-

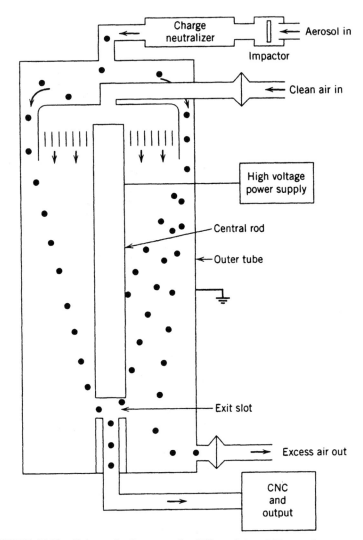

FIGURE 15.12   Schematic diagram of a differential mobility analyzer.

tribution can be obtained by stepping or continuously scanning through the voltage range. The lower size limit and maximum and minimum concentrations that can be measured are determined by the capabilities of the CNC that is used. DMAs can size particles from 0.005 to 1.0 μm at concentrations from 10^6 to $10^{14}/m^3$ [$1–10^8/cm^3$]. One version of this instrument, the TSI, Inc., Model 3934, scans up to 147 size channels under computer control in 1 to 8 minutes.

Two DMAs can be used in series or tandem (TDMA) to study processes that change particle size, such as evaporation, growth by condensation, or chemical reac-

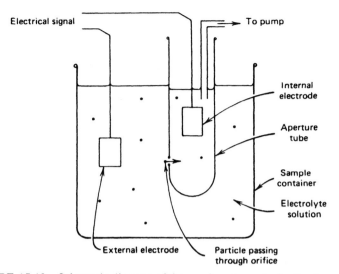

FIGURE 15.13   Schematic diagram of the sensing element of a Coulter counter.

tions. A monodisperse submicrometer aerosol of known size is produced by the first DMA. The aerosol is subjected to a growth or shrinkage process, and then its size is measured by the second DMA. By careful operation and data analysis, particle size changes as small as 0.3% can be measured. This method works best for particles in the 0.01-to-0.2-μm size range.

Although not an aerosol-measuring instrument, the Coulter counter, an instrument manufactured by Coulter Electronics, of Hialeah, Florida, uses the electrical resistance principle to measure particle size distributions in liquid suspensions. The instrument is suitable only for insoluble particles, and the particles must be captured directly in a conductive liquid, such as with an impinger, or transferred to such a liquid before measurement. As shown in Fig. 15.13, the liquid flows from one section to the other through a small orifice 10–400 μm in diameter. A current is established through the orifice by platinum electrodes in the liquid on either side of the orifice. Because the particle has an electrical resistance different from that of the liquid, there is a momentary change in the current as the particle passes through the orifice due to the displaced electrolyte. This change, which is directly proportional to the volume of the particle, is sensed electronically and converted to particle size, and these data are accumulated to produce a distribution of particle size weighted by volume. The instrument can be used for particle sizes ranging from 0.3 to 200 μm, but for a given orifice, the range is limited to 2–40% of its diameter. Size calibration is done with polystyrene spheres of known size. (See Section 21.2.) The maximum counting rate is 3000/s. A sample size of about 1 mg is required to determine a size distribution. Wetting agents and ultrasonic energy are used to ensure that all particles are fully dispersed in the liquid.

PROBLEMS

15.1 Determine V_{TE} for a 1.0-μm-diameter sphere of standard density positioned between two parallel plates 1 cm apart maintained at a 9000-V potential difference. Assume that the particle is charged to the average Boltzmann equilibrium charge. What is the ratio of this electrostatic velocity to its gravitational settling velocity?
ANSWER: 0.0023 m/s [0.23 cm/s], 66.

15.2 Raindrops falling from thunderclouds may carry charges as high as 3.3×10^{-11} C/drop [0.1 stC/drop]. According to the Rayleigh limit, what is the minimum size of such drops?
ANSWER: 280 μm.

15.3 A 1.6-μm-diameter particle of standard density is charged by diffusion charging at an ion concentration of $10^{12}/m^3$ [$10^6/cm^3$] for 10 s. What V_{TE} will this charged particle have in an electrostatic field of 300 kV/m [3 kV/cm]?
ANSWER: 0.016 m/s [1.6 cm/s].

15.4 Show that the collision velocity of two equal-sized particles carrying equal and opposite charge is proportional to the inverse of the cube of their diameter, d_p^{-3}. Assume that motion is in the Stokes region. Neglect slip correction.

15.5 An aerosol neutralizer has a volume of 1.7 L and contains a 1.0-mCi Kr-85 source that produces an ion concentration of $3 \times 10^{12}/m^3$ [$3 \times 10^6/cm^3$] inside the cylinder. At a flow rate of 50 L/min, what fraction of the original charge will remain on an aerosol after it has passed through the device?
ANSWER: 6.1×10^{-8}.

15.6 For a uniform electric field of 1200 kV/m [12 kV/cm] and an ion concentration of $10^{14}/m^3$ [$10^8/cm^3$], what is the saturation charge for a 2.0-μm-diameter particle? How long does it take to charge this particle to 95% of its saturation charge? ($\varepsilon = 10$.)
ANSWER: $2080e$, 0.28 s.

15.7 What is the maximum velocity a 2-μm particle can have in an electrostatic field of 100 kV/m [1 kV/cm]? [*Note*: Motion may be outside the Stokes region.]
ANSWER: 250 m/s [25,000 cm/s].

15.8 What is the average charge on a 0.15-μm-diameter particle in equilibrium with bipolar ions at 293 K [20°C]? Assume that Eq. 15.32 holds.
ANSWER: 1.47×10^{-19} C [4.4×10^{-10} stC]. (*Note*: Equation 15.30 gives an answer that is 7% larger.)

15.9 Show that Rayleigh limit disintegration leads to the formation of charge-stable, smaller droplets.

15.10 Determine the electrostatic velocity and the ratio of this velocity to the settling velocity for a 0.6-μm particle (ε = 10) in a field strength of 1200 kV/m [12 kV/cm]. Assume that the particle is charged to the average Boltzmann equilibrium charge.

ANSWER: 0.0043 m/s [0.43 cm/s], 320.

15.11 Determine the electrostatic migration velocity for a 0.6-μm particle (ε =10) in a field strength of 1200 kV/m [12kV/cm]. Assume that the particle is charged in the same field for one second at an ion concentration of 10^{14}/m^3 [10^8/cm^3].

ANSWER: 0.44 m/s [44 cm/s].

15.12 What particle size receives equal numbers of charges in 1 s by diffusion charging and by field charging at 2000 kV/m [20 kV/cm]? Assume that each mechanism operates independently and that the ion concentration is 10^{14}/m^3 [10^8/cm^3] for both types of charging. ε = 1. [*Hint*: A trial-and-error spreadsheet approach may be the easiest way to do this problem.]

ANSWER: 0.15 μm.

15.13 What voltage is required to just balance a singly charged (negative) 0.8-μm-diameter particle in a Millikan cell? The plate spacing is 5 mm and particle density is 1000 kg/m^3 [1.0 g/cm^3].

ANSWER: 82 V.

REFERENCES

Cheng, Y.-S., Yeh, H-C., and Kanapilly, G. M., "Collection Efficiencies of a Point-to-plane Electrostatic Precipitator," *Am. Ind. Hyg. Assoc. J.*, **42**, 605–610 (1981).

Flagan, R. C., "History of Electrical Aerosol Measurements," *Aerosol Sci. Tech.*, **28**, 301–380 (1998).

Hoppel, W. A., and Frick, G. M., "Ion-Aerosol Attachment Coefficients and the Steady-State Charge Distribution on Aerosols in a Bipolar Environment," *Aerosol Sci. Tech.*, **5**, 1–21 (1986).

Lawless, P. A., "Particle Charging Bounds, Symmetry Relations, and an Analytic Charging Rate Model for the Continuum Regime," *J. Aerosol Sci.*, **27**, 191–215 (1996).

Liu, B. Y. H., and Kapadia, H., "Combined Field and Diffusion Charging of Aerosol Particles in the Continuum Regime," *J. Aerosol Sci.*, **9**, 227–242 (1978).

Liu, B. Y. H., and Pui, D. Y. H., "Electrical Neutralization of Aerosols," *J. Aerosol Sci.*, **5**, 465–472 (1974).

Liu, B. Y. H., Pui, D. Y. H., and Kapadia, H., "Electrical Aerosol Analyzer: History, Principle, and Data Reduction," in Lundgren, D. A. et al. (Eds.), *Aerosol Measurement*, University Presses of Florida, Gainesville, FL, 1979.

Loeb, L. B., *Static Electrification*, Springer-Verlag, Berlin, 1958.

Smith, W. B., Felix, L. G., Hussey, D. H., Pontius, D. H., and Sparks, L. E., "Experimental Investigations of Fine Particle Charging by Unipolar Ions," *J. Aerosol Sci.*, **9**, 101 (1977).

Whitby, K. T., and Liu, B. Y. H., "The Electrical Behavior of Aerosols," in Davies, C. N. (Ed.), *Aerosol Science*, Academic Press, London, 1966.

White, H. J., *Industrial Electrostatic Precipitation*, Addison-Wesley, Reading, MA, 1963.

Yeh, H.-C., " Electrical Techniques," in Willeke, K., and Baron, P. A. (Eds.), *Aerosol Measurement: Principles, Techniques and Applications*, Van Nostrand Reinhold, New York, 1993.

16 Optical Properties

The optical properties of aerosols are responsible for many spectacular atmospheric effects, such as richly colored sunsets, halos around the sun or moon, and rainbows. They also cause the degradation of visibility associated with atmospheric pollution. The interaction of aerosol particles with light forms the basis for an important class of instruments used for measuring aerosol particle size and concentration. Optical measurement methods have the advantages of being extremely sensitive and nearly instantaneous and of not requiring physical contact with the particles.

The optical phenomena covered in this chapter are all a direct result of the scattering and absorption of light by aerosol particles. Black smoke looks black because the particles absorb visible light. Dense rain clouds, on the other hand, look black even though the droplets have negligible absorption, because the droplet scattering is so complete that light cannot get through the clouds. Fair-weather clouds look white because of the extensive scattering of the incident light from the surface of the cloud. When we see an object, we are seeing primarily the scattered light from that object. The scattered light reaching our eyes contains information about the color, surface texture, and form of the object.

The scattering of light by very small particles—less than 0.05 µm in diameter—is described in relatively simple terms by Rayleigh's theory, the theory of molecular scattering. Scattering by large particles—greater than 100 µm—can be analyzed readily by geometric optics, the tracking of diffracted, reflected, and refracted rays of light through the particle. Between these sizes, an important range of interest for aerosol technology is the Mie scattering region. In this size range, where the particle size and the wavelength of light are the same order of magnitude, the scattering of light by aerosol particles is a complicated phenomenon.

The scientific study of light scattering by aerosols began in the late 1800s with experimental studies by Tyndall and theoretical analysis by Lord Rayleigh. Maxwell's theory of electromagnetic radiation enabled Gustav Mie to develop the general theory of scattering in 1908. Unlike most other areas of aerosol science, the theory of light scattering, although complicated, is exact for spherical, homogeneous particles and thus provides an accurate and reliable tool for measuring properties of aerosols.

The discussion that follows is limited to elastically scattered light—that is, scattered light whose frequency is the same as the frequency of the incident light. Processes such as fluorescence or Raman scattering are not covered here. Nearly all real aerosols are sufficiently dilute that each particle will scatter light independently of its neighbors. Consequently, we follow the microscopic single-particle approach used in the previous chapters.

16.1   DEFINITIONS

Light is defined as visible electromagnetic radiation having a *wavelength* λ between about 0.4 (violet) and 0.7 μm (red) and a *frequency* f between 8×10^{14} and 4×10^{14} Hz. This wavelength band is an extremely narrow portion of the electromagnetic spectrum, which covers a range that is greater than 10^{20}. The center of the visible-light region is green light, with a wavelength of approximately 0.5 μm. Wavelength and frequency are related by

$$c = f\lambda \tag{16.1}$$

where c is the velocity of light in a vacuum, equal to 3.0×10^8 m/s [3.0×10^{10} cm/s].

The ratio of the velocity of light in a vacuum to the velocity of light in a particular material, V_p, is the *index of refraction,* or refractive index, m of that material. For a given wavelength, the index of refraction depends only on the material, but it varies slightly with the wavelength of the light used. For nonabsorbing materials, it is given by

$$m = \frac{c}{V_p} \tag{16.2}$$

This is called the *absolute index of refraction* and is always greater than unity.

The index of refraction for absorbing materials—those having appreciable electrical conductivity—is expressed as a complex number

$$m = m'(1 - ai) = m' - m'ai \tag{16.3}$$

where i is $\sqrt{-1}$, m' is the real refractive index, and a is related to the absorption coefficient of the bulk material A by

$$A = \frac{4\pi a}{\lambda} \tag{16.4}$$

The imaginary part of the refractive index is zero for nonabsorbing particles. Refractive indices for various materials are given in Table 16.1.

For particles in a two-phase system, a *relative index of refraction* is used. This is defined as the ratio of the velocity of light in the suspending medium, V_m, to the velocity in a particle, V_p:

$$m_r = \frac{V_m}{V_p} = \frac{m_p}{m_m} \tag{16.5}$$

Air has a refractive index that is practically equal to that of a vacuum (see Table 16.1), so that the absolute and relative refractive indices are equal for aerosol particles. For particles suspended in liquids, m_r may be greater or less than unity; m_r is less than unity for gas particles (bubbles) in a liquid.

TABLE 16.1   Index of Refraction of Some Common Materials at λ = 0.589 μm, the Wavelength of Yellow Sodium Light[a]

Material	Index of Refraction	Material	Index of Refraction
Vacuum	1.0	Quartz (SiO_2)	1.544
Water vapor	1.00025	Polystyrene latex	1.590
Air	1.00028	Diamond	2.417
Water	1.3330	Urban aerosol (avg.)	1.56–0.087i
Glycerine	1.4730	Methylene blue	1.55–0.6i
Benzene	1.501	Soot	1.96–0.66i
Ice (H_2O)	1.305	Carbon[b]	2.0–1.0i
Glass	1.52–1.88	Iron	2.80–3.34i
Sodium chloride	1.544	Copper	0.47–2.81i

[a]Data taken primarily from *Handbook of Chemistry and Physics*, 74th ed., CRC, Boca Raton, FL (1993).
[b]λ = 0.491 μm.

The *intensity* of electromagnetic radiation arriving at any surface, such as the surface of a detector, is expressed in terms of the radiant power arriving per unit area. When power (energy per unit time) is measured in watts (W), intensity is stated in W/m^2 [W/cm^2]. Intensity of light emitted from a point source is expressed as radiant power emitted into a given solid angle, i.e., watts per steradian (W/sr); for visible light, this intensity can also be expressed in candela (cd), the SI base unit for luminous intensity. The steradian (sr) is the dimensionless solid angle equal to the cone from the center of an imaginary sphere that subtends $1/4\pi$ of the total spherical surface. Thus, there are 4π sr in a sphere. The light scattered from an aerosol particle can be considered as coming from a point source, and its intensity in a given direction may be expressed in any of the units just mentioned.

Although we are concerned primarily with the scattering of visible light, the principles presented here can be applied to electromagnetic radiation of any wavelength. Thus, the scattering of radio waves by artificial satellites, the scattering of microwaves by raindrops, and the scattering of light by aerosol particles are equivalent processes with wavelengths and object sizes that are the same order of magnitude. The scattering in each case is governed by the ratio of the particle size to the wavelength λ of the radiation. This ratio, which is dimensionless, is called the *size parameter* and is given by

$$\alpha = \frac{\pi d}{\lambda} \tag{16.6}$$

The factor π is introduced to simplify certain light-scattering equations and has the effect of making α equal to the ratio of the circumference of the particle to the wavelength. For visible light, the value of α is approximately equal to six times the particle diameter expressed in micrometers.

Light can be considered a stream of photons or a transverse wave. To describe light scattering by aerosol particles, it is convenient to consider light as the electric wave component of electromagnetic radiation. When the transverse oscillations of the electric vector occur in all directions perpendicular to the direction of propagation, the light is said to be *unpolarized*. Normal light, such as sunlight or incandescent light, is unpolarized. When the electric vector oscillates only in one plane, the light is said to be *polarized* in that plane. Laser light and light that has passed through a polarizing filter are polarized. Any light beam can always be resolved into two perpendicularly polarized components, often called vertical and horizontal polarizations, that may differ in intensity.

16.2 EXTINCTION

Aerosol particles illuminated by a beam of light scatter and absorb some of that light, thereby diminishing the intensity of the beam. This process is called *extinction* and addresses only the attenuation of light along an axis. While all aerosol particles will scatter light, only those made of absorbing material will absorb light. The extinction characteristics of aerosols are important for aerosol measurement and, as will be explained in Section 16.4, for evaluating the effect of aerosols on visibility. We can look directly at the attenuated light of the setting sun because the intensity is reduced by extinction over the long air path. We see a red sun because the extinction is strongest for blue light and weakest for red light.

For a parallel beam of light, such as that shown in Fig. 16.1, the ratio of the light intensity traversing the aerosol, I, to that incident on the aerosol, I_0, is given by Bouguer's law (also known as the Lambert–Beer law),

$$\frac{I}{I_0} = e^{-\sigma_e L} \tag{16.7}$$

where σ_e is the extinction coefficient of the aerosol and L is the path length of the light beam through the aerosol. The extinction coefficient is the fractional loss in

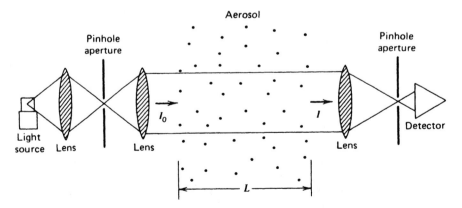

FIGURE 16.1   Schematic diagram of an extinction-measuring apparatus.

intensity per unit path length associated with an elemental thickness dL. The units of σ_e are (length)$^{-1}$, and σ_e and L must be expressed by the same unit of length to make the exponent of Eq. 16.7 dimensionless. The extinction coefficient σ_e is analogous to γ in Eq. 9.11 for penetration through a filter. (Note that Eqs. 16.7 and 9.11 have the same form.)

The aperture and lens arrangement shown in Fig. 16.1 is needed to ensure that the particles are illuminated with a parallel beam of light (i.e., to ensure that all rays are parallel) and that only the attenuated parallel light reaches the detector. Failure to include the lens and aperture at the detector allows forward-scattered light from the particles to reach the detector, in which case Eq. 16.7 does not hold. The extinction produced by a particle is a function of the particle extinction efficiency Q_e.

$$Q_e = \frac{\text{radiant power scattered and absorbed by a particle}}{\text{radiant power geometrically incident on the particle}} \qquad (16.8)$$

The geometrically incident power is the amount of energy per second that intercepts the cross-sectional area A_p of the particle. The particle extinction efficiency gives a relative measure of a particle's ability to remove light from a beam compared with simple blocking or interception by the projected area of the particle. The extinction efficiency times the projected area of a particle is the cross-sectional area of light removed from a beam by the particle. A Q_e of 2.0 indicates that a particle removes twice as much light as it would by simple projected-area blocking. Values of Q_e range from 0 to about 5. The particles in a cubic centimeter of a monodisperse aerosol will each remove $A_p Q_e$ watts from a beam of unit intensity. From the definition of the extinction coefficient, a monodisperse aerosol of N particles per unit volume has an extinction coefficient of

$$\sigma_e = NA_p Q_e = \frac{\pi N d^2 Q_e}{4} \qquad (16.9)$$

Q_e is defined (Eq. 16.8) in the same way as the combined single-filter efficiency, given in Eq. 9.14. Also, Eqs. 16.7 and 16.9 are analogous to Eq. 9.19 for filters. Both relate macroscopic phenomena—light attenuation and filter penetration—to microscopic-scale particle properties, Q_e and E_Σ, respectively.

The extinction efficiency of a particle is the sum of its scattering efficiency Q_s and its absorption efficiency Q_a,

$$Q_e = Q_s + Q_a \qquad (16.10)$$

where Q_s and Q_a are defined by equations equivalent to Eq. 16.8. It follows from Eq. 16.10 that, for monodisperse particles

$$\sigma_e = \sigma_s + \sigma_a \qquad (16.11)$$

where σ_s and σ_a are defined by Eq. 16.9 with Q_e replaced by Q_s and Q_a, respectively. For nonabsorbing particles, $Q_e = Q_s$ and $\sigma_e = \sigma_s$. For polydisperse aerosols, Eq. 16.9 holds for each particle size, and the combined effect is given by the sum of the σ_e's for all the particle sizes.

$$\sigma_e = \sum_i \frac{\pi N_i d_i^2 (Q_e)_i}{4} \qquad (16.12)$$

where N_i is the number concentration of particles with diameter d_i. For those situations where $(Q_e)_i$ can be treated as a constant ($d > 4$ μm), we can write

$$\sigma_e = \frac{\pi Q_e}{4} \sum_i N_i d_i^2 = \frac{\pi Q_e}{4} N \overline{d^2} \qquad \text{for } d > 4 \text{ μm} \qquad (16.13)$$

where N is the total number concentration. For geometrically similar particles, the diameter of average surface area is equal to the diameter of average projected area, and when Q_e is a constant, extinction measurements can be used to determine the surface area concentration (area/m^3 [area/cm^3]) C_s. From Eqs. 16.7 and 16.13,

$$C_s = \frac{-4}{L Q_e} \ln\left(\frac{I}{I_0}\right) \qquad \text{for } d > 4 \text{ μm} \qquad (16.14)$$

and the diameter of average surface area (the second moment average, see Eq. 4.22) can be obtained as follows if N is known:

$$d_{\bar{s}} = \left(\frac{C_s}{N \pi}\right)^{1/2} \qquad (16.15)$$

The use of Eqs. 16.7 and 16.9 for aerosols is straightforward, provided that Q_e is known. There is no single equation that gives Q_e for all particle sizes. Q_e depends on the particle refractive index, shape, and size relative to the wavelength of light. For small particles less than 0.05 μm in diameter, Q_e can be calculated directly from

$$Q_e = \frac{8}{3}\left(\frac{\pi d}{\lambda}\right)^4 \left(\frac{m^2 - 1}{m^2 + 2}\right)^2 \qquad \text{for } d < 0.05 \text{ μm} \qquad (16.16)$$

In this size range Q_e is proportional to d^4 and, consequently, decreases rapidly with decreasing particle size. For gases,

$$\sigma_e = \frac{32 \pi^3 (m - 1)^2 f}{3 \lambda^4 n} \qquad (16.17)$$

where m is the bulk refractive index of the gas (see Table 16.1), n is the number concentration of gas molecules, given by $n = p N_A / RT$, and $f = 1.054$ for air at standard conditions (van de Hulst, 1957, 1981).

For large particles ($d > 4$ μm), Q_e approaches, with oscillations, its limiting value of 2.0. For particles between 0.05 and 4 μm, there is no simple equation for Q_e, and its value can be obtained from graphs such as Figs. 16.2 and 16.3 or by computer calculation (see Wilson and Reist, 1994). Nonabsorbing particles have a maximum value of Q_e of about 4 in the size range 0.3–1 μm. Absorbing or irregular particles have only slight maxima and approach their limiting value of $Q_e = 2$ without oscil-

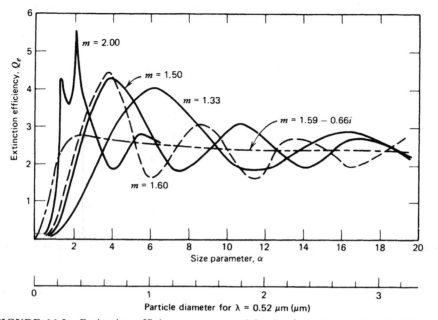

FIGURE 16.2 Extinction efficiency versus particle size for spheres (after Hodkinson, 1966).

lations. The curves for nonabsorbing particles in Fig. 16.2 have been smoothed to remove numerous small bumps and wiggles. These bumps are greatly reduced for absorbing particles. Extinction efficiency decreases rapidly with decreasing α in the region below $\alpha = 2$, as shown in Figs. 16.2 and 16.3.

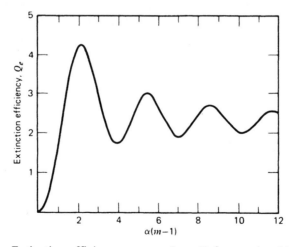

FIGURE 16.3 Extinction efficiency versus $\alpha(m - 1)$ for nonabsorbing spheres with refractive indices between 1.33 and 1.50. Data from Kerker (1969).

EXAMPLE

What is the fractional transmission of light through 10 km of air containing 0.5-μm fog droplets at a concentration of $10^8/m^3$? Assume that $\lambda = 0.5$ μm.

$$\alpha = \frac{\pi 0.5}{0.5} = 3.14 \quad \text{From Fig. 16.2, } Q_e = 2$$

$$\frac{I}{I_0} = e^{-\sigma_e L} = \exp(-\pi N d^2 Q_e L/4)$$

$$= \exp(-\pi \times 10^8 \times (0.5 \times 10^{-6})^2 \times 2 \times 10000/4) = e^{-0.392} = 0.68$$

$$[\exp(-\pi \times 100 \times (0.5 \times 10^{-4})^2 \times 2 \times 10^6/4) = 0.68]$$

The peculiarity of the quantity Q_e converging to a value of 2 for large values of α is called the *extinction paradox* and implies that large particles remove from a light beam an amount of light equal to twice their projected area. The explanation of this paradox is based on the condition that extinction must be observed at a long distance from the particles. As shown in Fig. 16.4, the distance between the object and the detector is such that all the diffracted light is removed from the beam before it reaches the detector. This distance must be large compared to $10d^2/\lambda$, the maximum extent of the shadow. This condition is easily met in the case of aerosol particles, but would require an observation distance greater than 100 km to evaluate the extinction of a coffee cup. Shadows observed in our normal experience seldom meet this criterion. Under proper conditions of observation, it follows from Babinet's principle [see Kerker (1969)] that the amount of light removed by diffraction is equal to that incident on the projected area of the particle. Thus, one projected area of light is removed by scattering and one by diffraction, to give a Q_e of 2.0 when $\alpha > 20$.

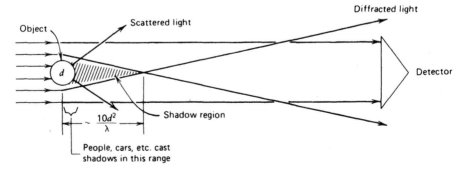

FIGURE 16.4   Diagram of extinction paradox.

It is of interest to express Bouguer's law, Eq. 16.7, in terms of the particle mass concentration C_m for monodisperse aerosols.

$$C_m = \frac{N\rho_p \pi d^3}{6}$$ (16.18)

Combining Eq. 16.18 with Eqs. 16.7 and 16.9 gives

$$\frac{I}{I_0} = \exp(-\sigma_e L) = \exp\left(-\frac{3 C_m Q_e L}{2\rho_p d}\right)$$ (16.19)

$$\frac{3\left(N\rho_p \frac{\pi d^3}{6z}\right)Q_e L}{4 Z \rho_p d}$$

For large particles, for which Q_e is a constant, extinction increases (scattering becomes greater) with *decreasing* particle size *for a constant mass concentration.* Figure 16.5 shows the relative scattering per unit mass for three materials as a function of particle size. Because Q_e decreases rapidly ($Q_e \propto d^4$) for small particles, there is a particle size for maximum extinction, as shown in the figure, and a range in which a reduction in size gives an increase in extinction for a given mass concentration. This fact explains why the visual impact of atmospheric aerosols and smokestack plumes is governed by the concentration of particles in the size range from 0.1 to 2 µm and why a reduction in concentration outside this range has little effect on extinction or visibility.

Table 16.2 gives the relative extinction for a given mass of water in the atmosphere in the form of molecules, aerosol particles, or raindrops. For a given mass concentration of water, aerosol particles in the range of 0.1 to 1.0 µm are a million times more effective than vapor molecules and a thousand times more effective than

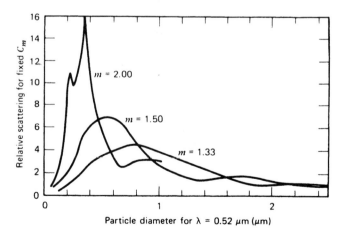

FIGURE 16.5 Relative scattering per unit mass of aerosol versus particle size for light at a wavelength of 0.52 µm.

**TABLE 16.2   Extinction and Visual Range in Air
Containing 18 g/m³ of Water[a] in Different Forms**

Particle Size (μm)	Extinction,[b] $1 - I/I_0$	Visual Range[c] (km)
Vapor	1.8×10^{-7}	220
0.01	3.8×10^{-5}	1.0
0.1	0.29	1.1×10^{-4}
1.0	0.64	3.8×10^{-5}
10	0.052	7.4×10^{-4}
1 mm (rain)	5.3×10^{-4}	0.074

[a]Equivalent to saturation at 293 K [20°C].
[b]$L = 1.0$ cm and $\lambda = 0.5$ μm.
[c]See Section 16.4.

raindrops in attenuating light and creating turbidity in the atmosphere. Thus, even though aerosols are usually present at low concentrations, their visual impact can be great.

The preceding discussion of extinction assumes that single scattering holds. If the aerosol concentration is sufficiently great, the light scattered by each particle will illuminate other particles in a direction that is not parallel to the incident light. This process allows multiple-scattered light, instead of attenuated incident light, to reach the detector, and Eq. 16.7 does not hold. Under favorable conditions, this effect is not significant for I/I_0 greater than 0.1.

16.3   SCATTERING

It is impractical to analyze the refracted and reflected rays of light for particles less than 50 μm, so the interaction of light with a particle in this size range is described in terms of the angular distribution of its scattered light. This angular scattering pattern depends strongly on the refractive index of the particle and the particle diameter, or, more correctly, the size parameter α. The scattered light from a particle is an extremely sensitive indicator of the particle's size, permitting size measurements of single submicrometer particles. Light scattering is responsible for the optical effects caused by aerosols and forms the basis for several types of aerosol measuring instruments. Light scattering also plays an important role in visibility and affects the earth's radiation balance. It is often helpful to think about light scattering as if the illuminated particle were reradiating the light it receives and thus acts as a light source with a unique angular distribution of light intensity.

Standard conventions have been adopted for describing the angular distribution of light scattered by an aerosol particle. As shown in Fig. 16.6, the plane formed by the incident beam and the direction of observation (the scattered beam) is called the *scattering plane*. The *scattering angle* θ is measured in the scattering plane from the direction of the incident beam to the scattered beam. Light that deviates only

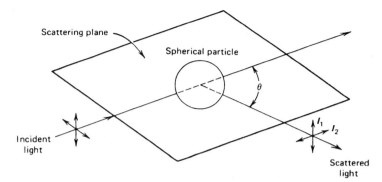

FIGURE 16.6 Diagram showing scattering angle, scattering plane, and the polarized components of scattered light.

slightly from its incident direction has a small scattering angle and is said to be *forward-scattering light.* Light that is reflected or scattered back to the source (i.e., $\theta = 180°$) is called *back-scattered light.* The incident and scattered beams can each be resolved into two independently polarized components, one whose electric vector is perpendicular to the scattering plane (designated by the subscript 1) and one whose electric vector is parallel to the scattering plane (subscript 2).

The general theory of light scattering by aerosols was developed in 1908 by Gustav Mie. The theory gives the intensity of light scattered at any angle θ by a spherical particle with known values of α and m illuminated by light of intensity I_0. The resulting Mie equations are very complicated for particles with sizes greater than the wavelength of light.

When particles are much smaller than the wavelength of light ($d < 0.05$ μm), the much simpler Rayleigh scattering theory can be used. For these particles (and gas molecules), the instantaneous electromagnetic field of the incident light is uniform over the entire particle, creating a dipole that oscillates in synchronization with the incident electromagnetic field. This oscillating dipole reradiates electromagnetic energy in all directions. The intensity and pattern of scattering is known as Rayleigh scattering and, for unpolarized illumination, is given by

$$I(\theta) = \frac{I_0 \pi^4 d^6}{8 R^2 \lambda^4} \left(\frac{m^2 - 1}{m^2 + 2} \right)^2 (1 + \cos^2 \theta) \quad \text{for } d < 0.05 \text{ μm} \qquad (16.20)$$

where $I(\theta)$ is the total intensity, at a distance R from the particle, of the light scattered in the direction θ.

Equation 16.20 indicates that the intensity of the scattered light at any angle will be proportional to d^6/λ^4. This is a strong function of diameter and a strong inverse function of wavelength. Rayleigh scattering depends on the particle volume squared and, as a result, is independent of the particle shape. The dependence of scattered light on wavelength (λ^{-4}) accounts for the blue color of the sky: Blue light is scattered (or reradiated) by the air molecules roughly 16 times more strongly than red light, so the predominant color of the sky is blue.

The $(1 + \cos^2\theta)$ term in Eq. 16.20 represents the two independent polarization components of the scattered light. For unpolarized incident light, the component of scattered light intensity that is polarized perpendicular to the scattering plane is given by

$$I_1 = \frac{I_0 \pi^4 d^6}{8 R^2 \lambda^4} \left(\frac{m^2 - 1}{m^2 + 2} \right)^2 \quad \text{for } d < 0.05 \ \mu\text{m} \qquad (16.21)$$

This component has the same intensity at all scattering angles. For polarization parallel to the scattering plane,

$$I_2 = \frac{I_0 \pi^4 d^6}{8 R^2 \lambda^4} \left(\frac{m^2 - 1}{m^2 + 2} \right)^2 \cos^2 \theta \quad \text{for } d < 0.05 \ \mu\text{m} \qquad (16.22)$$

Because of its dependence on $\cos^2\theta$, the I_2 component is equal to I_1 at scattering angles of 0 and 180° and goes to zero at 90°. The angular scattering pattern of both polarizations for Rayleigh scattering is customarily displayed in a polar diagram such as that shown in Fig. 16.7. In such a diagram the scattering angle is shown as the angle between a radial line from the particle and the direction of the incident light. The scattered intensity in that direction is proportional to the length of the line from the particle to the curve. The figure shows that the I_1 component is the same at all angles and the I_2 component is zero at $\theta = 90°$. Thus, light scattered at 90° by gas molecules and small particles will be completely polarized perpendicular to the scattering plane. Similarly, the light scattered from the sky on a clear day will be perpendicularly polarized when it is viewed 90° from the sun, a phenomenon that bees use for navigation. Rayleigh scattering is symmetrical in the forward and backward directions; that is, an equal amount of light is scattered in both directions.

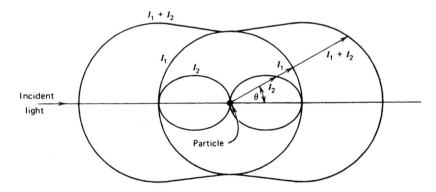

FIGURE 16.7 Polar diagram of Rayleigh scattered light. Scattering plane is parallel to the paper.

When Eq. 16.20 is integrated over the surface of a sphere, the result for non-absorbing particles is the total radiant power of the light scattered by the particle in all directions, which is equal to $I_0 \pi d^2 Q_e / 4$.

For particles larger than about 0.05 μm, the Mie equations must be used to determine the angular distribution of scattered light. Mie scattering theory provides a complicated but very general solution to the problem of scattering by spheres. The theory is valid for absorbing and nonabsorbing spheres from molecular dimensions to particles large enough to be treated by classical optics. The Mie theory is identical to the Rayleigh theory for particle sizes in the Rayleigh region.

The Mie solution is the complete formal solution of Maxwell's equations for the combined vector wave equations for the incident wave, the wave inside the particle, and the scattered wave, subject to a set of boundary conditions at the surface of the particle. A derivation of this theory and an explanation of the equations are presented in van de Hulst (1957, 1981) and Kerker (1969).

The Mie solution of Maxwell's electromagnetic equations is a solution to one of the classical problems of physics. Although the solution is exact, its utility was severely limited before the arrival of digital computers, because of the extreme complexity of the computations. With a mechanical calculator, a single solution for one value of m, α, and θ could take from several hours to several months. Efficient computer programs that provide nearly instant solutions to the Mie equations are now available. (See Dave, 1969, and Wilson and Reist, 1994.)

At a distance R in the direction θ from a spherical particle illuminated with unpolarized light of intensity I_0 (W/m² [W/cm²]), the scattered intensity is

$$I(\theta) = \frac{I_0 \lambda^2 (i_1 + i_2)}{8 \pi^2 R^2} \tag{16.23}$$

where i_1 and i_2 are the Mie intensity parameters for scattered light with perpendicular and parallel polarization, respectively. For polarized illumination, the scattered intensity is expressed by

$$I_1(\theta) = \frac{I_0 \lambda^2 i_1}{4 \pi^2 R^2} \tag{16.24}$$

for perpendicular polarization and by

$$I_2(\theta) = \frac{I_0 \lambda^2 i_2}{4 \pi^2 R^2} \tag{16.25}$$

for parallel polarization. The Mie intensity parameters i_1 and i_2 are functions of m, α, and θ and are defined by a complicated infinite series involving Legendre polynomials and Bessel functions. Because of the difficulty in calculating the Mie functions it is common to estimate scattered light intensities from values of i_1 and i_2, tabulated as functions of m, α, and θ or to calculate the functions directly using a computer. The use of power series approximations has been reviewed by Kerker (1969). In general, these approximations are restricted to a narrow range of values of α.

EXAMPLE

What is the ratio of backscattered intensity to incident intensity for 0.05-μm particles ($m = 1.5$) if 10^{12} particles are illuminated and observed at a distance of 0.3 m [30 cm]. Assume that $\lambda = 0.5$ μm.

$$\frac{I}{I_0} = 10^{12} \times \left(\frac{\pi^4 \times (0.05 \times 10^{-6})^6}{8 \times 0.3^2 \times (0.5 \times 10^{-6})^4} \right) \left(\frac{1.5^2 - 1}{1.5^2 + 2} \right)^2 (1 + \cos^2(180°)) = 5.9 \times 10^{-6}$$

$$\left[10^{12} \times \left(\frac{\pi^4 \times (0.05 \times 10^{-4})^6}{8 \times 30^2 \times (0.5 \times 10^{-4})^4} \right) \left(\frac{1.5^2 - 1}{1.5^2 + 2} \right)^2 (1 + \cos^2(180°)) = 5.9 \times 10^{-6} \right]$$

Figure 16.8 is a plot of i_1 and i_2 as a function of θ for $m = 1.33$ and α values of 0.8, 2, and 10. It is apparent that, as the particle size increases, the angular scattering pattern becomes more complicated. The smooth pattern for size parameter, $\alpha = 0.8$, changes to one that shows large variations in intensity, a factor of 100 or

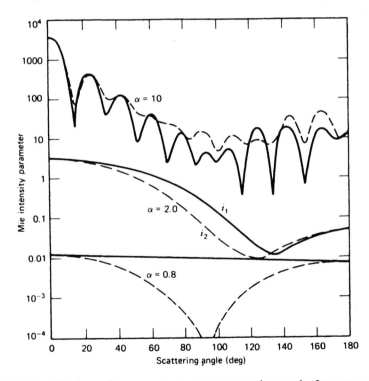

FIGURE 16.8 Mie intensity parameters versus scattering angle for water droplets ($m = 1.33$) having $\alpha = 0.8$, 2.0, and 10.0. Solid lines are i_1 and dashed lines are i_2.

more over a few degrees, for an α of 10 or more. As the particles get larger, the scattering in the forward direction becomes much stronger than at other angles. The segment of the curve for $\alpha = 10$ between 0 and about 20° is known as the *forward lobe* and consists primarily of diffracted light. It is absent in Rayleigh scattering. Agglomerated or very irregular particles and absorbing particles tend to smooth out the irregular scattering patterns shown. The irregular scattering patterns shown apply only to single particles or monodisperse aerosols. Most aerosols are polydisperse, and the patterns from many different sizes add together to give much smoother curves.

Figure 16.9 gives the combined Mie intensity parameter for water droplets as a function of the particle size parameter for scattering at $\theta = 30°$ and 90°. These curves

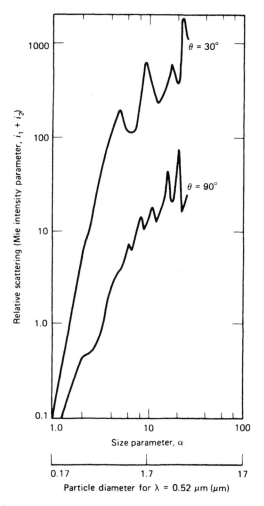

FIGURE 16.9   Relative scattering [Mie intensity parameter $(i_1 + i_2)$] versus size parameter for water droplets ($m = 1.33$) at scattering angles of 30 and 90°.

become smooth and approach a d^6 power curve as the particle size decreases to the Rayleigh region. Curves of scattered intensity versus particle size can be smoothed by accepting scattered light over a range of scattering angles, as is commonly done in light-scattering instruments.

16.4 VISIBILITY

An important application of the theory of light scattering and extinction is the study of visibility in the atmosphere, the most striking feature of particulate pollution. This subject has received considerable attention in the United States since the passage of the Clean Air Act of 1977, which mandated the protection of visibility in American national parks and wilderness areas.

Visibility is a subjective quantity that refers to the clarity with which distant objects can be seen. An exception to this is the use of the term visibility by the aviation industry to mean the maximum distance that objects can be seen throughout at least half the circle of the horizon. A more useful and scientific term is *visual range,* which expresses how far one can see in a given direction, or the distance at which an object is barely discernible. Both terms rely on the psychophysical concept of visual perception.

Atmospheric visual range is governed primarily by the scattering and absorption of light by aerosol particles present in the atmosphere. The scattering by air molecules usually has a minor effect, although it limits the maximum visual range to 100–300 km.

Visual range is limited by two factors: visual acuity and contrast. We cannot read a book at 10 m even though there is high contrast between the letters and the page, because of insufficient visual acuity. We can easily see stars on a clear night because of the extreme contrast between a bright star and the blackness of space. The same stars are invisible on a clear day, because the contrast is reduced to zero by the "air light"—the sunlight scattered by the air molecules. In most situations, it is the lack of apparent contrast between an object and its surroundings that limits how far we can see objects. Aerosol particles, primarily in the size range of 0.1–1.0 μm, reduce the apparent contrast by scattering light. Light from an object is scattered out of the sight path and does not reach our eyes. Sunlight is scattered *into* the sight path and makes dark objects appear lighter. The combined effect is to reduce the contrast between an object and its surroundings. The greater the distance between the observer and the object, the lower the contrast. The visual range is reached when the contrast is reduced to the point where the object is just discernible.

For an isolated object surrounded by a uniform and extensive background, the inherent contrast C_0 is

$$C_0 = \frac{B_0 - B'}{B'} \qquad (16.26)$$

where B_0 is the luminance of the object and B' is the luminance of the background. Luminance describes the brightness of a surface and is defined as the luminous intensity per unit solid angle per unit area of surface. The units of luminance are lumens/m² · sr, or candelas per square meter (cd/m²). The sky near the horizon has a luminance of 10^4 cd/m² on a clear day and 10^{-4} cd/m² on an overcast, moonless night. The luminance of a piece of white paper is 25,000 cd/m² in sunlight and 0.03 cd/m² in moonlight.

When the luminance of an object and its surrounding background are the same, the contrast is zero, and the object cannot be distinguished from the background. If the object is less luminous than its background, C_0 is negative, reaching a value of -1 for an ideal black object against a white (or lighted) background. When the object is brighter than its background, C_0 can have any positive value. For example, a light at night represents a case in which C_0 has a large value. Contrasts greater than about 10 seldom occur during daytime.

As we look at progressively more distant objects, such as the hills shown in Fig. 16.10, each more distant hill is lighter in tone. At the limit of visibility, the hills have the same brightness as the surrounding horizon. What lightens the appearance of the distant hills and mountains and reduces their contrast relative to the sky is the light scattered to the observer by the intervening aerosol. As the distance is increased, more aerosol is between the observer and the mountains, and they appear lighter.

The inherent contrast—that is, the contrast without the intervening aerosol—of forest-covered mountains against the sky is approximately -1. The apparent contrast C_R when the mountains are observed through the intervening aerosol is less than the inherent contrast because of the light scattered by the aerosol and is defined as

FIGURE 16.10 Photograph showing the increase in the apparent luminance of forest-covered hills with increasing distance. Photograph courtesy of W. L. Hinds.

$$C_R = \frac{B_R - B'_R}{B'_R} \qquad (16.27)$$

where B_R and B'_R are the observed luminances of the object and the background, respectively. The apparent contrast is equal to the inherent contrast if there is negligible aerosol (or air) scattering between the observer and the object.

In the common case, shown in Fig. 16.11, of an object viewed against the horizon, the change in luminance, dB, caused by a thin layer dx of a uniform aerosol between the observer and the object is

$$\frac{dB}{dx} = -\sigma_e B + B_a \qquad (16.28)$$

where B is the luminance of the object at the thin layer, $\sigma_e B$ is the loss of luminance due to scattering and absorption per unit thickness, and B_a is the luminance of a unit thickness of aerosol due to the skylight scattered to the observer. A similar equation can be written for the horizon sky (the background), but its luminance does not change with position, so dB/dx must equal 0, and

$$B_a = \sigma_e B' = \sigma_e B'_R \qquad (16.29)$$

Integrating Eq. 16.28 over the distance L between the object and the observer gives

$$\int_{B_0}^{B_R} \frac{dB}{B_a - \sigma_e B} = \int_0^L dx \qquad (16.30)$$

$$B_R = B'_R(1 - e^{-\sigma_e L}) + B_0 e^{-\sigma_e L} \qquad (16.31)$$

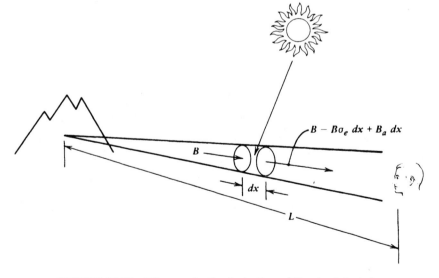

FIGURE 16.11 Diagram for the derivation of Koschmieder's law.

where B_a/σ_e has been replaced by B_R' according to Eq. 16.29. Substituting Eqs. 16.29 and 16.31 into 16.27 gives Koschmieder's equation, an expression for the apparent contrast of an object when viewed against the horizon as a function of the distance between the observer and the object:

$$C_R = \left(\frac{B_0 - B_R'}{B_R'} \right) e^{-\sigma_e L} \qquad (16.32)$$

$$C_R = C_0 e^{-\sigma_e L} \qquad (16.33)$$

A comparison of Eqs. 16.33 and 16.7 reveals that the exponential relationship between contrast and distance is the same as that for extinction. Equation 16.33 holds for horizontal viewing of an object with any value of inherent contrast when the objecct is viewed against the horizon. The extinction coefficient and the illumination of the aerosol must be uniform over the entire viewing distance. If the sight path distance is short or the aerosol sufficiently dilute, $\sigma_e L$ will be approximately zero, and the apparent contrast will equal the inherent contrast, giving nearly perfect viewing. With increasing distance (or increasing σ_e), the apparent contrast gets closer and closer to zero, as shown in Fig. 16.12.

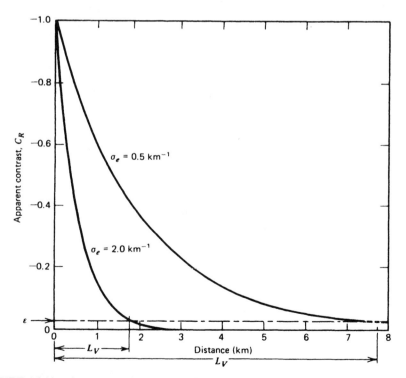

FIGURE 16.12 Apparent contrast versus distance for a dark object against the horizon for two values of extinction coefficient.

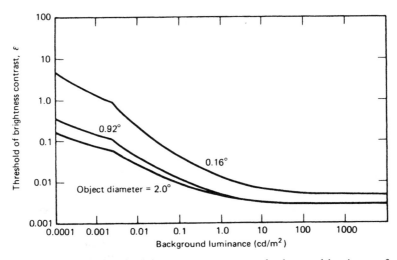

FIGURE 16.13 Threshold of brightness contrast versus background luminance for un-limited viewing time. Parameter is diameter of object, in degrees of arc. Data from Blackwell (1946).

The minimum contrast required to just distinguish an object from its background is called the *threshold of brightness contrast* and is given the symbol ε. The value of ε depends on many factors, the most important of which is the overall level of illumination as shown in Fig. 16.13. The values of ε given in the figure represent the 50% detection values and are valid for positive or negative contrast. Middleton (1952) discusses the factors that affect ε, such as the motivation of the observer, the angular size of the object, the state of dark adaption of the eyes, and the length of viewing time. The discontinuity in the curve at 0.002 cd/m² marks the transition from foveal vision (less sensitive cone vision for color) to parafoveal vision (more sensitive rod vision).

Equation 16.33 can be expressed in terms of the visual range L_V and ε as

$$\varepsilon = |C_0| \exp(-\sigma_e L_V) \tag{16.34}$$

for viewing any object against the horizon under any condition of illumination. For good daylight viewing conditions, a value for ε of 0.02 is commonly used. This value implies that we can just see a black object against the horizon when its apparent contrast is reduced to 2% of its inherent contrast—that is, a reduction in contrast of 98%. Figure 16.12 shows the relationship between ε, L_V, and σ_e. For a dark object viewed against the horizon during daylight, the visual range can be obtained by substituting $\varepsilon = 0.02$ into Eq. 16.34 and rearranging to get

$$L_v = \frac{-\ln(0.02)}{\sigma_e} = \frac{3.91}{\sigma_e} \tag{16.35}$$

For daylight viewing of a dark object against the horizon, the visual range is a simple inverse function of the atmospheric extinction coefficient. The visual range is given by Eq. 16.35 in the same units of length in which σ_e is expressed. The equation relates the visual range to σ_e and, consequently, to the aerosol properties of number or mass concentration and particle size by Eq. 16.9, 16.13, or 16.19. When the visual range is greater than 30 km, the effect of scattering by air molecules must be included by summing the extinction coefficients for air (Eq. 16.17) and aerosol.

In 1899, Lord Rayleigh used the preceding visibility principles and Eq. 16.17, a result of the Rayleigh theory, to determine Avogadro's number by visual observation. One clear day, while vacationing in Darjeeling, India, he observed Mount Everest at a distance of about 150 km from the terrace of his hotel. Through simple calculations, he arrived at a value of 3×10^{19} for n (adjusted to sea level), which corresponds to a value of N_a of 6.7×10^{23}. This is remarkably accurate considering the simplicity of the measurement and illustrates the power of light scattering as an atmospheric measuring tool.

EXAMPLE

What is the daylight visual range in polluted air containing 0.3-μm-diameter particles at a concentration of $2 \times 10^{10}/\text{m}^3$ [2×10^4 particles/cm³]? Assume that the particle refractive index is 1.5.

$$\sigma_e = \frac{N\pi d^2 Q_e}{4} \qquad \text{where } Q_e = 1.4 \text{ (from Fig. 16.2)}$$

$$\sigma_e = \frac{2 \times 10^{10} \times \pi \times (0.3 \times 10^{-6})^2 \times 1.4}{4} = 1.98 \times 10^{-3} \text{ m}^{-1}$$

$$\left[\frac{2 \times 10^4 \times \pi \times (0.3 \times 10^{-4})^2 \times 1.4}{4} = 1.98 \times 10^{-5} \text{ cm}^{-1} \right]$$

$$L_v = \frac{3.91}{\sigma_e} = \frac{3.91}{0.00198} = 1970 \text{ m} = 2.0 \text{ km} \qquad \left[\frac{3.91}{1.98 \times 10^{-5}} = 1.97 \times 10^5 \text{ cm} = 2.0 \text{ km} \right]$$

Equation 16.35 gives the true visual range, when the correct value of ε is 0.02. The latter depends on the level of illumination. The *meteorological range* is defined as the distance that yields a contrast threshold of 0.02, regardless of the illumination level or the uniformity of σ_e. Equation 16.35 is the definition of the meteorological range. In many situations the absorption by the atmospheric aerosol is negligible compared to scattering, and σ_e can be replaced by σ_s (Eq. 16.11). This assumption permits the meteorological range to be estimated by an instrument called an integrating nephelometer (see Section 16.5), which measures σ_s (also called b_{scat})

directly. In polluted urban air, σ_a may be as large as $0.5\sigma_s$, and the total extinction coefficient σ_e must be used.

A review of correlations between the aerosol mass concentration C_m and nephelometer measurements of σ_s by Lodge et al. (1981) found that

$$\frac{C_m}{\sigma_s} = 0.5 \text{ g/m}^2 \qquad (16.36)$$

for total mass concentration and

$$\frac{C'_m}{\sigma_s} = 0.3 \text{ g/m}^2 \qquad (16.37)$$

for C'_m equal to the mass concentration in the fine-particle mode. The correlation coefficients are 0.6–0.9 for total mass concentration and about 0.9 for the fine-particle mode, suggesting the importance of the latter mode in controlling visibility. At relative humidities greater than 70%, σ_s increases relative to C_m because of droplet formation on hygroscopic particles. Equations 16.36 and 16.37 are averages for many sites, and individual sites may differ significantly. Colored haze, usually brown or yellow, observed in urban areas is caused primarily by wavelength-dependent extinction of aerosols and secondarily by wavelength-dependent extinction of gases such as NO_2.

16.5 OPTICAL MEASUREMENT OF AEROSOLS

Light scattering provides an extremely sensitive tool for the measurement of aerosol concentration and particle size. A single particle as small as 0.1 μm produces a detectable scattered light signal. Light-scattering techniques have the advantage of minimally disturbing the aerosol and providing instantaneous information that is often suitable for continuous monitoring. A disadvantage of light-scattering instruments is that the scattering may be sensitive to small changes in refractive index, scattering angle, particle size, or particle shape, which can lead to confusing or misleading results. These optical measurement methods differ from the microscopy methods described in Chapter 20 in that the particles do not have to be captured and it is not necessary to form an image of them. Kerker (1997) gives a historical overview of light-scattering instruments for aerosol measurement.

A simple laboratory instrument called the *owl*, now primarily of historical interest, can be used to determine rapidly the size of *monodisperse* submicrometer test aerosols (Sinclair, 1950). The aerosol flows through a chamber 70 mm in diameter with a collimated beam of unpolarized, white light passing along a diameter. The aerosol is viewed through a microscope that is free to rotate so that it can view light scattered at any angle from 25 to 155°. The polarization (the ratio of I_2 to I_1) of the light scattered at 90° can be used to determine the diameter of particles between 0.1 and 0.4 μm by comparing the measured values with calculated values. A second method relies on the color of the scattered light as a function of angle to determine particle diameters from 0.2 to 1.5 μm. Monodisperse aerosols in this size range produce colored bands in their angular scattering pattern called *higher order Tyndall spectra*, or HOTS. The number or the angular positions of *red* bands can be used to determine the particle size by comparing the observed values with calculated values. The sharpness of the colored bands increases with increasing monodispersity.

Transmissometers, or light-attenuating photometers, are instruments that measure the attenuation (extinction) of light through a path length that may range from a few centimeters to a few kilometers. Stack transmissometers have a focused light source positioned on one side of a stack and directed across the stack to a detector. These instruments measure the extinction by the smoke in the stack when constructed according to Fig. 16.1, but often the detected beam is inadequately collimated, and such instruments detect significant amounts of forward-scattered light. Because of multiple scattering effects at high aerosol concentrations, the transmitted light intensity varies in an unpredictable way with concentration. This unpredictability is not a problem for pure extinction by aerosols. A gravimetric calibration can be used if the particle refractive index and size are constant. To eliminate forward scattering, the size of the pinhole before the detector must have an angular diameter, measured from the center of the preceding lens, of not more than $0.38/\alpha$ radians.

Three types of instruments rely on measurements of light scattered at fixed angles from polydisperse aerosols: photometers, nephelometers, and optical-particle counters. *Photometers* are used to measure relative concentration by measuring the combined light scattered from many particles at once. The aerosol flows continuously through the instrument at several liters per minute. The illumination and collection optics are arranged so that the light scattered at a fixed range of angles reaches the detector. Depending on the design, the instrument may measure the scattered light in the region of 90°, 45°, or less than 30°. Forward-scattering photometers (< 30°) are less sensitive to refractive index than 90° scattering instruments. Photometers are used with monodisperse test aerosols for filter penetration testing. Typical instruments can measure penetrations as low as 0.001%.

In the field of occupational hygiene, portable photometers are used as direct-reading instruments to measure mass concentration. The scattering angles and wavelengths used are selected to reduce the effect of particle size and refractive index on the response of the instrument. Electronic averaging is used to provide a digital readout of mass concentration over the range of 1 $\mu g/m^3$ to 100 mg/m^3. The output of such instruments is sensitive to aerosol material and particle size distribution, and errors in concentration of a factor of two or more are possible with variation in these parameters. Instruments should be calibrated with side-by-side filter samples if they are used for aerosols that differ appreciably from the manufacturer's calibration aerosol. These photometers provide continuous output and are well suited to monitoring concentration changes with time and location, provided that the particle size and chemical composition remain constant. Commercially available photometers and other light-scattering instruments are reviewed by Pui and Swift (1995).

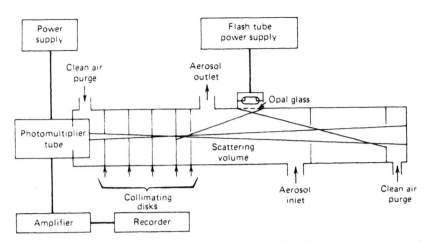

FIGURE 16.14 Diagram of an integrating nephelometer. Reprinted with permission from *J. Air Poll. Control Assoc.*, **17**, 467 (1967).

The *integrating nephelometer* is a photometer designed to measure the light scattered from an aerosol over as wide a range of angles as possible. The instrument shown in Fig. 16.14 measures the scattered light from 8 to 170^{v} and can measure scattering coefficients as small as 10^{-6}/m from a sensitive volume of about 1 L. This instrument provides a measurement of σ_s due to particles and gas molecules. If σ_a is zero, the nephelometer yields a direct measurement of σ_e that provides an instrumental estimate of meteorological range by Eq. 16.35. Integrating nephelometers can be used to estimate mass concentrations with a suitable calibration.

Optical particle counters (*OPCs*), such as that shown in Fig. 16.15, are similar to photometers. However, the aerosol flows through a focused light or laser beam as a thin stream surrounded by sheath air, so that only one particle at a time is illu-

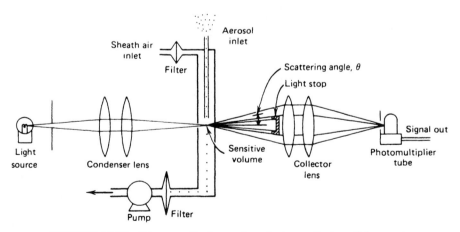

FIGURE 16.15 Diagram of a forward-scattering optical particle counter.

TABLE 16.3 Characteristics of Some Optical Particle Counters

Instrument	Particle Size Range (μm)	Number of Channels	Scattering Angle Range (deg)	Sample Flow Rate (cm³/min)	Maximum Concentration[a] (cm⁻³)
Climet[b]					
208[c]	0.3–>20	16	15–105	28,000	90
7400	0.1–>0.5	6	15–150	2800	14
7500	0.19–>5.0	6	15–150	28,000	14
Hiac/Royco[d]					
236	0.12–>6.0	16	35–120	280	3000
5100	0.25–>10	6	60–120	28,000	7
PMS, Inc.[e]					
HS-LAS	0.065–1.0	32	35–120	2800	8000
LAS-X	0.09–3	16[f]	35–120	60	17,000
LAS-X	0.12–7.5	16[f]	35–120	280	3000
LPC-525	0.2–>5.0	6	60–120	28,000	3.5

[a]Concentration for 10 % coincidence error.
[b]Climet Instruments Co., Redlands, CA.
[c]White light illumination; all others use laser light.
[d]Hiac/Royco, Menlo Park, CA.
[e]Particle Measuring Systems, Inc., Boulder, CO.
[f]Four ranges with 16 channels each.

minated and scatters light to the detector. OPCs have proven to be a valuable tool in aerosol research, air pollution studies, and clean-room monitoring because of their ability to rapidly provide information on aerosol size distributions. Commercial instruments cover the size range 0.05–20 μm, although 0.1–5 μm is most common, and provide measurements of number concentration in 5–16 size channels. As each particle passes through the focused light beam, it scatters a pulse of light to the detector, which is converted to an electrical signal. Electronic pulse height (or area) analysis is used to interpret the pulse and direct a count to the proper size channel, where the total counts in each size range are accumulated. The particle size distribution by count is obtained from the accumulated counts in each size channel. These instruments are based on the assumption that the scattered light intensity is a monotonic function of particle size, but this is not always the case, as is discussed shortly. Also, diluting the aerosol may be necessary so that only one particle passes through the beam at a time.

Characteristics of some commercially available optical particle counters are given in Table 16.3. The choice of scattering angle involves a trade-off of many factors. The forward direction provides stronger signals, but requires special care to reduce background noise, which is least at 90°. Forward scattering is primarily diffracted light, which is less sensitive to the refractive index and to whether the particle is absorbing. Some instruments use bowl-shaped mirrors to collect all the light scattered within a certain range of scattering angles. Instruments that use lasers as their light source have the lowest minimum detectable particle sizes. These instruments have counting efficiencies (ratio of indicated count to true count) of approximately 100% for particles > 0.1 μm. Monodisperse spherical particles of known size and refractive index are used to verify the size calibration of optical particle counters.

As shown in Fig. 16.16, the response of optical particle counters depends on the size and refractive index of the particles. When the refractive index is known, suitable calibration will permit an accurate measurement of the size distribution. For aerosol particles of unknown refractive index, the error in size estimation can be significant. For the response curves shown in Fig. 16.16 this error ranges from –50 to +140%. This large spread limits the usefulness of optical particle counters when a range of refractive indices is present or when the refractive index is not known. For some refractive indices, part of the response curve is not monotonic, usually in the range of 0.5–1.5 μm, and there is more than one size that corresponds to a particular signal strength. For example, a relative signal of 400 for iron in Fig. 16.16 could be due to particles having diameters of either 0.34 or 0.53 μm.

Optical particle counters are restricted by coincidence errors to the measurement of aerosols with relatively low number concentrations, usually less than $10^{10}/m^3$ [$10^4/cm^3$]. Aerosols with higher concentrations must be diluted to this level for unbiased measurement. Coincidence errors arise when two or more particles are in the sensitive volume simultaneously, causing a spurious signal that leads to an underestimation of the particle number concentration and an overestimation of the particle size. The sensitive volume is the region from which signals are generated, defined by the incident and scattered beams and the diameter of the aerosol stream. The ratio of the observed count N_o to the true count N_t is given by

$$\frac{N_o}{N_t} = \exp(-N_t Q \tau) \tag{16.38}$$

where Q is the flow rate through the sensitive volume and τ is the time for a particle to traverse the beam, plus the signal-processing time. Coincidence error can be minimized by reducing the flow rate and the dimensions of the beam; however, small beam dimensions require a small aerosol stream diameter so that nearly all particles will pass through the beam for a high counting efficiency. Some instruments use aerodynamic focusing of the aerosol stream to achieve a stream diameter of less than 0.1 mm. Table 16.3 gives maximum concentrations for a 10% error in count due to coincidence.

Care must be taken in interpreting size distribution data from optical particle counters. The distribution obtained is the size distribution by count or the number distribution, but the number of particles below the lower size limit is unknown and may be large. Plotting the results as counts per micrometer (number divided by channel width) gives an accurate representation of the data. Plotting the fraction per micrometer may be seriously in error if there is uncertainty in the total number of particles.

Another particle-sizing method, called photon correlation spectroscopy (PCS), dynamic light scattering (DLS) or quasi-electric light scattering (QELS), relies on both the light scattering and the Brownian motion of the particles. An ensemble of

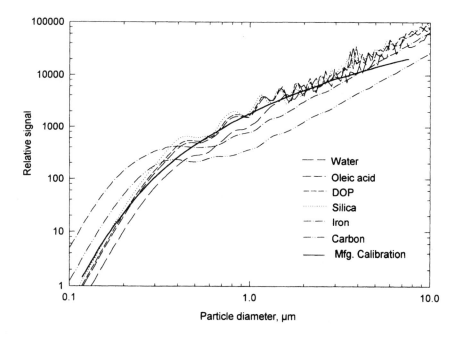

FIGURE 16.16 Calculated response curves for six materials and manufacturer's calibration curve for model LAS-X® (PMS, Inc., Boulder, CO) optical particle counter.

particles is illuminated with a laser beam, and the scattered light is detected at a specific angle. The incident light has a single frequency, but the scattered light has a narrow range of frequencies, due to the Doppler effect and the Brownian motion of the particles. This frequency broadening manifests itself as a low-frequency spectrum in the photomultiplier tube output signal. The signal is analyzed by a spectrum analyzer or a digital autocorrelator (PCS). The width of the spectrum is related to the average diffusion coefficient of the aerosol and thus to the particle size. This method requires a high number concentration, but does not require knowing the particle refractive index. It has been applied to soot formation in flames and to studies of diesel exhaust particles. (See Rader and O'Hern, 1993.)

PROBLEMS

16.1 A tabulation of Mie intensity parameters gives $i_1 = 4.89 \times 10^{-4}$ and $i_2 = 4.43 \times 10^{-8}$ for light scattered at 90° from 0.088-μm polystyrene latex spheres. What error in total scattered intensity at 90° would result if Rayleigh scattering were assumed to hold? $\lambda = 0.55$ μm; $m = 1.59$.
ANSWER: 280%.

16.2 Determine the aerosol number and mass concentration for which the particles and the air in a unit volume of aerosol scatter equal amounts of light. Assume that the particle diameter is 0.5 μm, $m = 1.5$, and $\rho_p = 1000$ kg/m^3 [1.0 g/cm^3].
ANSWER: 2.6×10^7/m^3 [26/cm^3], 1.7 μg/m^3.

16.3 Determine the fractional light transmission through 1 km of air containing 1.0-μm-diameter water droplets at a concentration of 10^8/m^3 [100/cm^3]. Assume that the wavelength of light is 0.52 μm.
ANSWER: 0.73.

16.4 What number concentration of 0.05-μm-diameter particles scatters as much light per cubic centimeter as 1 cm^3 of particle-free air? σ_e for air is 1.5×10^{-5}/m [1.5×10^{-7}/cm]. Assume that $m = 1.5$ for the particles and $\lambda = 0.5$ μm.
ANSWER: 3.4×10^{12}/m^3 [3.4×10^6/cm^3].

16.5 The California ambient air quality standard for visibility requires that the average σ_e be less than 0.23 km^{-1} when the relative humidity is less than 70%. To what visual range does this limit correspond? To what aerosol mass concentration does it correspond? Assume that the particle size is 0.18 μm (SMD for the accumulation mode), the particle refractive index is 1.5, and ρ_p is 1000 kg/m^3 [1.0 g/cm^3].
ANSWER: 17 km [10.6 mi], 110 μg/m^3.

16.6 One sunny day, while measuring the atmospheric aerosol concentration (25 µg/m³) on the top of the UCLA Center for Health Sciences building, you notice that you can *just* see Palos Verdes Hill 22 mi away. If you assume that the atmospheric aerosol is monodisperse, what is its particle size? Assume that Q_e = 1.0 and ρ_p = 1000 kg/m³ [1.0 g/cm³] and neglect extinction due to air molecules.

ANSWER: 0.34 µm.

16.7 Repeat problem 16.6, but include the effect of air molecules. Assume that λ = 0.5 µm.

ANSWER: 0.40 µm.

16.8 You observe the visual range in urban air. If you assume that the range depends solely on the fine-particle mode and that it can be represented by an equivalent nonabsorbing monodisperse aerosol, what is its particle size? Assume that $Q_e = Q_s$ = 1 and the particle density is 1500 kg/m³ [1.5 g/cm³]. [*Hint*: Use the empirical expression relating visual range to fine-particle mass concentration and Bouguer's law expressed in terms of mass concentration.]

ANSWER: 0.3 µm.

16.9 What is the meteorological range in particle-free air at 20°C at sea level? Assume that λ = 0.5 µm.

ANSWER: 220 km.

16.10 (a) What is the maximum distance at which one can see a black automobile on a foggy, moonlit night? Assume that the fog consists of 5-µm droplets at 10¹⁰/m³ [10⁴/cm³], that the background luminance is 10⁻³ cd/m², and that your eyes are completely dark adapted. (b) Repeat part (a) if the headlights are on and are directed toward you; assume that the inherent contrast is 10⁴.

ANSWER: (a) 6.1 m; (b) 30 m.

REFERENCES

Blackwell, H.R., "Contrast Thresholds of the Human Eye," *JOSA*, **36**, 624–643 (1946).

Cadle, R. D., *The Measurement of Airborne Particles,* Wiley, New York, 1975.

Cooke, D. D., and Kerker, M., "Response Calculations for Light Scattering Aerosol Particle Counters," *Appl. Optics,* **14**, 734–739 (1975).

Dave, J. V., "Scattering of Electromagnetic Radiation by a Large, Absorbing Sphere," *IBM J. Res. Develop.,* 302–313 (May 1969).

Dave, J. V., *Subroutines for Computing the Parameters of the Electromagnetic Radiation Scattered by a Sphere,* Rept. No. 320-3237, IBM Scientific Center, Palo Alto, CA, May 1968.

Fenn, R. W., "Optical Properties of Aerosols," in Dennis, R. (Ed.), *Handbook on Aerosols,* USERDA, 1976, TID-26608, Oak Ridge, TN, 1976.

Gebhart, J., "Optical Direct-Reading Techniques: Light Intensity Systems," in Willeke, K., and Baron, P. A. (Eds.), *Aerosol Measurement*, Van Nostrand Reinhold, New York, 1993.

Hodkinson, J. R., "The Optical Measurement of Aerosols," in Davies, C. N. (Ed.), *Aerosol Science*, Academic Press, New York, 1966.

Kerker, M., "Light Scattering Instruments for Aerosol Studies: An Historical Overview," *Aerosol Sci. Tech.*, **27**, 522–540 (1997).

Kerker, M., *The Scattering of Light and Other Electromagnetic Radiation*, Academic Press, New York, 1969.

Lodge, J. P., Waggoner, A. P., Klodt, D. T., and Crain, C. N., "Non-health Effects of Airborne Particulate Matter," *Atm. Env.*, **15**, 431–482 (1981).

Middleton, W. E. K., *Vision through the Atmosphere*, University of Toronto, Toronto, 1952.

Ozkaynak, H., Schatz, A. D., Thurston, G. D., Isaacs, R. G., and Husar, R. B., "Relationships between Aerosol Extinction Coefficients Derived from Airport Visual Range Observations and Alternate Measures of Airborne Particle Mass," *J. Air Pol. Control Assoc.*, **35**, 1176–1185 (1985).

Pui, D. Y. H., and Swift, D. L., "Direct-Reading Instruments for Airborne Particles," in Cohen, B. S., and Hering, S. V. (Eds.), *Air Sampling Instruments*, 8th ed., ACGIH, Cincinnati, 1995.

Rader, D. J., and O'Hern, T.J., "Optical Direct-Reading Techniques: In Situ Sensing," in Willeke, K., and Baron, P. A. (Eds.), *Aerosol Measurement*, Van Nostrand Reinhold, New York, 1993.

Sinclair, D., "Optical Properties of Aerosols," in *Handbook on Aerosols*, USGPO, Washington, DC (1950).

Van de Hulst, H. C., *Light Scattering by Small Particles*, Wiley, New York, 1957; republished by Dover, New York, 1981.

Whitby, K. T., and Willeke, K., "Single Particle Optical Counters: Principles and Field Use," in Lundgren, D. A. et al. (Eds.), *Aerosol Measurement*, University Presses of Florida, Gainesville, FL, 1979.

Wilson, W. E., and Reist, P. C., "A PC-based Mie Scattering Program for Theoretical Investigations of the Optical Properties of Atmospheric Aerosols as a Function of Composition and Relative Humidity," *Atm. Env.*, **28**, 803–809 (1994).

17 Bulk Motion of Aerosols

The analysis of aerosol motion in the preceding chapters has focused exclusively on the motion of individual particles. There are, however, some situations in which individual particle motion is negligible compared with motion on a larger scale. In one such situation, the gas phase of an aerosol has a density that differs from the surrounding gas. For example, the buoyancy of a stack gas plume causes the particles contained in it to rise many meters after leaving the stack. The movement of gases under these circumstances is a subject of fluid mechanics and meteorology and is not covered here. When there is no difference in gas density between an aerosol cloud and its surroundings, there are still situations in which the cloud as an entity can move much faster than the individual particles that make up the cloud. For the purposes of this chapter, an *aerosol cloud* is defined as a region of high aerosol concentration having a definite boundary in a much larger region of clean air. The mechanics of clouds is far more complicated than that of individual particles, and a complete description of the motion of clouds does not exist. In this chapter, we define the order of magnitude of the conditions under which cloud motion can occur.

Since cloud motion depends on the bulk properties of aerosols, we evaluate first the viscosity of an aerosol cloud relative to that of pure air. Consider a monodisperse aerosol of 1-μm spheres of standard density at a mass concentration of 100 g/m^3. This is a very high mass concentration and corresponds to the densest smokes that can be produced. The number concentration of such an aerosol is 2×10^8/cm^3. The average spacing between particles is 17 μm, or 17 times the diameter of the particles, so the aerosol is still mostly air, and the particles are far enough apart to have little effect on each other. The viscosity of such a two-phase system η_c is

$$\eta_c = \eta(1 + 2.5C_v) \qquad (17.1)$$

where C_v is the volume concentration—that is, the volume of particles per unit volume of aerosol. In this example, $C_v = 0.0001$ and the aerosol has a viscosity 0.025% greater than that for pure air, a negligible difference. The density of a cloud exceeds that of particle-free air by an amount equal to the mass concentration of the particles. At a mass concentration of 100 g/m^3, an aerosol has a density 8.3% greater than that of clean air. It is this difference in density that causes bulk motion of aerosols. When such motion is caused by gravity, it is called cloud settling or mass subsidence.

Cloud settling occurs when the aerosol concentration is sufficient to cause the entire cloud to move as an entity at a velocity significantly greater than the individual particle settling velocity. This effect can be caused solely by the particles and need not depend on differences in gas density inside and outside the cloud. Why such an effect occurs can be best understood by considering a spherical array of particles—a cloud—with each particle *fixed in space*. Let air be blown at the cloud with a uniform velocity U_0. The air can either pass through the cloud or pass around the cloud, depending on the relative resistance of the two paths. At a low particle concentration, the air will pass through the cloud, and each particle will experience a relative velocity of U_0. The total resistance presented by the particles will be the sum of the drag force on each particle. At a sufficiently high particle concentration, the resistance to airflow through the cloud will be so great that the air will flow around the cloud, that being the path of least resistance. In this case the relative velocity of the particles inside the cloud is zero, and the resistance is caused by the drag force of the air flowing around the spherical cloud. Between these extremes are intermediate conditions under which both mechanisms operate. A completely analogous situation exists for the gravitational settling of a free cloud which may push through the air as a cloud or as individual particles, depending on the aerosol concentration, cloud size, and particle size. These two types of aerosol motion are shown schematically in Fig. 17.1.

Consider a spherical cloud of diameter d_c having N particles per unit volume, each with a diameter d_p and density ρ_p. For the cloud to move as an entity, the gas in which it is suspended and the particles must move together. In a practical sense, the two can be thought of as coupled or completely entrained, since there is negligible relative motion between the gas and the particles within the cloud. In this case, the downward force of gravity on the cloud is equal to the mass of the gas plus the mass of the particles, and the buoyant force is equal to the mass of displaced gas. It is convenient to approach cloud settling in a manner equivalent to particle settling and define a density for the cloud. For a gas density ρ_g and no difference in gas density between the cloud and the surrounding clean air, the *net cloud density* (density minus buoyancy) ρ_c of a spherical cloud is the mass of the cloud minus the mass of the displaced volume of gas, divided by the volume of the cloud, v_c:

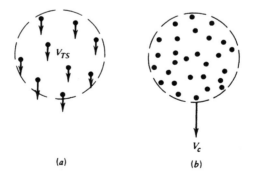

              (a)   (b)

FIGURE 17.1   (a) Individual particle settling. (b) Cloud settling.

$$\rho_c = \frac{\rho_g v_c + C_m v_c - \rho_g v_c}{v_c} = C_m \qquad (17.2)$$

Thus, the net cloud density is simply the particle mass concentration C_m expressed in appropriate units—kg/m^3 in SI units and g/cm^3 in cgs units. Because clouds are much larger than particles, their Reynolds numbers may be large, and their motion is likely to be outside the Stokes region. The settling velocity V_c of a spherical cloud is obtained by equating the force of gravity to Newton's drag

$$\frac{\rho_c \pi d_c^3 g}{6} = C_D \frac{\pi}{8} \rho_g d_c^2 V_c^2 \qquad (17.3)$$

and solving for the cloud settling velocity.

$$V_c = \left(\frac{4 \rho_c d_c g}{3 C_D \rho_g} \right)^{1/2} \qquad (17.4)$$

Equation 17.4 holds for rigid spheres, but settling clouds develop an internal circulation pattern which reduces the drag force that they experience and increases their settling velocity by a factor of $(6/5)^{0.5}$, or about 10%. For simplicity, this effect has been ignored. Equation 17.4 is equivalent to Eq. 3.29 with ρ_p replaced by ρ_c and d_p by d_c. The procedures for determining the settling velocities of particles at high Reynolds numbers, covered in Section 3.7, can be used to determine the settling velocities of clouds with these substitutions.

EXAMPLE

A spherical cloud of water droplets 5 m in diameter has a mass concentration of 1 g/m^3. What is the terminal settling velocity of the cloud?

$$\rho_c = 0.001 \text{ kg/m}^3 \quad [10^{-6} \text{ g/cm}^3]$$

$$C_D \text{Re}^2 = \frac{4 d_c^3 \rho_c \rho_g g}{3 \eta^2} = \frac{4 \times 5^3 \times 0.001 \times 1.2 \times 9.81}{3 \times (1.81 \times 10^{-5})^2} = 5.99 \times 10^9$$

$$= \left[\frac{4 \times 500^3 \times 10^{-6} \times 1.2 \times 10^{-3} \times 981}{3 \times (1.81 \times 10^{-4})^2} = 5.99 \times 10^9 \right]$$

This value is beyond the range of values given in Table 3.4 and is in the region where C_D has a constant value of 0.44. Hence,

$$\text{Re} = \left(\frac{5.99 \times 10^9}{0.44}\right)^{1/2} = 1.17 \times 10^5$$

$$V_c = \frac{\text{Re}}{66,000d_c} = \frac{1.17 \times 10^5}{66,000 \times 5} = 0.35 \text{ m/s} = \left[\frac{1.17 \times 10^5}{6.6 \times 500} = 35 \text{ cm/s}\right]$$

Thus, this cloud has a settling velocity that is equivalent to the settling velocity of a sphere of standard density 360 μm in diameter.

The settling velocity of an individual particle V_p is given by Eq. 3.21:

$$V_p = \frac{\rho_p d_p^2 g C_c}{18\eta} \tag{17.5}$$

Whether or not cloud settling is significant depends on the quantity G, the ratio of cloud settling velocity (Eq. 17.4) to particle settling velocity (Eq. 17.5). This ratio is

$$G = \frac{V_c}{V_p} = \frac{12\eta}{\rho_p d_p^2 C_c}\left(\frac{3\rho_c d_c}{C_D \rho_g g}\right)^{1/2} \tag{17.6}$$

where C_c is the slip correction factor of the particles. When $G \gg 1$, cloud settling is the predominant motion; when $G \ll 1$, only particle settling occurs; and in between, both mechanisms operate. For simplicity, a value of $G = 1$ will be considered the minimum for cloud settling; it is likely that values of $G \gg 1$ are required for prolonged cloud motion, because of the factors, described later in this section, that reduce G with time.

Equation 17.4 holds only for a cloud in the shape of a sphere. Although a cloud may be spherical initially, soon after it begins to move the uneven pressure on its surface will cause it to deform to a nonspherical shape. This deformation will slow the cloud's motion and may cause it to break up into smaller clouds. The size of a settling cloud is gradually reduced with time as its surface is eroded by air currents. Both effects reduce cloud settling velocities. Most cloud aerosols are polydisperse, and the differential settling velocities tend to separate the particles, diluting the cloud and thereby impeding cloud settling. Because of these effects, Eq. 17.4 correctly predicts only the initial motion of a cloud.

The significance of the different variables in Eq. 17.6 is obscured by the drag coefficient C_D, which depends on ρ_c and d_c. For the *special case where cloud motion is in the Stokes region*, $C_D = 24/\text{Re}$, and Eq. 17.6 can be written as

$$G = \frac{\rho_c d_c^2}{\rho_p d_p^2 C_c} = \frac{\pi C_N d_p d_c^2}{6 C_c} \quad \text{for } \text{Re}_c < 1 \tag{17.7}$$

where C_c is the slip correction factor of the particles. Equation 17.7 holds only when Re < 1 for the cloud, which usually implies a cloud size smaller than a few centimeters. Cloud settling in the Stokes region is favored by large cloud size, high mass concentration, and small particle size. A mass concentration of only 9.8 mg/m³ is required for a G value of unity for a cloud 1 cm in diameter with particles 1 μm in diameter.

Table 17.1 gives the minimum mass concentrations required for cloud settling ($G = 1$). The values were calculated using Eqs. 17.4 and 17.6 and the procedure given in Section 3.7. The table shows the importance of the cloud size and particle size in determining whether cloud settling occurs. It is apparent that most aerosol clouds will, at least initially, have cloud motion, although a value of $G > 100$ is probably required for sustained motion.

Because of the dilution, breakup, and erosion mechanism described earlier, the lifetime of clouds less than a few meters in diameter is brief. Furthermore, a value of $G = 1$ means only that the cloud settling velocity is equal to the particle settling velocity, a velocity that is very small, particularly for submicrometer-sized particles. Consequently, it is of greater practical significance to identify the concentration required for a cloud to have an appreciable settling velocity, such as 0.01 m/s [1 cm/s], as has been done in Table 17.2. In this case the required mass concentration does not depend on the particle size, although the number concentration does. Note that the mass concentration values are micrograms per cubic meter in Table 17.1 and milligrams per cubic meter in Table 17.2.

When the particle number concentration in a cloud is high, coagulation will rapidly change both the number concentration and the particle size, although the mass concentration will remain constant. As can be seen from Eq. 17.6, an increase in particle size with a constant cloud diameter and mass concentration leads to a rapid decrease in the value of G. Thus, coagulation inhibits cloud settling. In sum, the several effects described here suggest that aerosol clouds often move initially as an entity, but quickly revert to individual particle motion because of the combined effect of erosion, breakup, size distribution, dilution, and coagulation.

Generally, the gas composition and temperature inside a cloud and in the surrounding air will be different. This difference in density is usually much greater than the net cloud density ρ_c, due to the weight of the particles. For example, in natural atmospheric clouds, a temperature difference of 0.2°C produces a buoyant force

TABLE 17.1 Minimum Concentration for Cloud Settling ($G = 1$) at Standard Conditions

Particle Diameter (μm)	Mass Concentration (μg/m³)			
	$d_c = 0.01$ m	$d_c = 0.1$ m	$d_c = 1$ m	$d_c = 10$ m
0.04	82	0.82	0.0082	8.2×10^{-5}
0.1	240	2.4	0.024	2.4×10^{-4}
0.4	1900	19	0.19	0.0028
1.0	9800	98	1.3	0.023

TABLE 17.2 Concentration Required for a Cloud Settling Velocity of 0.01 m/s
[1 cm/s] at Standard Conditions

Particle Diameter (μm)	Mass Concentration (mg/m³) and [Number concentration (cm⁻³)]			
	$d_c = 0.01$ m	$d_c = 0.1$ m	$d_c = 1$ m	$d_c = 10$ m
0.04	4400 $[1.3 \times 10^{11}]$	100 $[3.1 \times 10^9]$	3.8 $[1.1 \times 10^8]$	0.34 $[1.0 \times 10^7]$
0.1	4400 $[8.5 \times 10^9]$	100 $[2.0 \times 10^8]$	3.8 $[7.3 \times 10^6]$	0.34 $[6.4 \times 10^5]$
0.4	4400 $[1.3 \times 10^8]$	100 $[3.1 \times 10^6]$	3.8 $[1.1 \times 10^5]$	0.34 $[1.0 \times 10^4]$
1.0	4400 $[8.5 \times 10^6]$	100 $[2.0 \times 10^5]$	3.8 $[7.3 \times 10^3]$	0.34 $[6.4 \times 10^2]$

equivalent to the force of gravity on a cloud with a particle mass concentration of 1 g/m³. A similar effect is produced by an absolute humidity difference of 130 Pa [1 mm Hg]. The weight of the particles plays a secondary role in controlling the vertical movement of atmospheric clouds. Some combustion clouds rise initially because of their high gas temperature, and after cooling they settle because of their high CO_2 content. Cloud motion has been observed in experiments with cigarette smoke, an aerosol of high concentration with a gas phase denser than air. (See Martonen, 1992, Chen and Yeh, 1990, and Phalen et al., 1994.)

The special case of an aerosol cloud in the form of a layer overlying clean air can be analyzed in terms of *Rayleigh–Taylor instability*. This phenomenon exists when a denser fluid layer overlays a lighter fluid in a gravitational or other type of force field. Such a configuration can occur naturally in the atmosphere, in dust clouds from volcanos, in the oceans, and in cultures of microorganisms that are denser than water and exhibit negative geotaxis (swim against gravity). Rayleigh–Taylor instability is characterized by an abrupt breakthrough of the heavier layer through the underlying lighter layer. The spacing between the breakthrough points and the rate of fall has been characterized by Plesset and Whipple (1974). Differences in gas density play a major role with this phenomenon, as they do with cloud settling. Small differences in gas density, which have little effect on individual particle settling, can dramatically modify particle motion if they cause Rayleigh–Taylor instability or cloud settling.

All elutriation devices that act as spectrometers have two gas streams: an aerosol stream and a clean carrier gas stream. Consequently, they are susceptible to cloud settling and instability effects. Aerosol centrifuges (see Section 3.9) are particularly susceptible because of the high centrifugal acceleration they exert on the particles and gas streams.

The cloud motion described in this chapter is to be distinguished from the situation in which a cloud entirely fills a container. In this situation, there can be no cloud settling and only a very slight reduction of particle settling velocity, due to the increase in viscosity of the aerosol at high concentrations and the slight upward velocity of the suspending gas caused by the displaced particle volume. These two effects have only a minor influence (< 0.1%) on the settling velocity of particles in contained aerosols, even extremely dense aerosols.

PROBLEMS

17.1 A cigar smoker blows a smoke ring that falls with a noticeable velocity. If the ring is considered to be a sphere 0.05 m [5 cm] in diameter, what is its cloud settling velocity? Assume that the smoke gas is air at room temperature and the smoke particle concentration is 30 g/m³. If the smoke particles are 0.4 μm in diameter and of standard density, what is the value of G?
ANSWER: 0.17 m/s [17 cm/s], 25,000.

17.2 Repeat the first part of Problem 17.1 for a smoke gas-phase density that is 2% greater than that of air.
ANSWER: 0.24 m/s [24 cm/s].

REFERENCES

Chen, B. T. and Yeh, H. C., "Effects of Cloud Behavior on a Cigarette Smoke Aerosol," Lovelace Foundation Report LMF–129, Albuquerque, NM, 1990.

Fuchs, N. A., *The Mechanics of Aerosols*, Pergamon, Oxford, 1964.

Kaiser, G. D., and Griffiths, R. F., "The Accidental Release of Anhydrous Ammonia to the Atmosphere: A Systematic Study of Factors Influencing Cloud Density and Dispersion," *J. Air Poll. Control Assoc.*, **32**, 66–71 (1982).

Martonen, T. B., "Deposition Patterns of Cigarette Smoke in Human Airways," *Am. Ind. Hyg. Assoc. J.*, **53**, 6–18 (1992).

Maude, A. D., and Whitmore, R. L., "A Generalized Theory of Sedimentation," *Brit. J. Appl. Phys.*, **9**, 477–482 (1958).

Phalen, R. F., Oldham, M. J., Mannix, R. C., and Schum, G. M., "Cigarette Smoke Deposition in the Tracheobronchial Tree: Evidence for Colligative Effect," *Aerosol Sci Tech*, **20**, 215–226 (1994).

Plesset, M. S., and Whipple, C. G., "Viscous Effects in Rayleigh–Taylor Instability," *Phys. Fluids*, **17**, 1–7 (1974).

Stöber, W., Martonen, T. B., and Osborne, S., Jr., "On the Limitations of Aerodynamical Size Separation of Dense Aerosols with the Spiral Duct Centrifuge," in Shaw, D. T. (Ed.), *Recent Developments in Aerosol Science*, Wiley, New York, 1978.

Wen, C. S., *The Fundamentals of Aerosol Dynamics*, World Scientific, Singapore, 1996.

18 Dust Explosions

The most spectacular property of aerosols is their ability to explode violently at high concentrations, as shown in Fig. 18.1. On average, there are more than 40 severe dust explosions in the United States each year (Eckhoff, 1977; Factory Mutual, 1998). About 20 of these are grain dust explosions. Together these accidents account for several fatalities and millions of dollars in property damage every year. Because of the importance of dust explosions, their characteristics and control methods are briefly reviewed here. The phenomenon is complicated, and, extensive empirical data are available, but no general theory of dust explosions exists. Although less common, mists of combustible liquids can explode. See Bowen and Shirvill (1994).

For a dust explosion to occur, dust and oxygen must be present at the proper concentrations, and there must be a source of ignition. Any oxidizable material, including all organic materials and some inorganic compounds and metals, will burn if present as an aerosol at a sufficiently high concentration. The minimum concentration required is usually 20–200 g/m^3. When combustion takes place in an enclosed space, the result is a dust explosion. In the process, heat is generated much faster than it is dissipated, causing a runaway reaction with rapid gas expansion and the generation of gaseous combustion products. This leads to the rapid increase in pressure that causes the explosion damage. Typically, a severe dust explosion in an occupied space is initiated by the compression wave of a small explosion that resuspends dust from the floor, walls, and beams to produce a dust concentration sufficient to cause a second, much larger, explosion. The process may continue with a series of explosions spreading through the plant or mine. It is this ability to generate, by resuspension, explosive dust concentrations in large spaces that accounts for the destructiveness of dust explosions. Explosive gases, on the other hand, are not usually permitted to build up to explosive concentrations in such large volumes. In coal mines, the initial explosion may be a small methane gas explosion or blasting. In a plant, the initial explosion may be caused by the rupture of a piece of equipment under pressure, or the dust may be resuspended by local mechanical action in the presence of a source of ignition. Some explosions are initiated by a small fire in a settled dust layer that gets resuspended by the action of a water hose or fire extinguisher.

In some ways, dust explosions are similar to gas explosions. For example, with both types of explosion, there is a *lower explosive limit*, the minimum concentration required for a flame to propagate. On the other hand, gases have a well-defined upper explosive limit, but such a limit for dusts is poorly defined and difficult to

FIGURE 18.1   A demonstration coal dust explosion in a Bruceton, Pennsylvania, experimental mine (Verakis and Nagy, 1987). Copyright ASTM. Reprinted with permission.

determine. It occurs when there is insufficient oxygen reaching the particles, usually at a concentration of 1500 to 3000 g/m³ for organic dust. This is shown in Figure 18.2: Methane has well-defined minimum and maximum explosive concentrations, but brown coal dust has only a well-defined lower limit. Minimum explosive concentrations and other explosive properties of representative dusts are given in Table 18.1. The data shown for concentration and pressure were obtained with a 1-m³ or 20-L spherical chamber. The dust is sized and tested as received or sieved through a 63-μm (No. 230) sieve. It is dispersed in a chamber by compressed air and ignited by a 10-kJ chemical igniter. These larger test chambers have replaced the traditional 1.2-L cylindrical Hartmann apparatus, improving the reproducibility and representativeness of the data obtained (ISO, 1985 and ASTM, 1988). The data show the characteristics of different dusts on a relative scale, but differ, usually by less than a factor of 2, from results obtained with other types of laboratory-scale test apparatus.

The *pressure rise index* K_{st} is a measure of explosive violence. It is a constant for a given dust material, particle size distribution, and moisture content and is related to the maximum rate of pressure increase $(dp/dt)_{max}$ by

$$K_{st} = (V_{ch})^{1/3}\left(\frac{dp}{dt}\right)_{max} \quad \text{for } V_{ch} > 0.002 \text{ m}^3 \qquad (18.1)$$

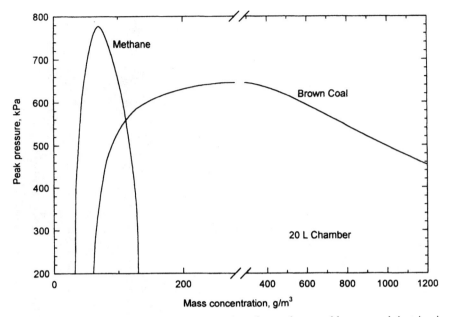

FIGURE 18.2 Range of explosive concentrations for methane and brown coal dust in air. Data from Hertzberg and Cashdollar (1987) and Eckhoff (1997).

where V_{ch} is the volume of the chamber in m³. K_{st} is numerically equal to $(dp/dt)_{max}$ for a 1-m³ chamber. The minimum explosive concentrations given in Table 18.1 are all very high concentrations and are not normally found in areas where people are working. Instead, these concentrations exist only in ducts or enclosed equipment or during the transient resuspension caused by a blast or other aerodynamic disturbance.

The minimum explosive concentration depends on the particle size and material, as shown in Fig. 18.3. For many materials, the shape of the curve relating the minimum explosive concentration to particle size is controlled by the rate of volatilization (pyrolysis or thermal decomposition) of the heated particles, which controls the combustion. In the region where the minimum explosive concentration does not vary with particle size (0–80 μm for polyethylene dust), the particles are fully volatilized so quickly that the rate of volatilization does not affect the flame propagation process. In the intermediate region (80–120 μm for polyethylene dust), volatilization is incomplete, occurring in only a surface layer, and thus requires a higher concentration to achieve flame propagation. For particles larger than an upper limit (120 μm for polyethylene dust), the surface-to-volume ratio is so low that volatilization is insufficient for flame propagation in the time that is available.

Unlike the combustion of gases, combustion during an unconfined dust explosion does not reach a *detonation* condition, where the flame speed equals or exceeds the speed of sound. During detonation, a transient high-pressure shock wave accompanies the flame front. A dust explosion is characterized as a *deflagration* in which

TABLE 18.1 Explosive Characteristics of Selected Dusts[a]

Material	MMD (μm)	Minimum Explosive Concentration (g/m³)	Maximum Pressure[b] (kPa)	Pressure Rise Index, K_{st}[c] (kPa · m/s)	Minimum Ignition Temperature (°C)
Cotton	44	100	720	2,400	560
Wood (chipboard)	43	60	920	10,200	490
Maize (corn) starch	16	60	970	15,800	520
Wheat grain	80	60	930	11,200	—
Bituminous coal	38	125	860	8,600	610
Epoxy resin	26	30	790	12,900	510
Polyvinyl chloride (PVC)	25	125	820	4,200	750
Ascorbic acid	39	60	900	11,000	460
Organic dyestuff (red)	<10	50	11,200	24,900	520
Aluminum	22	30	11,500	110,000	500
Zinc	<10	250	670	12,500	570
Sulphur	20	30	680	15,100	280

[a]Data from Eckhoff (1997).
[b]For maximum pressure in bar, divide value in kPa by 100.
[c]For K_{st} in bar · m/s, divide value in kPa · m/s by 100.

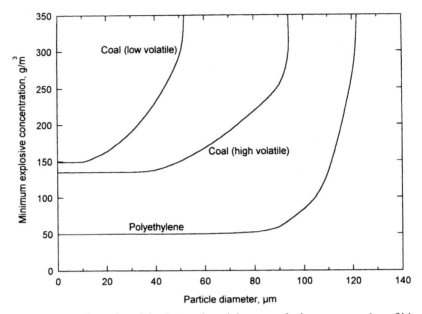

FIGURE 18.3   Effect of particle size on the minimum explosive concentration of bituminous coal and polyethylene dust. Data from Hertzberg and Cashdollar (1987).

the flame speed, typically a few meters per second, is much less than the speed of the compression wave generated by the explosion—about 340 m/s, the speed of sound. The compression wave precedes the flame and resuspends settled dust, providing an explosive concentration when the flame arrives. In a dust explosion, it is the pressure produced by the heat of combustion and the generation of gaseous combustion products that causes the damage, rather than the transient shock wave typical of a detonation. The speed of the flame front, or flame speed, of a dust explosion depends on the laminar burning velocity (about 0.2–0.3 m/s), the speed of the gas due to expansion and buoyancy, the turbulent intensity at the flame front, and the geometry of the structure in which the explosion takes place. Flame speed increases up to a maximum value with increasing dust concentration above the minimum explosive concentration. When a dust explosion is confined to a long tunnel or a tube that is open at one end with a dust layer on the floor, the gas expansion can accelerate the flame front along the tube to sonic velocity. For particles larger than a few micrometers, flame speed decreases with increasing particle size at constant concentration. Only a thin layer of dust on a floor is required to produce an explosive concentration when the dust is resuspended. In typical industrial spaces, a layer of dust on the floor less than 1 mm thick will produce an explosive concentration if the dust is fully resuspended.

The severity of a dust explosion depends on the maximum pressure developed, p_{max}, and the rate of pressure buildup, dp/dt. Values for these quantities are given in Table 18.1 for representative dusts. For natural organic dusts, the maximum pressure obtained in a sealed apparatus is about 6–10 times the preignition pressure and occurs at a concentration of about 500 g/m^3, somewhat above the stoichiometric concentration and well above the lower explosive limit. The maximum pressure is reached rapidly, in a fraction of a second within a piece of equipment and in a few seconds in a building. Pressure dies away at a rate that depends on the leakage of the building (which may be dramatically altered by the explosion) or the cooling of the test apparatus. The rate of pressure rise is important for the design of automatic venting equipment to limit damage due to explosions.

Ignition may be caused by a spark, flame, or hot surface. The minimum temperature required to ignite selected dusts is given in Table 18.1. The minimum ignition temperature increases with increasing particle size ($\propto d_p^3$) and with increasing relative humidity and moisture content of the dust. There is some evidence that an ignition source with a large area can initiate dust explosions at a lower temperature than can smaller sources of ignition. Sparks caused by electricity, friction, or welding are above the minimum ignition temperature, but are confined to a very small volume. The key parameter for spark ignition is the energy contained in the spark. The minimum energy required for spark ignition ranges from 1 to 4000 mJ and varies significantly with the type of apparatus used for testing.

The *hazard* associated with dust explosions depends on the dispersibility (see Section 21.3) of the dust layer, the ease with which ignition occurs, and the severity of the explosion. Ignition is more likely to occur if the dust has a low ignition temperature, a low ignition energy, or a low minimum explosive concentration. The severity of an explosion increases with increasing values of p_{max} and K_{st}.

The primary control methods are limited to preventing explosions and to mitigating the spread and consequences of those that do occur. Ignition sources such as hot surfaces, flames, sparks from friction and welding should be controlled. Electrical equipment and light fixtures should be enclosed, and ducts, blowers, and air-cleaning equipment should be well grounded to prevent the buildup of static charge. Housekeeping is extremely important, because it is the settled dust that, when resuspended, permits the explosion to propagate. Frequent cleaning is required to prevent any buildup of dust, particularly on beams, ledges, and other out-of-the-way horizontal surfaces. For example, the U.S. Occupational Safety and Health Administration (OSHA) requires that no more than 1/8 in. (3mm) of settled dust be allowed to accumulate in grain-handling facilities. Equipment for grinding, crushing, and transferring dust should be enclosed in dust-tight enclosures. Potentially dangerous operations can be segregated so that an explosion cannot spread. These operations can be placed in heavily constructed rooms fitted with automatic vents to prevent the buildup of destructive pressures. Guidelines for the size of the vents are given by safety organizations such as the National Fire Protection Association (NFPA, 1994).

Ventilation and monitoring are required in mines to prevent the accumulation of explosive concentrations of methane. In many coal mines, it is impractical to achieve the cleanliness necessary to prevent dust explosions, so a different approach is used. The walls, ceiling, and floors are dusted with an inert dust such as limestone. In the event of an explosion, the inert dust gets resuspended along with the coal dust, but it absorbs so much heat, that it prevents flame propagation and the dust explosion. The amount of inert dust required depends on the volatile content of the coal and may be as much as four times the weight of the coal dust. At critical points, larger amounts of inert dust may be placed in special overhead bins so that it will be resuspended automatically by the pressure wave and prevent the explosion from spreading. Enclosed processes, such as grinding or milling, can be done in a low-oxygen atmosphere by adding sufficient CO_2 to reduce the oxygen concentration below the explosive minimum. Except for metal dusts, the minimum oxygen concentration required for flame propagation and dust explosion ranges from 10 to 15%.

PROBLEMS

18.1 What are the number concentration and visual range (see Section 16.4) for a cornstarch aerosol at its minimum explosive concentration? Assume that $d_p = 16$ μm (from Table 18.1) and $\rho_p = 1000$ kg/m^3 [1.0 g/cm^3].
ANSWER: 2.8×10^{10}/m^3 [28,000/cm^3], 0.35 m [35 cm].

18.2 A bituminous coal mine tunnel has a 2 m × 2 m-square cross section. How thick must a layer of coal dust on the floor be to give an explosive concentration when 10% of the dust is resuspended? Assume that the bulk density of the coal dust layer is 500 kg/m^3 [0.5 g/cm^3].
ANSWER: 5.0 mm.

18.3 A cornstarch explosion produces its maximum pressure. What force (in N and lb) would the explosion exert on a 2-m^2 door?
ANSWER: 1.9 MN, 440,000 lb.

REFERENCES

ASTM, "ASTM Standard E-1226-88 for Pressure and Rate of Pressure Rise for Combustible Dust," *Annual Book of ASTM Standards*, *14.02*, American Society for Testing and Materials, Philadelphia, 1988.

Bartknecht, W., *Dust Explosions: Course, Prevention, Protection*, Springer-Verlag, Berlin, 1989.

Bowen, P. J. and Shirvill, L. C., "Combustion Hazards Posed by the Pressurized Atomization of High Flash Point Liquids," *J. Loss Prev. Process Ind.*, **7**, 233–241 (1994).

Cross, J., and Farrer, D., *Dust Explosions*, Plenum, New York, 1982.

Eckhoff, R. K., *Dust Explosions in the Process Industries*, 2d ed., Butterworth-Heinemann, Oxford, 1997.

Factory Mutual Engineering Corp., *Property Loss Prevention Data Sheet 7-76, Prevention and Mitigation of Combustible Dust Explosions and Fires*, Factory Mutual System, Norwood, MA, 1998.

Hertzberg, M., and Cashdollar, K. L. "Introduction to Dust Explosions," in Cashdollar, K. L., and Hertzberg, M. (Eds.), *Industrial Dust Explosions*, American Society for Testing and Materials, Philadelphia, 1987.

ISO, *Explosion Protection Systems. Part 1: Determination of Explosion Indices of Combustible Dusts in Air*, ISO/DIS 6184/1, International Standardization Organization, Geneva, 1985.

National Fire Protection Association, *NFPA 68, Guide for Venting of Deflagrations*, 1994 ed., National Fire Protection Association, Quincy, MA, 1994.

Verakis, H. C., and Nagy, J., "A Brief History of Dust Explosions" in Cashdollar, K. L., and Hertzberg, M. (Eds.), *Industrial Dust Explosions*, American Society for Testing and Materials, Philadelphia, 1987.

19 Bioaerosols

Bioaerosols are aerosols of biological origin. They include viruses, living organisms, such as bacteria and fungi; and parts or products of organisms, such as fungal spores, pollen, and allergens from dogs, cats, and insects. Bioaerosols are a special category of aerosols because their biological properties can affect the health of humans, animals, and plants. This aspect demands special analysis, which imposes limitations on collecting and handling samples. This chapter presents a summary of bioaerosol characteristics and sampling.

19.1 CHARACTERISTICS

Bioaerosols are found nearly everywhere, indoors and outdoors, but usually at low concentrations. Only minute amounts of inhaled bioaerosol are needed to cause disease—as little as one bacterium, weighing less than 1 pg. Particle size and the usual range of outdoor concentrations for the most common bioaerosols are shown in Table 19.1. The sources of bioaerosols are plants, animals (including humans), soil, and water. The health effects caused by bioaerosols include (1) infectious disease, (2) sensitization reactions, such as asthma, and (3) reactions to toxins or irritants such as endotoxins (components of the cell walls of certain bacteria) and mycotoxins from fungi. Because of the nature of their growth and aerosolization, bioaerosol particles often occur as agglomerates, as clusters of organisms in droplets, or attached to other airborne debris. Bioaerosols can be subdivided into two main groups: viable and nonviable. Particles of the former group are living organisms that can be identified and quantitated by growing individual organisms with suitable nutrients into visible clusters or colonies. Nonviable bioaerosols, including dead organisms, pollen, animal dander, and insect excreta, require other types of analyses. The most common bioaerosols are bacteria and fungal spores. The concentration of bioaerosols, like that of all aerosols, decays with time due to settling and deposition on surfaces. (See Sections 3.8 and 7.4.) Also, viable bioaerosols show a decline in biological activity with time that depends on the relative humidity, oxygen content, and the concentrations of trace gases in the air. Characteristics of bioaerosols are covered in Burge (1995) and Nevalainen et al. (1993).

Bacteria are single-celled organisms with sizes from 0.3 to 10 μm. They are mostly water with a density that ranges from 1000 to 1500 kg/m³ [1–1.5 g/cm³], depending on their degree of hydration. Usually, bacteria are spherical or rod shaped, but they may occur as clusters or chains. More than 1700 species of bacte-

TABLE 19.1 Particle Size and Natural Background Concentration of Bioaerosols[a]

Type of Bioaerosol	Size (μm)	Concentration (number/m³)
Viruses	0.02–0.3	—
Bacteria	0.3–10	0.5–1000
Fungal Spores	0.5–30	0–10,000
Pollen	10–100	0–1000

[a]Data primarily from Jacobson and Morris (1976).

ria have been identified. Those that cause human disease are described as human pathogens. Some examples of infectious diseases known to be caused by aerosolized bacteria are tuberculosis, legionellosis, and anthrax. Opportunistic pathogens are not harmful to healthy people, but can infect those with compromised immune systems. Many environmental bacteria colonize water or soil and are released as aerosols when the water or soil is disturbed. A gram of soil can contain 10^9 bacteria. In indoor environments, bacteria may colonize accumulations of moisture in ventilation systems and become aerosolized by air currents or vibration. Some bacteria release spores called endospores. These are hardy, dormant versions of the bacteria, 0.5 to 3 μm in size, that are easily carried by air currents. Included in this category are actinomycetes associated with hypersensitivity pneumonitis.

Fungi represent a unique group of organisms that are found everywhere in the outdoor environment and in most indoor environments. They may occur as single-celled organisms such as yeast or, more commonly, as microscopic, multicellular branching structures called hyphae. Masses of hyphae are easily visible. Molds and mildew are visible fungal growths on surfaces. Only about 70,000 of the estimated 1.5 million fungi have been identified. The majority of fungi are saprophytes, organisms that get their nutrients from dead organic material and facilitate its decomposition. They are found primarily in soil, in damp locations, and on decaying vegetation. Most fungi disperse by releasing spores into the air. These *fungal spores*, particles of 0.5 to 30 μm in size, are resistant to environmental stresses and adapted to airborne transport. They may be single celled or fragmented hyphae, typically 2–4 μm, or fruiting bodies that form at the end of hyphae. They are aerosolized by air currents or mechanical disturbances. Yeasts, by contrast, do not produce spores, but become airborne by aerosolization of the liquid in which they grow. Inhalation of fungi or spores can cause infectious disease, such as histoplasmosis, or opportunistic infection, such as aspergillosis. Most fungi are associated with allergic diseases (e.g., asthma).

Viruses are intracellular parasites that can reproduce only inside a host cell. They consist of encased RNA or DNA and little else. Although naked viruses range from 0.02 to 0.3 μm, most airborne viruses are found as part of droplet nuclei or attached to other airborne particles that can have a wide range of sizes. Viruses infect humans, other animals, and plants, although most are specific to one type of host. Infected humans represent the primary reservoir and source for human viruses. Viruses are transmitted by direct contact, through contaminated food or water, or by inhalation of aerosolized viruses. Aerosolization can occur by coughing, sneezing, or talking. Under favorable conditions, they can survive for weeks on fabric or carpets. Viruses cause infectious diseases such as colds, flu, chicken pox, measles, and hantavirus pulmonary syndrome.

Pollen grains are relatively large, near-spherical particles produced by plants to transmit genetic material to the flowers of other plants of the same species. The grains range in size from 10 to 100 µm, with most between 25 and 50 µm. The average density of pollen grains is 850 kg/m³ [0.85 g/cm³]. Anemophilous (wind-pollinated) plants produce abundant bioaerosol pollen; insect-pollinated plants produce sticky pollen that is not readily aerosolized. Pollen has a tough outer coating to protect it from environmental stresses. Wind-borne pollen is produced in the plants' flowers on the end of protruding stalks called anthers. This arrangement facilitates the entrainment of pollen in the wind. The production and release of pollen is seasonal and controlled by wind and weather. Pollen causes allergic diseases of the upper airways, such as hay fever.

Other bioaerosols associated with allergic diseases include aerosolized algae, insect fragments, excreta from dust mites and cockroaches, and saliva and dander from dogs, cats, and birds. Viable aerosols have variable survivability (viability) in air during transport. Some do better at high humidity while others do better at low humidity. Many are affected by temperature and ultraviolet radiation. Oxygen is toxic to some, and trace gases probably affect many others.

All types of bioaerosols are likely to occur in farm and agricultural environments. Bioaerosols in the outdoor, nonfarm environment are dominated by fungal spores with smaller numbers of bacteria and pollen, but concentrations are highly variable and influenced greatly by wind, weather, and local sources. Typical outdoor concentrations of culturable bioaerosols are 100–1000 cfu/m³ (cfu = colony-forming units; see Section 19.2). In the absence of indoor sources, indoor environments with natural ventilation resemble outdoor environments, but have lower bioaerosol concentrations. In mechanically ventilated office spaces bioaerosol concentrations are low, typically less than 100 cfu/m³. High concentrations of bioaerosols, 10^4–10^{10} cfu/m³, can occur in specific work environments, such as farm buildings, food-processing plants, textile plants, and sawmills. Currently, there are no health-based environmental standards for culturable or countable bacteria or fungi or for infectious agents. Nor are there guidelines for acceptable levels of bioaerosols. Information about dose-response relationships for bioaerosols is limited.

19.2 SAMPLING

In sampling for bioaerosols, the objective is to determine the presence or the number concentration of specific species or the total bioaerosol concentration. Sampling for bioaerosol particles employs many of the methods used for nonbioaerosol particles, but the analytical methods impose certain limitations on the sampling process. There are physical and biological aspects to bioaerosol sampling. The physical aspects are the same for bioaerosols and nonbioaerosols. The biological aspects may require aseptic handling to prevent contamination and special attention to the viability of the organisms during the process of sampling and analysis. Viability, which depends on the organism, the environment, and the sample collection and analysis methods used, introduces significant uncertainty into the quantitation of airborne concentrations of bioaerosols. Estimates of the fraction of viable particles that survive the sampling and analysis process, grow, and are counted range from 0.1 to 0.001 (Macnaughton et al., 1997). Other important factors in bioaerosol sampling are high variability in number concentration, wide particle size range of interest, analytical sensitivity to under- and overloading, and interference by other particles. Comprehensive reviews of bioaerosol sampling methods are given by Nevalainen et al. (1993), Macher (1995), and Willeke and Macher (1998).

There are three stages of bioaerosol sampling. The first concerns the inlet efficiency, which is the same as that for nonbioaerosol particles. This stage is covered in Sections 10.1 and 10.2 on isokinetic and still air sampling. The second stage is the collection or deposition of particles onto glass slides, a semisolid culture medium, or water. This is accomplished by adapting the basic methods used for nonbioaerosols to bioaerosols. The third stage is biological analysis to identify and quantitate the bioaerosol particles present. This stage is unique to bioaerosol sampling. Most bioaerosol analysis involves either microscopic examination or growth in culture medium followed by counting colonies. A discussion of the second stage of bioaerosol sampling follows.

The most common collection methods rely on impaction, the same process as described in Section 5.5 for nonbioaerosols. *Slit impactors* impact particles directly onto a culture medium. For bacteria and fungal spores, the culture medium, called agar, is a semisolid material containing water and nutrients that foster the growth of the viable particles that are collected. For viruses, cell or tissue culture media are used. Typically, the agar fills a 100-mm or 150-mm disposable petri dish, called a culture plate, that is slowly rotated under the slit to provide a history of bioaerosol concentration. Rotating slit impactors have flow rates of 28–50 L/min and cutoff diameters of about 0.5 μm.

Multijet impactors have 100–500 jets impacting directly onto agar culture plates. This action spreads the collected particles to many locations to prevent overloading. Single and multistage impactors are available with cutoff diameters from 0.6 to 8 μm. Shear forces in the high-velocity jet and during impaction may cause loss of viability.

Other types of impactors are used to collect pollen. The pollen grains are impacted directly onto glass slides or adhesive-coated transparent tape for microscopic examination and counting. Culturing is not used for pollen. One type of sampler (the Rotorod® sampler) impacts pollen grains directly onto adhesive-coated polystyrene rods. The rods are transparent for direct microscopic particle counting. The rods are square in cross section, are 60 mm long and 0.5 to 1.6 mm wide, and are attached to an exposed rotating frame. The cutoff size for collection can be estimated by Eq. 5.23, the definition of the Stokes number, as

$$d_{50} = \left(\frac{18\eta d_r(\mathrm{Stk}_{50})}{\rho_p V_T} \right)^{1/2} \tag{19.1}$$

where d_r is the width of the rod, V_T is the tangential velocity of the rods, Stk_{50} is approximately 0.3. The Stokes number for 85% collection, Stk_{85}, is approximately 1.5.

Impingers (see Section 10.6 and Fig. 10.13) are frequently used to collect bioaerosols in a liquid, such as water, buffered saline solution, or nutrient broth. The two most commonly used impingers are the all-glass impingers, AGI-30 and AGI-4. The numerals refer to the distance in mm from the glass nozzle to the base of the impinger body. Liquid collection prevents desiccation, but the shear forces in the jet and in the turbulent liquid cause loss of viability. The AGI-30 has a cutoff diameter of 0.3 µm; the AGI-4 has a slight lower cutoff, but has a greater loss of viability. Water containing the collected organisms can be applied directly to the culture plates or can be diluted and applied to give the desired surface density of colonies.

The *centrifugal sampler* has a rotating "paddle wheel," open to the environment on one end, with a stationary removable agar-coated plastic strip around the rim of the wheel. Air enters along the axis, and the particles contained in it are driven by centrifugal force onto the collecting strip. Sample flow rates are 40 to 50 L/min.

Membrane filters are also used to collect bioaerosol particles. The collected particles can be examined by an optical microscope (see Section 20.2), or they can be cultured by placing the filter, particle side up, on a culture medium. The filtration process causes significant desiccation of collected microorganisms and loss of viability. This method has been used for bioaerosols in highly contaminated environments.

Bioaerosols can be sampled by allowing them to settle directly onto culture plates, a simple, inexpensive method that allows one to determine the presence of a specific microorganism. The technique, however, is not useful for quantifying airborne concentrations because, after culturing, the size of the initial particle cannot be determined, so V_{TS} and the flux are not known.

Respirable sampling (see Section 11.5) is appropriate for some infectious agents. Inhalable particle sampling (see Section 11.4) is appropriate for all bioaerosol sampling, especially of pollen and other large particles.

An important consideration in bioaerosol sampling is the appropriate surface density for counting organisms or colonies. If the loading is insufficient there will be few counts and high statistical uncertainty. In microscopic counting of organisms, overloading causes overlap with nonbioaerosol particles, making identification difficult. Environmental samples usually have far more nonviable particles than viable particles. In counting colonies, overloading leads to three types of underestimation. First, if two of the same species are sufficiently close to each other, they will be counted as a single colony. Second, if two different species are sufficiently close, one may inhibit the growth of the other. Finally, due to its chemical nature, a nonbioaerosol particle may inhibit the growth of a nearby viable particle. For a given sampler, the sampling time t to achieve a desired surface density s (colonies or particles per unit area) is given by

$$t = \frac{sA}{\overline{C}_N Q} \tag{19.2}$$

where A is the area onto which particles are deposited, $\overline{C}_N$ is the average number concentration of bioaerosol particles, and Q is the sampler flow rate. For microscopic counting, a surface density of $10^8/m^2$ [$10,000/cm^2$] gives good results; for counting colonies, a surface density of $10^4/m^2$ [$1/cm^2$] is desired.

A different approach is taken for multijet impactors. There is a deposition site directly below each jet. After culturing, one cannot tell whether just one or more than one organism has been deposited at a particular site. Regardless of the number of organisms involved, a site with a colony is called a *filled site* or a *positive hole*. Plates are analyzed by counting the number of filled sites. For a given multijet impactor, the more filled sites, the greater the proportion of those filled sites that have multiple organisms. Willeke and Macher (1998) and Macher (1989) give tables with correction factors to estimate the total number of viable organisms collected, based on the number of filled sites. Also given is the statistical uncertainty of these estimates. As shown in Table 19.2, the correction is less than 5% when less than 10% of the sites are filled sites. The correction factor is 2.0 when 80% of the sites are filled sites. For an impactor with N_j jets, the following empirical equation gives

TABLE 19.2 Correction Factors for Multiple Particle Collection in Multijet Impactors[a]

Filled Fraction[b]	Correction Fraction[c]	Filled Fraction	Correction Factor
0.05	1.026	0.55	1.452
0.10	1.054	0.60	1.527
0.15	1.084	0.65	1.615
0.20	1.116	0.70	1.720
0.25	1.151	0.75	1.848
0.30	1.189	0.80	2.012
0.35	1.231	0.85	2.232
0.40	1.277	0.90	2.559
0.45	1.329	0.95	3.154
0.50	1.386	1.00	>5.878

[a]Calculated from Macher (1989).
[b]Fraction of deposition sites with colonies.
[c]Total number of viable particles collected equals the number of filled sites times the correction factor.

the total number of viable organisms collected n_c in terms of the number of filled sites—i.e., those with colonies, n_f,

$$n_c = n_f \left(\frac{1.075}{1.052 - f} \right)^{0.483} \quad \text{for } f < 0.95 \tag{19.3}$$

where, n_f is the number of filled sites with colonies and $f = n_f/N_j$ is the fraction of sites with colonies. Equation 19.3 agrees with tabulated values within 1% for filled fractions less than 0.95.

The only commercially available, direct-reading instrument for bioaerosols is the TSI, Inc. (St. Paul, MN), Model 3312 UV-APS®. This instrument combines time-of-flight measurement of particle size (see Section 5.7) with light-scattering particle detection (see Section 16.5) and single-particle ultraviolet fluorescence measurement. The last of these involves illuminating the particle with an ultraviolet laser beam and measuring the emitted fluorescent intensity at certain wavelengths. This fluorescence is associated with certain molecules produced by living cells. The instrument samples at 1.0 L/min, measures particles from 0.5 to 15 μm, and measures concentrations up to $10^9/m^3$ [$1000/cm^3$].

PROBLEMS

19.1 An adjustable-speed, rotating slit impactor has a total deposition area of 50 cm^2 for one full rotation. For how long should a sample be taken to get the desired colony surface density if bioaerosol concentration is 1000 cfu/m³? Assume the sample flow rate is 28 L/min, and that sampling takes place for one full rotation.
ANSWER: 107 s (1.8 min).

19.2 A 400-jet impactor samples a bioaerosol at 28 L/min for 20 minutes. If 344 colonies are counted, what is the average airborne number concentration of viable particles. Assume no loss of viability during sampling.
ANSWER: 1410 cfu/m³.

REFERENCES

Burge, H. A., Feeley, J. C., Kreiss, K., et al., *Guidelines for the Assessment of Bioaerosols in the Indoor Environment*, ACGIH, Cincinnati, 1989.

Burge, H. A., (Ed), *Bioaerosols*, Lewis, Boca Raton, FL, 1995.

Jacobson, A. R., and Morris, S. C., "The Primary Air Pollutants—Viable Particles, Their Occurrence, Sources, and Effects," in Stern, A. C. (Ed.), *Air Pollution*, 3d ed., Academic Press, New York, 1976.

Macher, J. M., "Positive-Hole Correlation of Multiple-Jet Impactors for Collecting Viable Microorganisms," *Am. Ind. Hyg. Assoc. J.*, **50**, 561–568 (1989).

Macher, J. M., "Sampling Airborne Microorganisms and Aeroallergens," in Cohen, B. C., and Hering, S. V. (Eds.), *Air Sampling Instruments*, 8th ed., ACGIH, Cincinnati, 1995.

Willeke, K., and Macher, J. M., "Air Sampling," in Macher, J. M. (Ed.), *Bioaerosols: Assessment and Control*, ACGIH, Cincinnati, 1998.

Macnaughton, S. J., Jenkins, T. L., Alugupolli, S., and White, D.C., "Quantitative Sampling of Indoor Air Biomass by Signature Lipid Biomarker Analysis: Feasibility Studies in a Model System," *Am. Ind. Hyg. Assoc. J.*, *58*, 270–277 (1997).

Nevalainen, A., Willeke, K., Liebhaber, F., and Pastuszka, J., "Bioaerosol Sampling" in Willeke, K., and Baron, P. A., (Ed.), *Aerosol Measurement*, Van Nostrand Reinhold, New York, 1993.

20 Microscopic Measurement of Particle Size

Microscopic observation of aerosol particles permits direct measurement of particle size. This is in contrast to indirect methods such as sedimentation, impaction, mobility analysis, and light scattering, wherein the particle size is estimated from the measurement of a property related to size. Microscopy also provides the opportunity to observe particle shapes, and it requires only an extremely small amount of sample. Linear measurements made with a microscope can be very accurate and, often serve as a primary measurement for the calibration of other aerosol-sizing methods. However, microscopic methods for determining particle size distributions are, in general, tedious and require consistency, skill, and careful preparation.

20.1 EQUIVALENT SIZES OF IRREGULAR PARTICLES

In measuring particle size with a microscope, it is necessary to assign to each particle a size based on its two-dimensional projected image, or silhouette. For spheres, this is simply the diameter of the circular silhouette observed in the microscope, but for the more common case of irregular particles, we must use equivalent diameters such as those shown in Fig. 20.1. The equivalent diameters shown are based solely on the geometry of the silhouette, in contrast to the equivalent diameters defined in Section 3.5, which are based on the aerodynamic behavior of the particles. The smallest diameter shown is Martin's diameter d_M, the length of the line parallel to a given reference line that divides the projected area (silhouette) of the particle into two equal areas. This diameter is often referred to as a "statistical" diameter, because its value depends on particle orientation, and only its mean value for all particle orientations is unique for a given particle. In practice, orientation averaging is rarely done; instead, in sizing particles, it is more common to measure a single Martin's diameter for each of many particles randomly oriented with respect to a reference line. This procedure has the practical effect of averaging over all orientations. Martin's diameter always passes through the centroid of the particle silhouette, and the average Martin's diameter of a particle represents the mean of all chords through the centroid of the silhouette.

Another "statistical" diameter is Feret's diameter d_F, shown in Fig. 20.1. Feret's diameter is defined as the length of the projection of a particle along a given reference line or the distance between the extreme left and right tangents that are perpendicular to the reference line. It is most convenient to use Feret's diameter when

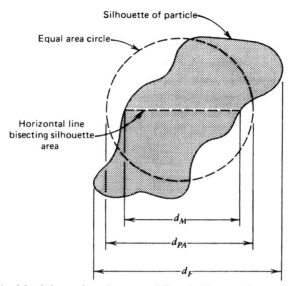

FIGURE 20.1 Martin's, projected area, and Feret's diameter for an irregular particle.

a scale along a given axis is available, such as that in a microscope equipped with a filar micrometer (Section 20.3).

The most commonly used equivalent diameter is the projected area diameter d_{PA}, also shown in Fig. 20.1. The projected area diameter is defined as the diameter of the circle that has the same projected area as the particle silhouette. This diameter has the advantage of providing a unique size for a given silhouette, regardless of its orientation. It is difficult to make an accurate measurement of d_{PA} for a single particle because doing so requires determining the area of an irregular shape. Nevertheless, the projected area diameter is widely used because it can be determined by simple visual comparison between the silhouette area of a particle and standard circles of known area. Such comparisons allow particles to be grouped rapidly into size intervals based on their projected area and the size distribution determined by the methods described in Chapter 4.

In general, $d_M \leq d_{PA} \leq d_F$, with the equality holding only for spheres. For irregular particles of most crushed materials, d_M is only slightly less than d_{PA} (< 10% difference), and they are often assumed to be equal. Feret's diameter is generally about 20% larger than d_{PA} for such materials.

In optical microscopy, the projected area diameter is measured using a set of circles of standard area printed on an eyepiece graticule or reticule. The graticule, a glass disk about 2 cm in diameter with the pattern printed on it, is inserted into the eyepiece of a microscope so that its pattern is superimposed on the magnified image of the particles. A common type is the Porton graticule shown in Fig. 20.2. Each circle on the Porton graticule has twice the area of the next smaller circle; that

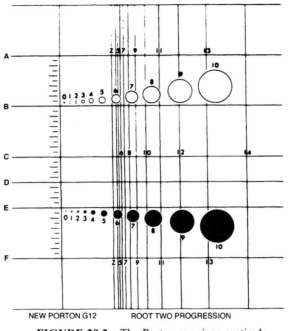

NEW PORTON G12   ROOT TWO PROGRESSION

FIGURE 20.2   The Porton eyepiece graticule.

is, the circle diameters are in a square-root-of-two geometric progression, and every
other diameter differs by a factor of two. The diameter of the nth circle d_n is

$$d_n = d_0(2)^{n/2} \tag{20.1}$$

where d_0 is the base diameter, or the diameter of the zeroth circle. The spacing of
the numbered lines from the centerline (labeled Z) follows the same progression
as the numbered circles. On the version, shown in the figure, the inside diameter
of the open circles follows the same progression. For each microscope, the apparent size of the circles must be calibrated with a stage micrometer, a miniature ruler
1 mm long mounted on a glass slide. As shown in Fig. 20.3, the 1-mm distance is
divided into 100 equal segments (numbered by tens), each 10 μm wide. The stage
micrometer is placed in the microscope at the location of the sample and is observed
with the pattern of the graticule superimposed on it. The vertical distance in Fig.
20.2 between horizontal lines A and F is 200 times the diameter of the zeroth circle.
This distance is measured with the stage micrometer, and the sizes of the circles
calculated by Eq. 20.1.

EXAMPLE

The apparent distance between lines A and F on a Porton graticule is measured with
a stage micrometer in a microscope having a 43× objective and a 10× eyepiece
and is found to be 98 μm. The diameter of the zeroth circle is 98/200, or

0.49 μm, and the number-9 circle is $0.49 \times 2^{9/2}$, or 11.1 μm. This result could have been obtained by measuring the number-9 circle directly with the stage micrometer (a less accurate procedure) or by measuring the distance between the center line (Z) and the number-14 line, 63 μm. For the latter measurement, the apparent diameter of the number-9 circle is $63 \times 2^{-5/2}$, or 11.1 μm. The diameters of the other circles can be calculated quickly using the fact that the diameter of every other circle (or line) changes by a factor of two. Thus, circles 7 and 5 are 5.5 and 2.8 μm in diameter, respectively, and circles 8 and 6 are 7.8 and 3.9 μm, respectively.

Sizing is usually done by grouping particles into ranges defined by successive circle sizes and plotting the results on probability graph paper as a cumulative percent for each group versus the upper size limit of that group. The same sizing procedure is applicable to electron microscopy: One uses a photograph of the magnified particles and a transparent overlay of an enlarged Porton graticule to compare the various areas. In practice, particles are grouped into size ranges easily and rapidly because of the factor-of-two difference in area between successive circles. The geometric progression of the circles also provides an even spacing of data points along the size axis of a log-probability graph.

Most aerosol particle size distributions are skewed, so that relatively few large particles are present. To count a meaningful number of large particles involves counting unnecessarily large numbers of small particles. This problem can be

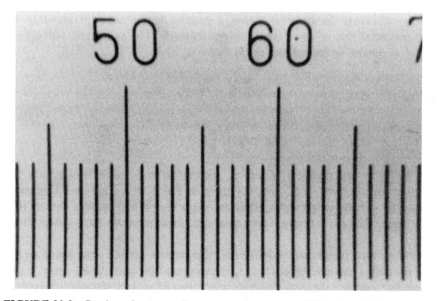

FIGURE 20.3 Portion of a stage micrometer as it appears at 400×. Actual distance between the smallest lines is 10 μm.

avoided by *stratified counting*. As illustrated in Table 20.1, sizing is done by using fields—defined areas such as the box that surrounds the Porton graticule lines or some fraction of it. All particles are sized in the first field. In subsequent fields, only those particles in size ranges having fewer than 10 particles (or another predetermined number) in them are counted. When an adequate number of particles is obtained in each size range of interest, the data are normalized in terms of counts per field, and the normalized results are used to calculate percentages for cumulative or differential distributions. Without stratified counting, the example (Table 20.1) would require sizing about 1250 particles to achieve the same accuracy in the largest size range.

The utility of microscopic size measurements of irregular particles often depends on the ability to convert the measured sizes to other equivalent diameters that describe the behavior rather than the geometry of the particles. The most useful of these diameters is the equivalent volume diameter, which can be combined with the dynamic shape factor (Section 3.5) to describe the aerodynamic properties of a particle. The volume shape factor α_v relates the volume of a particle, v_p, to one of the silhouette diameters described before and is defined, for the projected area diameter, by

$$\alpha_v = \frac{v_p}{d_{PA}^3} \qquad (20.2)$$

Similar definitions give α_v in terms of d_F or d_M, but for simplicity, we consider only α_v, defined in terms of d_{PA}. Table 20.2 gives volume shape factors for geometric shapes and mineral dusts. The volume shape factor based on projected area diameter has a maximum value of $\pi/6 = 0.52$, for a sphere. For regular geometric shapes, α_v can be calculated; for irregular shapes, it must be determined experimentally by a combination of two or more measurement methods. The equivalent volume diameter d_e is related to the volume shape factor α_v by

$$v_p = \frac{\pi}{6} d_e^3 = d_{PA}^3 \alpha_v$$

$$d_e = d_{PA} \left(\frac{6\alpha_v}{\pi} \right)^{1/3} \qquad (20.3)$$

Except for fibers and platelets, d_e and d_{PA} do not differ by more than a factor of two. Values of the ratio d_a/d_{PA} are also given in Table 20.2. This useful quantity combines the volume shape factor, dynamic shape factor, and particle density into a single factor. Except for very heavy materials, most minerals have values close to unity.

A similar shape factor can be defined for the particle surface area, based on the ratio of the actual particle surface area to the square of the particle diameter. It can be shown by Cauchy's theorem that the surface area of an irregular convex particle is equal to four times its projected area, averaged over all possible particle orienta-

TABLE 20.1 Example of Stratified Counting Procedure

Field Number	Number of Counts in Indicated Size Range						
	>1.4 μm	1.4–2.0 μm	2.0–2.8 μm	2.8–4.0 μm	4.0–5.7 μm	5.7–8.0 μm	Total
1	44	81	75	42	6	2	250
2					9	3	12
3						3	3
4						1	1
5						3	3
Total	44	81	75	42	15	12	269
Number per field	44	81	75	42	15	12	251.9
Percent in size range	17.5	32.1	29.8	16.7	3.0	1.0	100.0
Cumulative percent	17.5	49.6	79.3	96.0	99.0	100.0	100.0

TABLE 20.2   Volume Shape Factor and the Ratio d_a/d_{PA} for Geometric Shapes and Mineral Dusts[a]

Particle	Volume Shape Factor, α_v	d_a/d_{PA}
Geometric shapes		
Sphere	0.52	1.0[b]
Cube	0.38	0.89[b]
Prolate spheroid (axial ratio = 5)	0.33	0.76[b]
Cylinder (axial ratio = 5)	0.30	
Mineral dusts		
Anthracite coal	0.16	0.70
Bituminous coal	0.24	0.88
China clay		0.92
Glass		1.08–1.34
Limestone	0.16	
Quartz	0.21	0.97–1.16
Sand	0.26	1.00
Talc	0.16	0.75

[a]Based on projected area diameter determined for particles lying in the most stable position. Data for mineral dusts from Davies (1979) and α_v for geometric shapes from Mercer (1973).
[b]Standard density.

tions. For convex particles having no preferred orientation, the second moment average (rms) of the projected area diameter is equal to the diameter of average surface, $d_{\bar{s}}$.

Liquid aerosol particles are spherical when airborne, but when collected on a glass slide, they spread out to form a lens-shaped (plano-convex) droplet on the surface of the slide. The diameter of the lens is greater than the diameter of the original airborne droplet. The extent of the spreading depends on the properties of the specific liquid and surface involved. Coating a glass slide with an oleophobic surfactant reduces the spreading, as well as the influence of different liquids on spreading. The spreading factor B relates the observed size of a droplet (lens), d_o, on a glass slide to the size of the original airborne droplet, d_p.

$$B = \frac{d_p}{d_o} \tag{20.4}$$

Table 20.3 gives values of B for different liquids and surfaces.

20.2   FRACTAL DIMENSION OF PARTICLES

The shape of an aerosol particles having a complex structure, such as an agglomerated metal fume or soot particle, can be characterized in terms of a fractal dimen-

TABLE 20.3   Spreading Factor for Liquids on Glass Slides[a]

Liquid	Surface Coating[b]	Particle Size Range (μm)	Spreading Factor
DOP[c]	None	2–50	0.30
DOP	L-1429	2–5	0.58
DOP	L-1428	2–50	0.69
DOP	FC-721	10–20	0.74
DOP	L-1083	50–150	0.69
Oleic acid	none	2–50	0.42
Oleic acid	L-1428	2–50	0.70
Oleic acid	FC-721	3–29	0.75
Sulfuric acid	L-1429	0.02–0.07	0.71

[a]Data from Liu, Pui, and Wang (1982); Cheng, Chen, and Yeh (1986); Olan-Figueroa, McFarland, and Ortiz (1982); and John and Wall (1983).
[b]L-1083, L-1428, L-1429, and FC-721 are fluorocarbon oelophobic surfactants manufactured by 3M, Inc., St. Paul, MN.
[c]Di (2-ethylhexyl) phthalate.

sion. Fractals are structures that have geometrically similar shapes at different levels of magnification. This characteristic is called *self-similarity* or *fractal morphology*. The *fractal dimension* relates a property such as the perimeter or surface area of an object to the scale of the measurement.

For example, the length of a coastline can be represented by a continuous series of equal straight-line segments. The size of each segment represents the step size, or "ruler" length, from one contact point to the next. The total length depends on the step size. The smaller the step size, the finer is the detail one can measure and the longer the measured coastline. The relationship between step size and measured length is shown in Figure 20.4 for the coastline of Great Britain. This log–log graph, called a Richardson plot, of the measured coastline length L versus the step size λ approximates a straight line with a negative slope m. The equation for this straight line is

$$L = k\lambda^m = k\lambda^{1-D_f} \qquad (20.5)$$

where k is a constant and D_f is the fractal dimension.

$$D_f = 1 - m \qquad (20.6)$$

In this situation, the fractal dimension, 1.24, is a measure of the complexity of the coastline. Its value lies between 1 and 2. If the coastline were a perfectly smooth straight line, there would be no change in length with the measurement scale, so m would be zero and D_f would be 1.0. As the self-similar complexity of the coastline increases, the fractal dimension increases, approaching its maximum value of 2.0, for this situation. Figure 20.5 shows line segments of increasing complexity and increasing fractal dimension. Since they are all line segments, their topological

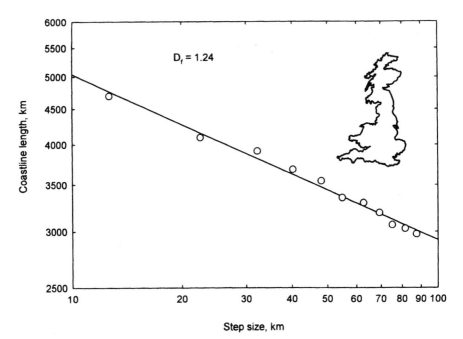

FIGURE 20.4 Measured length of the coastline of Great Britain versus step size (13–88 km). Data from Kaye (1989).

dimension is 1; that is, they would be straight lines if stretched out. The fractal dimension can be thought of as a fractional dimension, in this case intermediate between a straight line with a dimension of 1 and a surface with a dimension of 2.

EXAMPLE

For the data shown in Fig. 20.4, *calculate* the fractal dimension of the coastline of Great Britain and the length of the coastline for a step size of 30 km.

$$m = \frac{\log(y_1/y_2)}{\log(x_1/x_2)} = \frac{\log(5050/2920)}{\log(10/100)} = -0.238$$

$$D_f = 1 - (-0.238) = 1.238$$

$$k = \frac{L}{\lambda^m} = \frac{5050}{10^{-0.238}} = 8740 \text{ km}$$

For $\lambda = 30$ km,

$$L = k\lambda^{1-D_f} = 8740 \times 30^{-0.238} = 3890 \text{ km}$$

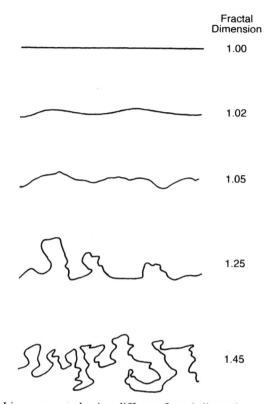

Fractal
Dimension

1.00

1.02

1.05

1.25

1.45

FIGURE 20.5   Line segments having different fractal dimensions. All have topological dimensions of 1.0. After Kaye (1989). Reprinted with permission of Wiley-VCH Verlagsgesellschaft, Weinheim, Germany.

The concept and approach given here for calculating the fractal dimension and the length of the coastline of Great Britain can be applied to the perimeter of any particle. With suitable scaling, the data shown in Fig. 20.4 could apply equally well to an aerosol particle with the same shape. The fractal dimensions would be the same. Graphs such as Fig. 20.4 are usually put in dimensionless (normalized) form by dividing L and λ by the maximum Feret's diameter of the profile of the object, be it Great Britain or an aerosol particle. The appearance of the graph remains the same. Figure 20.6 (Kaye, 1989) shows simulated particle profiles, their fractal dimensions, and normalized graphs of the particle perimeter, P/d_F, versus the normalized step size, λ/d_F.

Fume particles have been measured microscopically and found to have fractal dimensions of 1.3–1.5 (Baron and Willeke, 1993). Soot particles analyzed by Sorenson and Feke (1996) displayed fractal morphology over the range 0.05–400 μm. This range corresponds to agglomerates of 10 to 10^8 primary particles. These same authors found the relationship between the number of primary particles in an

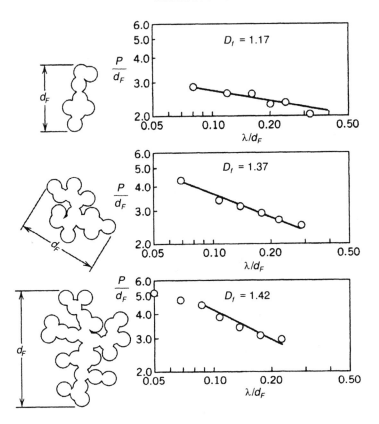

FIGURE 20.6 Profiles of simulated particle agglomerates and their Richardson plots. After Kaye (1989). Reprinted with permission of Wiley-VCH Verlagsgessellschaft, Weinheim, Germany.

agglomerate, N_{pp}, the diameter of the agglomerate, $2R_g$, and the size of the primary particles, d_{pp}, to be

$$N_{pp} = 1.7\left(\frac{2R_g}{d_{pp}}\right)^{1.8} \tag{20.7}$$

where 1.8 is the fractal dimension. R_g is the radius of gyration of the agglomerate, a size measure equal to the rms average distance the particle's mass is from its center of mass. A liquid sphere composed of N_{pp} primary particles of diameter d_{pp} would have a fractal dimension of 3.0 (for $\lambda < 0.5d_p$) and a constant of 3.95 in Eq. 20.7.

Other quantities in the field of aerosol science can also be described as fractals. For example, tracing the path of a particle undergoing Brownian motion by straight-line segments between the positions of the particle at fixed time intervals yields a fractal. As the interval gets shorter, more details are resolved and the path gets longer, but retains its self-similarity. Other examples of quantities that can be described as fractals are the borders of clouds and smoke plumes, turbulence, and the geometry of the airways of the lungs.

20.3 OPTICAL MICROSCOPY

Because the optical microscope is a common instrument used in many areas of science, only those characteristics of microscopes that are important to particle sizing are covered here. Optical microscopes are most suitable for counting and sizing solid particles having diameters of 0.3–20 μm. While any professional-grade compound microscope can be used to size particles, those with multiple objectives mounted in a nosepiece and a mechanical stage are the most convenient. Figure 20.7 shows the major components of such a microscope. Other desirable features, not shown, are a built-in substage illuminator and binocular eyepieces.

The light path for an optical microscope is shown in Fig. 20.8. Light from an external source, usually a microscope lamp, is reflected up into the condenser by a mirror. The condenser concentrates the light onto the sample from below. The particles to be measured are mounted on a glass slide held in the mechanical stage. A magnified image of the particles is formed by the objective lens positioned just above the slide. The image is located at the eyepiece, which further magnifies it to create a virtual image at a distance comfortable for viewing. In this situation, the

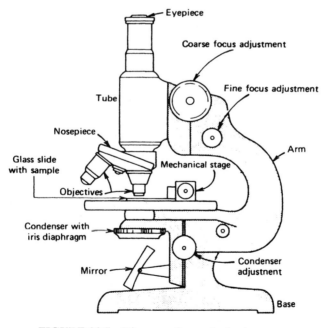

FIGURE 20.7 Diagram of an optical microscope.

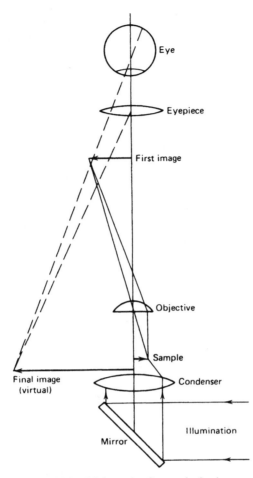

FIGURE 20.8   Light path of an optical microscope.

eyepiece functions like a simple magnifying glass. The total magnification—the ratio of the image size to the particle size—is approximately equal to the product of the magnifications of the objective and eyepiece lenses. Objective lens magnifications range from 10 to 100; eyepiece magnifications range from 5 to 25, with 10, 15, and 20 being the most useful. Eyepieces slide into the microscope tube and are easily interchanged. Focusing is accomplished by moving the tube-and-lens system up or down relative to the sample with a rack-and-pinion mechanism.

At high magnifications it is necessary to use the fine-focusing adjustment, a 500:1 gear system that permits the tube to be positioned to a precision of 1 μm. The mechanical stage holds a glass slide for viewing and provides micrometer adjustment of the position of the sample along two axes. This feature is particularly useful for traversing a nonuniform sample and also enables particle images to be easily positioned for comparison with the standard circles or line spacings of a graticule. The nosepiece, shown in Fig. 20.7, permits 2–5 objectives to be mounted on the microscope and rotated into place as needed. The objectives are positioned so that, when the microscope is exactly in focus for one objective, the others will also be in focus.

Particles can be captured directly on a glass slide by sedimentation, electrostatic precipitation, or thermal precipitation and viewed directly in a microscope. Powders or bulk dust samples can be spread out on a slide by gentle smearing with a toothpick. The procedure for viewing particles captured on a filter is to cut out a small pie-shaped piece of the filter and place it on a clean glass microscope slide. A drop of immersion oil is allowed to saturate the filter to make it transparent. The filter is covered with a glass coverslip and examined in the microscope. The immersion oil must have the same refractive index as the filter material. The most common type of filter for optical microscope examination of aerosol particles is a cellulose ester membrane of 0.45-μm pore size. These filters have a refractive index of 1.510. Unfortunately, particles having exactly this refractive index are also made transparent and cannot be observed by this method.

Although a total magnification of about 2500× can be obtained with standard objectives and eyepieces, the size of the smallest observable particle is governed not by magnification, but by the limit of resolution, which is controlled by the wavelength of light and the characteristics of the objective lens. The *limit of resolution*, or, more simply, the *resolution*, is the smallest scale of detail that can be observed in the magnified image. Although greater magnification produces a larger image, it does not increase the observable detail beyond the limit of resolution. Resolution is usually defined as the minimum separation distance required for two dots to be observed as two separate dots rather than as one elongated dot. Resolution is characterized by a property of the objective lens called the *numerical aperture*, NA, defined as

$$NA = m \sin(\theta/2) \tag{20.8}$$

where m is the refractive index of the viewing medium between the objective and the sample and θ is the angular aperture—the angle shown in Fig. 20.9. The limit of resolution, L_R, obtained with an objective having a numerical aperture NA is

$$L_R = \frac{1.22\lambda}{2(NA)} \tag{20.9}$$

where λ is the wavelength of the light used, 0.55 μm for white light.

Characteristics of the three objectives shown in Fig. 20.9 are given in Table 20.4. It is apparent that the maximum value of θ is 180°, so the numerical aperture can exceed unity only when $m > 1$. This is accomplished by filling the space between the objective and the sample with an oil of high refractive index. Such an approach allows numerical apertures as great as 1.4 and yields resolutions of about 0.2 μm for white light. In general, the condenser should be adjusted so that the angle of illumination equals the angular aperture of the objective. For the best resolution, immersion oil should be used to couple the top of the condenser lens to the bottom of the glass slide. A rule of thumb for the selection of eyepieces and objectives is

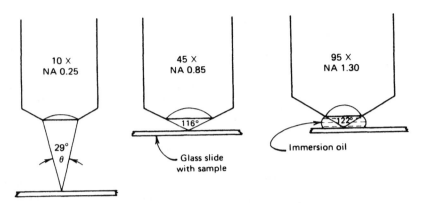

FIGURE 20.9   Comparison of microscope objectives, showing angular aperture.

that the maximum useful magnification is equal to 1000 times the numerical aperture of the objective. The depth of field or vertical resolution is the thickness (depth) of the sample region that is simultaneously in focus. This thickness is approximately equal to $L_R/\tan(\theta/2)$. At high magnifications, the small depth of field is a limitation of the optical microscope that makes it difficult to completely observe large particles and to see all particles deposited at different depths in a filter. It is desirable to use a small-pore membrane filter, such as the Millipore HA, for microscopic counting and sizing, because all particles are deposited on or near the surface and can be observed simultaneously.

The filar micrometer and the image-splitting eyepiece are microscope attachments that can be used for particle sizing. Both replace the regular eyepiece. The filar micrometer has a movable crosshair in the field of view that is positioned by a calibrated micrometer screw. Particle size is measured by positioning the crosshair on the left edge of the particle, reading the micrometer setting, moving the crosshair to the right edge of the particle, and reading the setting again. The particle's Feret's diameter is calculated from the difference in the readings. The micrometer is calibrated with a stage micrometer. An image-splitting eyepiece has a prism assembly that creates two images of a particle. When the micrometer screw is in the zero position, both images are superimposed. As the micrometer screw is turned, the images shift, and the micrometer is read when the left edge of one image just touches the right edge of the other. Particle size is calculated from this setting and a calibration obtained with a stage micrometer. Both devices are slower and more tedious, but more accurate, than the circle comparison method. Special types of microscopes used for particle sizing and analysis, such as dark field, phase contrast (used for asbestos counting; see Section 20.5), and polarizing microscopes, are described in Silverman et al. (1971) and Cadle (1975).

20.4   ELECTRON MICROSCOPY

To examine particles smaller than the limit of resolution of optical microscopes, one must use an electron microscope. Of the two common types, scanning electron

TABLE 20.4 Characteristics of Typical Microscope Objectives Used for Particle Sizing

Objective and Nominal Magnification	Focal Length (mm)	Numerical Aperture	θ (deg)	Depth of Field (μm)	Limit of Resolution, λ = 0.55 μm (μm)
10× apochromatic	16	0.25	29	5.19	1.34
45× apochromatic	4	0.85	116	0.25	0.39
95× apochromatic (oil immersion)	1.8	1.30	122	0.14	0.26

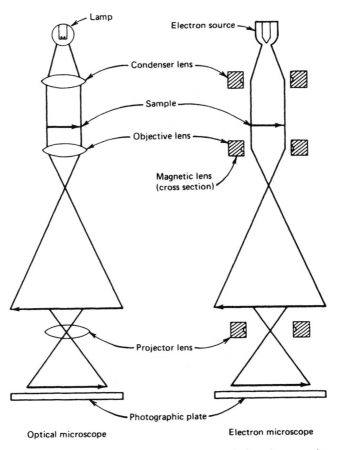

FIGURE 20.10   Comparison of optical and transmission electron microscopes.

microscopes (SEMs) and transmission electron microscopes (TEMs), the latter is most analogous to the light microscope and is covered first. As shown in Fig. 20.10, transmission electron microscopes use electrons in the same way that optical microscopes use light (photons). For purposes of comparison, the optical microscope is shown in the projection mode rather than the direct viewing mode described in Section 20.3. In the TEM, the electrons are generated by thermionic emission from a heated tungsten filament and are focused by magnetic lenses. These serve the same function as optical lenses and are given the same names. Magnetic lenses, however, have the advantage that their focal length can be adjusted by controlling the current through them. The interior of an electron microscope, including the sample area, must be under a high vacuum (less than 0.01 Pa [10^{-7} atm]), to prevent the scattering of the electron beam by air molecules. This is accomplished by a combination of mechanical and oil diffusion vacuum pumps. Particles in the electron beam absorb and scatter electrons to produce a two-dimensional silhouette image similar to that produced in optical microscopy. The magnified image is viewed by projecting the electron image onto a fluorescent screen or a photographic plate.

The principles of resolution and depth of field discussed in Section 20.3 hold for electron microscopes, but the wavelength of the radiation used in electron microscopes is much less than the wavelength of light. The wavelength of an electron is given by

$$\lambda_e = \frac{0.0012}{\sqrt{V}} \qquad (20.10)$$

where λ_e is in μm and the microscope acceleration voltage V is in volts. For a typical acceleration voltage of 70 kV, the wavelength of the electron beam is 4.5×10^{-6} μm, or 100,000 times smaller than the wavelength of light. The resolution, however, is *not* 100,000 times smaller than that of an optical microscope, because of the small value of angular aperture used in electron microscopes to minimize distortion. The limit of resolution for a typical TEM is less than 0.001 μm. The small value of aperture angle used has the advantage of providing a large depth of field, more than 100 times the resolution of the instrument. Table 20.5 compares the magnification, resolution, and depth of field of different types of microscopes.

Transmission electron microscopes permit the smallest aerosol particles to be measured, but these instruments can be used only for solid particles that do not evaporate or degrade under the combined effects of high vacuum and heating by the electron beam. Particles must be deposited on specially prepared 3-mm-diameter grids mounted in a grid holder and inserted into the microscope. The grids are 200 mesh electrodeposited screen with a thin film of carbon or Parlodion (1–4% nitrocellulose in amyl acetate) covering the screen openings. The film is sufficiently thin, compared to the particles, that it causes only slight attenuation of the electron beam. The particles, on the other hand, extensively scatter and absorb the electron beam to form a high-contrast silhouette image. Particles can be deposited directly onto the film surface of a grid by electrostatic precipitation using the miniature type of precipitator described in Section 15.8. In some cases, replication or transfer procedures can be used to deposit particles or their replicas from another surface onto a grid.

Sizing of particles is usually done from photographs made by enlarging a negative exposed to the electron image in the microscope. Special fine-grained, blue-sensitive photographic film is used for this purpose. The sizing procedures are the same as those described in Section 20.1 for optical microscopy, except that a transparent overlay of the Porton graticule or other standard circles is used. The exact magnification of the photographs and the calibration of the overlay are determined by photographing a carbon film replica of a diffraction grating under the same conditions of magnification as that for the particles. The spacing between the grating lines, usually less than 1 μm, is accurately known from independent measurements.

Although the scanning electron microscope (SEM) also uses electron beams, magnetic lenses, and a high vacuum, it operates on a different principle than the TEM does and creates images having a three-dimensional appearance. As shown

TABLE 20.5   Magnification, Resolution, Depth of Field, and Sample Environment for Different Types of Microscopes

Type of Instrument	Magnification	Limit of Resolution (μm)	Ratio of Depth of Field to Resolution	Sample Environment
Eye	1	~200		Air
Magnifying glass	2–10	25–100		Air
Compound optical microscope 43× objective (dry)	200–850	0.3	3	Air
Compound optical microscope 95× objective (oil immersion)	500–1300	0.2	2	Oil
Scanning electron microscope (SEM)	20–100,000	0.01	>1000	Vacuum
Transmission electron microscope (TEM)	1000–1,000,000	<0.001	>100	Vacuum

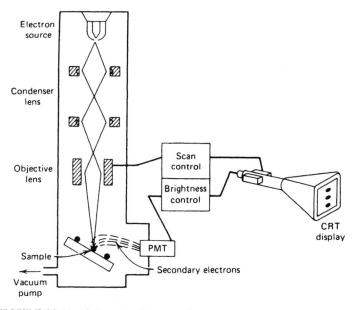

FIGURE 20.11   Schematic diagram of a scanning electron microscope.

in Fig. 20.11, the electron beam is focused to a spot about 0.01 μm in diameter and passed back and forth across the sample in a regular pattern. The electron beam causes secondary electrons to be emitted from the surface of the sample. These low-voltage electrons are attracted to a detector maintained at a positive voltage with respect to the sample. They can follow a curved path to reach the detector (a scintillator plus a photomultiplier tube, or PMT), where they are converted to an electrical signal. The number of secondary electrons emitted and reaching the detector depends on the topography of the surface; high points emit many electrons and valleys emit few. A cathode ray tube (CRT) is scanned in synchronization with the scanning pattern of the electron beam, and the brightness of the picture at any point is made proportional to the number of secondary electrons reaching the detector at that instant. This creates a reconstructed image of the sample surface, with high points appearing bright and valleys appearing dark. By tilting the sample with respect to the electron beam, realistic shadows are produced that give a remarkable three-dimensional effect to the picture. SEM photographs such as those shown in Sections 1.1 and 9.1 are photographed directly from the CRT screen. The sample can be tilted and rotated by external controls while being viewed to reveal all sides of a particle.

The resolution of a typical SEM is about 0.01 μm, not as good as a TEM, but much better than an optical microscope and adequate for most particle studies. Like the TEM, the SEM is limited to solid, nonvolatile particles. Unlike the TEM, the electron beam, in a SEM does not have to penetrate the sample support, so SEM samples are usually mounted on the end of an aluminum cylinder 10–30 mm in diameter. Precautions must be taken to prevent the buildup of electric charge on the sample surface, which would deflect the electron beam and cause distortion of the image. This is usually done by applying a very thin (< 0.01 μm) coating of gold or carbon to the surface of the particle with a sputter-coating apparatus. Particles can be deposited onto any surface that can be attached to the sample cylinder and made conductive. The smooth surface of a capillary pore membrane filter provides a suitable surface for viewing particles, and particles larger than its pore size can be sampled directly onto the filter surface. Figure 1.4(b) is an example of an SEM sample taken on a capillary pore membrane filter.

Scanning electron microscopes have a depth of field that is more than 300 times greater than that of optical microscopes and a wide range of magnification, 20–100,000×. To measure and evaluate particles, a magnification range of 100–10,000× is most commonly used. The great depth of field and the accurate rendition of surface features make the SEM well suited to particle sizing and morphology studies.

An electron microscope can be used as a microprobe to determine the elemental composition of a particle. The electron beam is focused on a single particle, and the X rays that are emitted are detected by a solid-state X-ray detector. The energy of the X rays is related to the Z number of the elements in the particle and provides an elemental analysis of the particle for $Z > 12$. Microprobe analyses of single particles for elemental or molecular composition are reviewed by Fletcher and Small (1993).

20.5   ASBESTOS COUNTING

The mineral asbestos possesses some unique properties that cause lung injury and necessitate microscopic methods for evaluating its hazard. The characteristic fibrous shape of asbestos particles [see Fig. 1.4(b)] enables them to get past respiratory defense mechanisms and cause asbestosis (scarring of lung tissue), mesothelioma (cancer of the lining of the lung), and lung cancer. Microscopy is used to identify those asbestos fibers whose size and shape enable them to cause these diseases.

Asbestos originates as fiber bundles that break longitudinally into finer and finer fibers during milling. Because of the high strength of the material, fibers of long aspect ratio (length to diameter) are formed. The aerodynamic properties of fibers are such that fibers as long as 50 μm or as big as 3 μm in diameter can reach the alveolar region and are considered respirable. The reason they can do this is that the fibers align themselves with the streamlines and "snake" their way through the narrow airways to the alveolar region. Once lodged in an alveolus, large asbestos fibers cannot be removed by normal clearance mechanisms. They are insoluble in lung fluids, too long to be engulfed by macrophages, and only slightly able to migrate to lymph nodes. It is thought that when macrophages try to engulf long fibers, they leak enzymes that cause fibrosis, the scarring and thickening of alveolar surfaces associated with asbestosis. Asbestos fibers less than about 5 μm in length are cleared by the normal clearance mechanisms. Each of the asbestos-related diseases is associated with fibers in a specific size range. Fibers 0.15 to 3 μm in diameter and 2–100 μm long are associated with asbestosis, fibers less than 0.1 μm in diameter and 5–100 μm long with mesothelioma, and fibers 0.15–3 μm in diameter and 10 to 100 μm long with lung cancer (Lippmann, 1991). Because the risk of disease depends critically on the shape of the particles, microscopy must be used to evaluate the concentration of those asbestos particles whose size and shape make them hazardous.

The U.S. occupational exposure standard for asbestos is 100,000 fibers/m^3 [0.1 f/cm^3] greater than 5 μm in length and having an aspect ratio greater than 3:1, as determined by the membrane filter method using phase contrast microscopy (PCM). The membrane filter method, or "NIOSH method," is an optical microscope technique used for the quantitative determination of the concentration of asbestos fibers in air. The method consists of collecting breathing zone samples at 0.5–16 L/min on 25-mm membrane filters (mixed cellulose ester membrane filters with a pore size of 0.45–1.2 μm). Open-face filter holders with a 50-mm protective cowl (inlet extension) are used. The sample volume is adjusted to produce a fiber surface density in the optimal range of 100–1300 f/mm^2. A sample volume of 1.0 m^3 [1000 L] is suitable for concentrations of 40,000–500,000 f/m^3 [0.04–0.5 f/cm^3]. The filter is "melted" to a solid, transparent film by hot acetone vapor and is sealed in triacetin. Fibers meeting the size and shape criteria are counted at 400–450× magnification (45× objective, NA = 0.65–0.75, and 10× eyepiece) by phase contrast optical microscopy. (Phase contrast microscopy is used to enhance the contrast of the fibers because the refractive index is near that of the mounting medium.)

Any particle having a length greater than 5 μm and an aspect ratio of 3:1 (length:diameter) or greater is considered a fiber, whether or not it is asbestos. Standard practice in the United States calls for using a Walton–Beckett graticule, shown in Figure 20.12. The Walton–Beckett graticule defines a 100-μm-diameter counting field with standard fiber shapes and sizes around the perimeter for comparison. Fibers meeting the size and shape criteria that are completely within the field are counted as one fiber, while those that are partially within the field are counted as one-half fiber. Sample fibers are also shown in Figure 20.12. Because of the limit of resolution of optical microscopes, this method cannot detect fibers less than 0.25 μm in diameter. Consequently, a significant number of hazardous fibers are too small to be detected by PCM, so the count that is obtained should be considered an index of asbestos concentration, and not the true concentration. Because of the difficulty of sizing particles near the limit of resolution of the microscope, novice counters can underestimate fiber concentrations by a factor of two or more.

More advanced methods of optical and electron microscopy may be used to determine whether a fiber that meets the size and shape requirements is in fact an asbestos fiber. This can be done most definitively by transmission electron microscopy (TEM), which allows observation and measurement of fiber size and shape for even the smallest asbestos fibers. TEM also provides the opportunity for positive identification through elemental analysis by energy-dispersive X-ray analysis

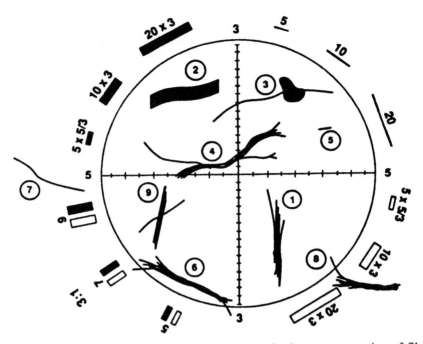

FIGURE 20.12 Walton–Beckett graticule for optical microscope counting of fibers. Particles 1–4 are counted as one fiber, 5–7 are not counted, and 8 and 9 are counted as one-half and two fibers, respectively. From Baron, 1993. Reprinted by permission of John Wiley & Sons, Inc.

(EDXA) and crystal structure by diffraction pattern analysis. TEM samples can be taken on mixed cellulose ester membrane filters, but require special preparation. (See Baron, 1993.)

A portable direct-reading fibrous aerosol monitor, the FAM-1® (MIE, Inc., Bedford, MA), provides rapid measurement of airborne fiber concentrations. Aerosol is sampled at 0.12 m³/hr [2 L/min] and passes through a 10-mm-diameter sensing tube with laser illumination along its axis. The flow rate through the sensing volume is 10 cm³/min. The fibers in the aerosol stream line up with an oscillating electric field. The light scattered by fibers pulses at the oscillation frequency while that scattered by compact particles does not. The sharpness of the pulses is related to the fiber length. Fibers meeting the criteria of length and aspect ratio have a characteristic signature that is sensed electronically. Fibers are counted for a preset sampling period and the average number concentration displayed. The correlation between results obtained using the FAM-1 and those obtained using manual counting methods is reasonable.

20.6 AUTOMATIC SIZING METHODS

The visual comparison methods for particle sizing described in Section 20.2 are satisfactory for most particle size analysis, but are slow, tedious, and subject to operator error or bias. Automated image analysis attempts to overcome these problems by using computer analysis of particle images. First, the image of many particles is scanned, creating a digital image in the form of an array of millions of picture elements (pixels). The image may be from an optical or electron microscope or a photograph. The computer interprets particle size and shape from the arrangement of contiguous dark pixels on a light background or vice versa. The computer can then calculate the projected area diameter, maximum chord, Feret's diameter, Martin's diameter, perimeter, and various shape factors. Mean and standard deviations and moments of these quantities for all particles in the field can be calculated automatically. Measurements can be made on all or selected particles in a field. Image analysis can be combined with elemental analysis in an electron microscope. Here, the microscope is operated under computer control to obtain information about particle size, shape, and elemental composition of many particles automatically. Analysis of hundreds or thousands of particles yields more elaborate information, such as size distributions for selected elemental composition or average concentrations of a specific element versus particle size.

The accuracy, speed, versatility, and ability to make measurements without the possibility of operator bias are the primary virtues of these automatic instruments. They work best with a high-contrast image in which the smallest particles have an image size of several millimeters. When particle images have low contrast with fuzzy, gray edges, the size of the particles measured is controlled by instrument settings that define what gray level constitutes the edge of the particle. In this situation, the entire sizing process, microscopy, photography, and image analysis must be calibrated to yield accurate results that are free from bias. The procedure can be further complicated by overlapping particles and a "noisy" background. Care must be taken that the background texture, such as the holes of a capillary pore membrane (Nuclepore) filter, is not included in the analysis. Such elements are instinctively screened out when we view an image, but the computer does not have our instincts.

PROBLEMS

20.1 Consider an aerosol composed entirely of spherical doublets (particles that consist of two spheres joined at a point). Assume that all spheres are identical with a diameter d. If the particles are viewed perpendicular to the line between their centers:

(a) What is the projected area diameter for these particles?

(b) What is their volume shape factor?

(c) What is the maximum and minimum ratio of Feret's diameter to the projected area diameter?

ANSWER: $1.41d$, 0.37, 1.41, 0.71.

20.2 A straight fiber has an aspect ratio of 5:1. If the fiber is viewed perpendicular to its axis, what is the maximum ratio of Feret's diameter to projected area diameter?

ANSWER: 2.0.

20.3 For particles having a square projected cross section, what is the ratio of Feret's diameter (averaged over all orientations) to the projected area diameter?

ANSWER: 1.13.

20.4 A cylindrical fiber has a diameter of 1 μm and a length three times its diameter. What are the projected area diameter and volume shape factor of the fiber when it is viewed perpendicular to its axis?

ANSWER: 2.0 μm, 0.32.

20.5 A particle has a smooth, circular cross section of diameter d_p. The fractal dimension of the particle's perimeter is to be estimated by fitting a four-sided and a six-sided regular inscribed polygon to the circle. Based only on this information, what fractal dimension is estimated for this step size range. [*Hint*: The length b of the side of a regular inscribed polygon with n sides is $b = d_p \times \sin(180/n)$.]

ANSWER: 1.17.

20.6 What is the theoretical limit of resolution of a microscope having a 90× objective (angular aperture of 120°) and a 15× eyepiece used with blue light ($\lambda = 0.4$ μm) and with immersion oil having a refractive index of 1.51?

ANSWER: 0.19 μm.

20.7 What is the ratio of the maximum to the minimum size of circle provided by the Porton graticule as shown in Fig. 20.2?

ANSWER: 32.

20.8 Between what two circle numbers on the Porton graticule will the image of a 1-μm particle fall when viewed with a 43× objective and a 15× eyepiece? Circle number 10 has an image diameter of 9.9 mm.

ANSWER: 2 and 3.

REFERENCES

Baron, P. A., "Measurement of Asbestos and Other Fibers," in Willeke, K., and Baron, P. A. (Eds.), *Aerosol Measurement*, Van Nostrand Reinhold, New York, 1993.

Baron, P. A., and Willeke, K., "Aerosol Fundamentals" in Willeke, K., and Baron, P. A. (Eds.), *Aerosol Measurement*, Van Nostrand Reinhold, New York, 1993.

Cadle, R. D., *The Measurement of Airborne Particles*, Wiley, New York, 1975.

Cheng, Y. S., Chen, B. T., and Yeh, H. C., "Size Measurement of Liquid Aerosols," *J. Aerosol Sci.*, **17**, 803–809 (1986).

Davies, C. N., "Particle–Fluid Interaction," *J. Aerosol Sci.*, **10**, 477–513 (1979).

Fletcher, R. A., and Small, J. A., "Analysis of Individual Collected Particles," in Willeke, K., and Baron, P. A. (Eds.), *Aerosol Measurement*, Van Nostrand Reinhold, New York, 1993.

John, W., and Wall, S. M., "Aerosol Testing Techniques for Size-Selective Samplers," *J. Aerosol Sci.*, **14**, 713–727 (1983).

Kaye, B. H., *A Random Walk through Fractal Dimensions*, Wiley-VCH Verlagsgesellschaft, Weinheim, Germany, 1989.

Lippmann, M., "Industrial and Environmental Hygiene: Professional Growth in a Changing Environment," *Am. Ind. Hyg. Assoc. J.*, **52**, 341–348 (1991).

Liu, B. Y. H., Pui, D. Y. H., and Wang X-Q., "Drop Size Measurement of Liquid Aerosols," *Atm. Environ.*, **16**, 563–567 (1982).

Mandelbrot, B. B., *Fractals, Form, Chance and Dimension*, Freeman, San Francisco, 1977, 1983.

McCrone, W. C., and Delly, J. G., *The Particle Atlas*, 2d ed., Ann Arbor Science, Ann Arbor, MI, 1973.

Mercer, T. T., *Aerosol Technology in Hazard Evaluation*, Academic Press, New York, 1973.

NIOSH, "Asbestos and Other Fibers 7400," in *NIOSH Manual of Analytical Methods*, U.S. Gov. Printing Office, Washington, DC, 1989.

Olan-Figueroa, E., McFarland, A. R., and Ortiz, C. A., "Flattening coefficients for DOP and oleic acid droplets deposited on treated glass slides," *Am. Ind. Hyg. Assoc. J.*, **43**, 395 (1982).

Silverman, L., Billings, C. E., and First, M. W., *Particle Size Analysis in Industrial Hygiene*, Academic Press, New York, 1971.

Sorenson, C. M., and Feke, G. D., "The Morphology of Macroscopic Soot," *Aerosol Sci. Tech*, **25**, 328–337 (1996).

21 Production of Test Aerosols

An important element of aerosol technology is the production of test aerosols for calibrating instruments, conducting aerosol research, and developing and testing air-cleaning and air-sampling equipment. Monodisperse aerosols are used to calibrate particle-size measuring instruments and to determine the effect of particle size on sampling devices. Polydisperse aerosols may be used for calibration or to simulate the actual use of equipment under controlled laboratory conditions. A monodisperse aerosol is usually defined as an aerosol that has a geometric standard deviation of less than 1.2. In aerosol research, monodisperse test aerosols of known size, shape, and density are highly desirable because most aerosol properties depend strongly on particle size, which is best controlled by using monodisperse aerosols. Tests made with a series of monodisperse aerosols, each having a different particle size, permit the evaluation of the effect of particle size on aerosol properties or the performance of an instrument. For example, the cutoff curves for impactors presented in Sections 5.5 and 5.6 were determined with monodisperse test aerosols. Test aerosols also are used for various types of inhalation studies, such as studies of animal exposure to toxic substances, respiratory deposition in humans or animals, and the administration of therapeutic aerosols.

The characteristics of an *ideal* aerosol generator are a constant and reproducible output of monodisperse, stable, uncharged, solid, spherical aerosol particles whose size and concentration can be easily controlled. In this chapter, we outline the characteristics of common types of aerosol generators, focusing on devices and methods that provide continuous and steady output suitable for calibrating instruments.

21.1 ATOMIZATION OF LIQUIDS

Atomization is the general name for the process of disintegrating a liquid into airborne droplets. Atomizers are classified by the type of energy used to break up the liquid. *Pressure atomizers* are the simplest; they convert liquid pressure into kinetic energy that causes the liquid to break into droplets. There are two types of pressure atomizers, jet atomizers and swirl atomizers. The *jet atomizer* ejects the liquid in a stream at sufficiently high velocity that it disintegrates into relatively large droplets some distance from the nozzle. *Swirl atomizers* cause the liquid to spin as it exits the nozzle, forming a hollow cone that facilitates the breakup of the liquid. Another type is the *pneumatic atomizer*, which uses the energy from compressed air to break up a liquid stream. This type of atomizer produces the smallest droplets and can be used to aerosolize viscous liquids. The nebulizers discussed shortly are examples of pneumatic atomizers. A third type is the *rotary atomizer*, which uses a rotating element to impart kinetic energy to the liquid, thereby causing its disintegration. These atomizers are often used for spray drying. Other types of atomizers include *ultrasonic* and *electrostatic atomizers*. The general topic of atomization is reviewed by LeFebvre (1989) and Bayvel and Orzechowski (1993).

The most common way to generate droplet aerosols is by *compressed-air nebulization*. A nebulizer is a type of atomizer that produces an aerosol of small particle size by removing larger spray droplets by impaction within the device. As a class, nebulizers produce aerosols at concentrations of 5–50 g/m³ with MMDs of 1–10 µm and GSDs of 1.5–2.5. The operating principle of most compressed-air nebulizers is similar to that of the DeVilbiss model 40 shown in Fig. 21.1. Compressed air at a supply pressure of 35–340 kPa (5–50 psig) exits from a small tube or orifice at high velocity. The low pressure created in the exit region by the Bernoulli effect causes liquid to be drawn from a reservoir into the airstream through a second tube. The liquid exits the tube as a thin filament that is stretched out as it is accelerated in the airstream until it breaks into droplets. The spray stream is directed onto an impaction surface, where large droplets are deposited and drain back to the liquid reservoir. The operating principles are the same for most nebulizers, but the geometry of the components differs with each device. Median particle size decreases with increasing air velocity and decreasing viscosity and surface tension. Properties of some common nebulizers are given in Table 21.1. Most nebulizers produce a maximum particle number concentration of 10^{12}–10^{13}/m³ [10^6–10^7/cm³].

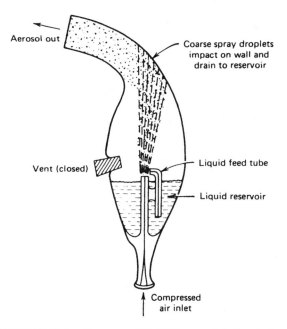

FIGURE 21.1   Diagram of DeVilbiss model 40 nebulizer.

TABLE 21.1 Characteristics of Some Compressed-Air and Ultrasonic Nebulizers[a]

Nebulizer	Operating Pressure (kPa [psig])	Flow Rate (L/min)	Output Concentration[b] (g/m³)	Droplet Size Distribution[b] MMD (μm)	GSD	Reservoir volume (mL)
Compressed Air						
Collison[c]	100[15]	2.0	8.8	2.5–3		20–100
	140[20]	12		2.1–3[d]	2.7–3.4[d]	
	170[25]	2.7	7.7	1.9–2		
DeVilbiss[e] D-40	100[15]	12	16	4.2	1.8	10
	140[20]	16	14	3.2	1.8	
	210[30]	21	12	2.8	1.9	
DeVilbiss D-45	100[15]	9.4	23	4.0		
	210[30]	14	23	3.4		
TDA-4B (Laskin)[f]	140[20]	84	27	0.7	2.1	500–5000
Lovelace[g]	140[20]	1.5	40	5.8	1.8	4
	340[50]	2.3	27	2.6	2.3	
Retec[g] X-70/N	140[20]	5.0	46	5.7	1.8	10
	340[50]	9.7	47	3.2	2.2	
Ultrasonic						
DeVilbiss 880	(1.35 MHZ)	(2)[h] 41	54	5.7	1.5	5
		(4)[h] 41	150	6.9	1.6	

[a]Data primarily from Cheng and Chen (1995).
[b]For water or liquid with a density of 1000 kg/m³ [1.0 g/cm³].
[c]BGI, Inc., Waltham, MA.
[d]For DOP.
[e]DeVilbiss Co., Somerset, PA.
[f]ATI, Inc., Owings Mills. MD. Author's data for DOP.
[g]In-Tox Products, Albuquerque, NM.
[h]Power Setting.

The *Laskin nebulizer* differs from the other compressed air nebulizers listed in the table in that the operating part of the nebulizer is submerged in the liquid. The aerosol stream impacts on the liquid surface as it exits the jet. For some liquids, such as di(2-ethylhexyl) phthalate (DOP), mineral oil, and corn oil, the agitation of the air jet creates a dense foam of microscopic bubbles in the liquid. Additional aerosol particles are created as these bubbles burst at the surface of the liquid. The final size distribution is influenced by both mechanisms of particle formation. Properties of some low-vapor-pressure liquids used for aerosol generation are given in Appendix A13.

When compressed-air nebulizers are used with liquids of low volatility, such as DOP (vapor pressure $< 10^{-5}$ Pa [$< 10^{-7}$ mm Hg] at 293 K [20°C]), they produce droplet aerosols whose size is stable for hundreds of seconds. (See Table 13.3.) When a volatile solvent containing dissolved solid material is used, the solvent evaporates rapidly after droplet formation, to form smaller solid particles. This is a simple way of producing solid-particle aerosols by the nebulization of liquids. The size of the final aerosol particle, d_p, depends on the volume fraction of solid material, F_v, and the droplet diameter, d_d, according to the equation

$$d_p = d_d(F_v)^{1/3} \qquad (21.1)$$

Nebulizing a 0.1% solution of sodium chloride in water will produce a solid particle when the droplets are dried completely which has a volume that is 0.001 of the original droplet volume and, therefore, an equivalent volume diameter one-tenth of the droplet diameter. Since all particles are reduced in size by the same factor, the GSD of the particle size distribution does not change after drying. The nebulization of a low-vapor-pressure liquid dissolved in a volatile solvent can also be used to generate small liquid-droplet aerosols. The final particle size can be calculated by Eq. 21.1. In either case, care must be taken to ensure that drying is complete. Diffusion drying columns in which the aerosol passes through a screen tube surrounded by desiccant granules are useful for this purpose. A problem with generating solid particles by nebulization from liquid solutions is the evaporative loss of solvent from the reservoir during the operation of the nebulizer. Loss of solvent increases the concentration of solute in the reservoir and causes the particle size to increase with time. Presaturation of the supply air and cooling of the nebulizer reduce these evaporative losses.

Impactors can be used with compressed-air nebulizers to narrow and control the output particle size distribution (Pilacinski et al., 1990). The aerosol exiting the nebulizer passes through a conventional single-stage impactor (Section 5.5) to remove the large particles and then through a virtual impactor (Section 5.7) to remove the small particles. By selecting the appropriate cutoff sizes, aerosols with narrow size distributions (GSD < 1.4) can be produced with CMDs of 0.5–5 μm. The technique is somewhat analogous to size classification by the differential mobility analyzer (DMA) to produce monodisperse submicrometer aerosols as described in Section 15.9.

Ultrasonic nebulizers produce aerosol droplets in the size range of 1–10 μm (MMD) without the use of a compressed air jet. Ultrasonic waves generated by a piezoelectric crystal are focused near the surface of a small volume of liquid. The ultrasonic energy creates intense agitation of the liquid and forms a conical fountain above the surface of the liquid. The action of the compression waves in the liquid causes capillary waves to form on the surface of the fountain, and these waves shatter to create a dense aerosol. The particle size produced is related to the frequency of excitation by the empirical expression (Mercer, 1973)

$$\text{CMD} = \left(\frac{\gamma}{\rho_L f^2}\right)^{1/3} \qquad (21.2)$$

where γ and ρ_L are, respectively, the surface tension and density of the liquid and f is the excitation frequency in Hz. This proportionality has been found to hold for excitation frequencies of 0.012–3 MHz. A gentle airflow is directed over the surface of the liquid to carry away the aerosol. Compared with compressed-air nebulizers, ultrasonic nebulizers produce a larger volume of aerosol at a higher concentration with a narrow distribution (typically, GSD = 1.4–1.6; see Table 21.1). Most of the ultrasonic energy is dissipated as heat in the liquid, and the resulting increase in temperature may increase evaporative losses.

Compressed-air nebulizers form particles by breaking a liquid filament with an air jet. Two other methods of aerosol generation also rely on breaking a liquid filament, but use different methods to form and break it. Both instruments produce monodisperse droplets and are used primarily in aerosol research. The *vibrating-orifice aerosol generator* shown in Fig. 21.2 forms a thin filament of liquid by pumping the liquid through a small orifice (5–50 μm in diameter) with a syringe pump. The orifice is oscillated along its axis, the axis of the filament, by means of a piezoelectric crystal such that each oscillation breaks the filament and forms a particle. With a constant liquid flow rate and a constant oscillation frequency, the resulting droplets are monodisperse, and their size can be calculated directly from these two quantities:

$$d_d = \left(\frac{6Q_L}{\pi f}\right)^{1/3} \qquad (21.3)$$

where Q_L is the liquid flow rate in m³/s or cm³/s and f is the frequency in Hz. Initial droplet diameters range from 15 to 100 μm. When first formed, the droplets are very close together and can coagulate rapidly, so it is necessary to direct a jet of air at the exit point to dilute and disperse the particle stream. The output concentration is 3×10^7–4×10^8/m³ [30–400/cm³] after dilution and 8×10^8–1.2×10^{10}/m³ [800–12,000/cm³] before dilution. Any low-viscosity liquid free from impurities that might clog the orifice can be used. Volatile solvents containing dissolved solutes can be used in this type of generator to reduce the particle size and produce solid-particle aerosols by the procedures described in this section, using Eq. 21.1.

A second type of monodisperse atomizer is the *spinning-disk aerosol generator*. A horizontal disk a few centimeters in diameter is rotated at speeds up to 70,000

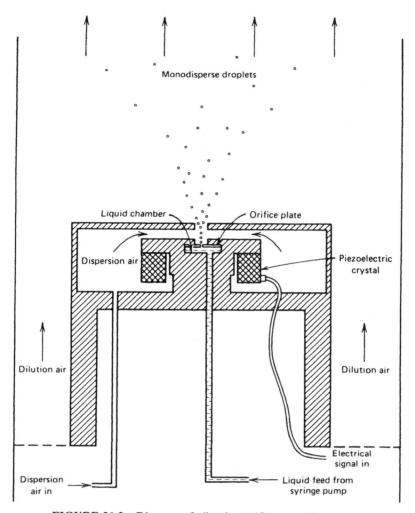

FIGURE 21.2 Diagram of vibrating-orifice aerosol generator.

rpm. Liquid is introduced at the center of the disk at a constant rate. Centrifugal force causes the liquid to travel to the edge of the disk as a thin film. The film forms filaments at the edge that break into particles. The droplet size produced depends on the disk radius R and the rotational speed ω in revolutions/s (Mercer, 1973).

$$d_d \cong \frac{0.47}{\omega} \left(\frac{\gamma}{\rho_L R} \right)^{1/2} \tag{21.4}$$

Careful control of all variables yields aerosols with initial droplet sizes of 20–100 μm and GSDs of about 1.1. To achieve this degree of monodispersity, it is necessary to remove the smaller satellite droplets that form as the filaments break. The satellite droplets are about one-fourth the size of the primary droplets. Aerodynamic separation of the two sizes uses the fact that their stopping distances differs by a factor of about 16. A flow of air at the disk periphery sweeps away the smaller particles, while the inertia of the primary particles carries them into the mainstream airflow.

The ubiquitous *aerosol spray can* is another type of atomizer. These cans contain a mixture of dissolved or suspended material and a liquid propellant. The propellant has a high vapor pressure at room temperature that pressurizes the can and forces the mixture out of the exit nozzle when it is activated. The exit stream is fractured by the rapid evaporation of the propellant as the mixture exits the nozzle and encounters atmospheric pressure. The resulting aerosol has a wide range of sizes with droplets as large as 100 μm. A significant percentage of the final aerosol can be in the respirable size range.

A special version of the aerosol spray can is the *metered-dose inhaler* (MDI), used for the administration of therapeutic drugs. This device operates on the same principle as the aerosol spray can, but uses a dosing valve to dispense a fixed amount of propellant-drug mixture with each activation. The drug is usually in the form of a fine powder suspended in the liquid propellant. The aerosol produced is inhaled directly. The desired final particle size depends on the targeted region of the respiratory system. Particles larger than about 7 μm impact on the back of the throat and are thus not available for uptake in the lung airways or alveolar regions. Some devices use a spacer tube to provide additional distance and time for the droplets to slow and evaporate completely.

Another type of atomizer is the *electrostatic atomizer*, or *electrospray* device. A liquid is fed slowly ($\approx$3 mm^3/s [10 mL/hr] through a hollow needle facing downward. A high voltage ($\approx$10 kV) is established between the needle and a coaxial ring a few centimeters below the needle to create a strong electrostatic field near the tip of the needle. In cone-jet mode, the liquid exiting the needle forms a cone that emits large numbers of droplets from its tip. The droplets are charged and initially repel each other, until they are neutralized. The particle size can be monodisperse and is controlled by the dielectric constant of the liquid, liquid flow rate, field strength, and current. By using volatile solvents, particles from nanometers to micrometers can be produced. This subject is reviewed by Grace and Marijnissen (1994).

All the nebulization methods described here are likely to produce moderately charged particles for which neutralization (see Section 15.7) may be required, especially if the particle size is reduced by evaporation of solvent.

21.2 ATOMIZATION OF MONODISPERSE PARTICLES IN LIQUID SUSPENSIONS

A simple way to generate monodisperse solid-particle aerosols, and a common way to check the size calibration of instruments, is by nebulizing a liquid suspension containing monodisperse solid particles of known size. After nebulization, the liquid is removed by drying to produce a solid-particle aerosol. Liquid suspensions of monodisperse polystyrene (PSL) and polystyrene-divinylbenzene (PS-DVB) latex spheres are used for this purpose. The spheres have relative standard deviations of a few percent, are perfect spheres, and have homogeneous properties. For both materials, the density of the spheres is close to standard density—1050 kg/m^3 [1.05 g/cm^3]—so their aerodynamic diameters are only slightly (2.5%) larger than their physical diameters.

These spheres are available in a wide range of sizes. For example, Duke Scientific, of Palo Alto, California, sells spheres from 20 nm to 1 mm, with more than 100 sizes between 0.02 μm and 100 μm. About half are provided with National Institute of Standards and Technology (NIST) traceable particle-size data. The majority of these have uncertainty in their mean size of less than 1% and relative standard deviations for their size distributions of less than 2%. They are sold in 15-mL vials containing 0.5, 1, 2, or 10% solids in aqueous suspension. The water contains 0.02 to 0.2% stabilizer (surfactant and dispersant), to prevent coagulation in the liquid. Other particles are available that incorporate fluorescent dyes into the polymer matrix to aid in detection and analysis. NIST sells eight sizes of polystyrene spheres between 0.1 and 30 μm as standard reference materials (SRM). These have been extensively characterized in terms of their physical and chemical properties.

Three problems may arise in the generation of monodisperse aerosols using uniform spheres. First, the size reported by the manufacturer may differ from the size reported by other investigators who use more careful measurement techniques. One source of error is the tendency of the spheres to change size due to volatilization or decomposition in the beam of an electron microscope. Accurate measurements of these spheres are now made using low-intensity electron beams to prevent such effects. Also, particles may swell slightly in certain solvents.

The second problem arises when more than one sphere is present in a droplet when it is formed. When the droplet dries, the resulting particle is a cluster or chain of spheres. If two spheres are present, the resulting particle is called a doublet or twin, if three a triplet, and so forth. The probability $P(n)$ of n spheres occurring in a droplet of diameter d_d is given by

$$P(n) = \frac{(\bar{x})^n}{n!} \exp(-\bar{x}) \tag{21.5}$$

where $\bar{x}$, the average number of spheres per droplet, is given by

$$F_v v_d = \bar{x} v_p$$

$$\bar{x} = F_v \left(\frac{d_d}{d_p}\right)^3 \tag{21.6}$$

in which v_d and v_p are the respective volumes of the droplets and spheres and F_v is the volume fraction of spheres of diameter d_p. As shown in Table 21.2, as F_v is decreased (the suspension is diluted) below 10^{-3}, the ratio of singlets to doublets in-

TABLE 21.2   Fraction of Droplets Containing 0, 1, 2, or 3 Spheres when Droplet Diameter is Ten Times the Sphere Diameter

Volume Fraction of Spheres, F_v	Fraction of Droplets with Indicated Number of Spheres				
	$n = 0$	$n = 1$	$n = 2$	$n = 3$	$n > 3$
0.1	0	0	0	0	1.000
0.01	5×10^{-5}	5×10^{-4}	0.002	0.007	0.990
10^{-3}	0.368	0.368	0.184	0.061	0.019
10^{-4}	0.905	0.090	0.005	1.5×10^{-4}	4.3×10^{-6}
10^{-5}	0.990	0.010	5.0×10^{-5}	1.6×10^{-7}	0
10^{-6}	0.999	0.001	5.0×10^{-7}	0	0
10^{-7}	0.999	1.0×10^{-4}	0	0	0

creases, but the ratio of "empties"—that is, droplets with no particles—to singlets also increases. This means that for a fixed droplet generation rate, the concentration of singlets decreases as the proportion of doublets decreases. The real situation is, more complicated than that suggested by Table 21.2 because of the distribution of droplet sizes produced by a nebulizer. Figure 21.3 shows the ratios of empties to total droplets, singlets to total droplets, and singlets to droplets containing particles obtained by numerical integration of Eq. 21.5 for lognormal droplet

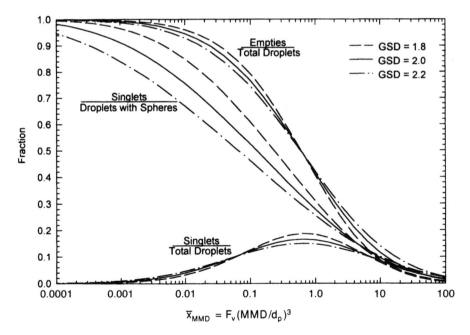

FIGURE 21.3   Fraction of empties to total droplets, singlets to droplets containing spheres, and singlets to total droplets for lognormal droplet size distributions having GSD = 1.8, 2.0 and 2.2 versus $\bar{x}_{MMD}$, the average number of spheres in an MMD-sized droplet.

distributions with the indicated geometric standard deviation. The abscissa, $\bar{x}_{MMD}$ is the average number of spheres in an MMD-sized droplet:

$$\bar{x}_{MMD} = F_v \left(\frac{MMD}{d_p} \right)^3 \tag{21.7}$$

For lognormal droplet size distributions (GSD = 1.8–2.2), the fraction of droplets containing only one sphere (singlets) peaks at $\bar{x}_{MMD} = 0.7$. This value of $\bar{x}_{MMD}$ produces the largest concentration of singlets after drying, but 64–71% of the particles are multiplets (particles consisting of two or more spheres). A value of $\bar{x}_{MMD}$ of about 0.001 yields 84–97% singlets in the dried aerosol, but has a low number concentration because more than 98% of the droplets contain no spheres.

The third problem is created by the empty droplets. Some PSL sizes have a significant percentage of stabilizer in the liquid. The stabilizer is present in the empty droplets, and when they dry, a residue particle of stabilizer is formed whose diameter is given by Eq. 21.1. The stabilizer also forms a coating that changes the size of small particles slightly.

A measurement of the ratio R' of singlets to doublets for PSL spheres of known size and volume fraction provides a simple way of estimating the diameter of average volume of the nebulizer droplets. Substituting $n = 1$ and $n = 2$ into Eq. 21.5 gives

$$R' = \frac{P(1)}{P(2)} = \frac{2}{\bar{x}} = \frac{2v_p}{F_v \bar{v}_d} \tag{21.8}$$

and

$$d_{\bar{v}} = \left(\frac{6\bar{v}_d}{\pi} \right)^{1/3} = d_p \left(\frac{2}{F_v R'} \right)^{1/3} \tag{21.9}$$

EXAMPLE

A DeVilbiss model 40 nebulizer at 140 kPa [20 psig] is used to aerosolize monodisperse 0.5-μm PSL spheres. What suspension concentration should be used to give a final aerosol that is 90% singlets? What median size would surfactant residue particles have if the nebulized liquid contained a surfactant volume fraction of 5×10^{-6}?

From Table 21.1 and Fig. 21.3, MMD = 3.2 μm, GSD = 1.8, and $\bar{x}_{MMD} = 0.005$. Substituting these values in Eq. 21.7 gives,

$$F_v = \bar{x}_{MMD} \times (d_p/MMD)^3 = 0.005 \times (0.5/3.2)^3 = 1.9 \times 10^{-5}$$

residue $d_p = d_d(F_v)^{1/3} = 3.2(5 \times 10^{-6})^{1/3} = 0.055$ μm

21.3   DISPERSION OF POWDERS

The most widely used method for generating solid-particle test aerosols is the pneumatic redispersion of dry powders. This method, often called *dry dispersion*, can accommodate a wide range of powdered materials and dust feed rates. The output concentration of dry-dispersion generators ranges from milligrams per cubic meter to greater than 100 g/m^3. This method of aerosol generation finds application in filter testing, air-cleaning research, and animal inhalation toxicology studies.

The basic requirements for all dry-dispersion dust generators are (1) a means of continuously metering a powder into the generator at a constant rate and (2) a means of dispersing the powder to form an aerosol. The simplest dust-metering systems feed loose powder into an airstream by means of gravity, usually assisted by vibrators. These systems can give variable aerosol concentratration because of uneven delivery of powder and variations in the bulk density of the powder. A more stable metering method is the use of a cylinder of compressed powder that is eroded or scraped away at a constant rate. This method, used in the Wright and TSI 3410 dust feeders described shortly, is more reliable for long-term studies because the bulk density of the powder can be controlled during packing.

The *dispersibility* of a powder, the ease of separating powder particles, depends on the powder material, particle size and size range, particle shape, and moisture content of the powder. Incomplete dispersion results in an aerosol particle size distribution that is larger than the size distribution of the original powder particles. Hydrophobic materials such as talc are easier to disperse than hydrophilic materials such as quartz or limestone. To disperse a powder fully, it is necessary to supply sufficient energy to a small volume of the bulk powder to separate the particles by overcoming the attractive forces between them. Dispersibility increases rapidly with particle size, and there is a size below which particles cannot be satisfactorily dispersed with a given generator.

A related quantity is *dustiness*, a quantitative measurement of the fraction of a powder that becomes airborne during a simulated process. A powder's dustiness is used as a relative indication of the ease with which the powder becomes airborne and to estimate the strength of a source in various industrial operations. The two most common methods for testing dustiness are the single-drop and rotating cylinder tests. In the single-drop test, a known mass of powder (typically, 100–300 g) is dropped from a fixed height above the floor of a chamber. The airborne dust is measured by capturing it on a filter during a set time interval following the drop. In the rotating cylinder method, a known mass of powder (typically, 20–100 g) is placed in a horizontal cylinder with internal ribs. Air is drawn through the cylinder while it is slowly rotated for a fixed duration. The airborne dust is captured on a filter. The ratio of captured dust mass to the initial dust mass is the dustiness index, usually expressed in mg/kg. Typical results range from one to a few hundred mg/kg. Results are apparatus specific, and careful control of all operational variables is required to provide consistent results. Plinke et al. (1995) give predictive equations in terms of cohesive forces within the powder and impaction forces experienced in the drop test.

The most common method for dispersing dust is to feed the dust into a high-velocity airstream. The shear forces in the turbulent airstream disperse the powder and break up agglomerates. Another approach is to introduce powder into a fluidized bed of 100- to 200-μm beads to break up the agglomerates, suspend the powder particles in the airstream, and, by elutriation, prevent large agglomerates from escaping the fluidized bed. The compressed air used to disperse dust must be clean, dry, and free from compressor oil mist. Extremely dry air (relative humidity < 5%) can cause strong electrostatic forces between particles, reducing the dispersibility of the dust.

A problem common to all dry-dispersion aerosol generators is the buildup of charge on the particles during dispersion. This buildup, a result of contact charging of the dust particles as they touch and separate from surfaces in the generator, causes variations in output concentration due to variable loss to the walls of the system. The charging effect is greatest for powders with low moisture content. Although charging can be reduced by careful selection of generator materials, it is more common to discharge the aerosol to the Boltzmann equilibrium charge distribution as it leaves the generator. This is usually done by passing the aerosol through a chamber containing a radioactive source that produces a high concentration of bipolar ions in the chamber, as described in Section 15.7.

It is often necessary to include a size-selective classifier in the output stream of a dust generator, to ensure that agglomerates or large particles are removed from the exiting aerosol. Elutriation columns, cyclones, impactors, and centrifugal separators have been used for this purpose. Cyclones operated to match the ACGIH respirable cutoff curve (Section 11.5) have been used to eliminate nonrespirable particles from aerosols in animal exposure studies.

Table 21.3 summarizes the characteristics of six commercially available dust generators. The subject is reviewed by Cheng and Chen (1995) and Chen (1993). The most widely used dry-dispersion aerosol generator is the *Wright dust feed*, shown in Fig. 21.4. By packing the dust in a cylinder under controlled pressure and scraping away the surface of the dust cake at a constant rate, this device has eliminated the stability problems characteristic of loose-dust feeding systems. The dust cylinder containing the packed dust rotates about a stationary scraper head. Filtered, dry, compressed air passes up through an annular space in the spindle to the scraper head and through a groove at the outer edge of the scraper. The air flows radially along the scraper blade, entraining dust that is being cut away from the compacted mass of dust. Air and dust exit at high velocity through an axial hole to a partial impactor that breaks up any remaining agglomerated particles. Differential gearing causes a rotation between the cylinder and the threaded spindle that advances the dust cylinder onto the scraper blade at a constant rate, ensuring a constant feed rate of dust. The rate of this advance is controlled by the pitch of the spindle thread and the speed of the drive motor, which can be varied a thousandfold. Dust cylinders come in two diameters, 12.7 and 38 mm, to give an additional ninefold range in dust feed rate. The device works best for hard, dry materials with 90% of the particles less than 10 μm, such as silica, uranium-dioxide, or other mineral dusts. The dust generator has been found unsatisfactory for soft, sticky dusts such as coal and

TABLE 21.3   Characteristics of Some Dry-Dispersion Aerosol Generators

Model and Type of Generator	Recommended Particle Size Range (μm)	Airflow Rate (m³/hr [L/min])	Volumetric Feed Rate (mm³/min)	Output Mass Concentration[a] (g/m³)
Wright II[b] (packed cylinder with scraper)	0.2–10	0.6–2.4 [10–40]	0.24–210	0.012–12
NBS II[b] (gear feed)	1–100	3–5 [50–85]	1200–15,000	15–100
Palas RGB 1000[c] (packed cylinder with brush)	0.1–100	0.3–5 [5–85]	0.6–7200	0.1–200
TSI 3400[d] (fluidized bed)	0.5–40	0.3–1.2 [5–20]	1.2–36	0.13–4
TSI 3410[d] (packed cylinder with brush)	1–100	0.6–3 [10–50]	20–100	0.07–1.5
TSI 3433[d] (turntable with venturi)	1–50	0.7–1.3 [12–21]	0.05–1.5	0.0003–0.04

[a] For $\rho_b = 1000$ kg/m³ [1.0 g/m³].
[b] BGI, Inc., Waltham, MA 02154.
[c] Palas GmbH, Karlsruhe, Germany.
[d] TSI, Inc., St Paul, MN, 55164.

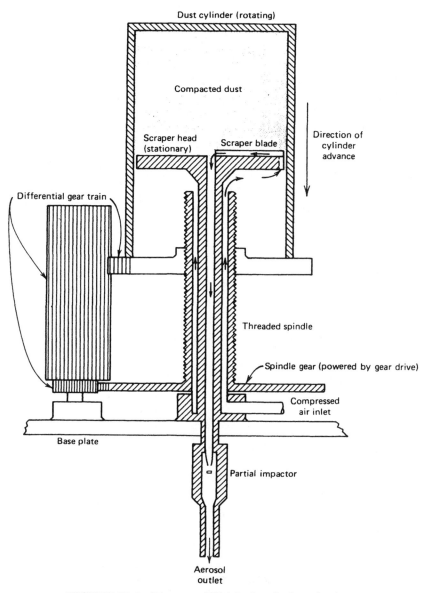

FIGURE 21.4 Diagram of Wright dust feed mechanism.

carbon black and for fine iron-oxide and zinc-oxide powders. A related device capable of handling a wider range of particles is the *rotating brush disperser* used in the TSI 3410. The powder is packed into a cylinder and pushed by a piston onto a rotating wire brush at a controlled rate. The brush transfers the particles into a stream of compressed air.

The *NBS dust generator* is most suitable when a high dust concentration and high flow rate are required. The dust flows from a hopper, assisted by a vibrator, into the spaces between the teeth of a metering gear. A contoured spreader plate removes the excess dust as the gear rotates at constant speed (adjustable over a 40:1 range). The dust is sucked up from between the teeth of the metering gear by a compressed-air ejector, which is positioned diagonally to extract dust from parts of three teeth to reduce pulsations. The *turntable dust feeder* is similar, except that the dust is deposited in a groove or on the surface of a slowly rotating turntable. The dust is aspirated from the turntable into a venturi disperser as the turntable rotates.

A *fluidized-bed aerosol generator*, based on the design of Marple et al. (1978) is shown in Fig. 21.5. This type of generator has a 51-mm-diameter fluidized-bed chamber filled with 180-μm bronze beads to a depth of 15 mm. The dust is eluted directly from the fluidized bed into a krypton-85 neutralizing chamber at 9 L/min.

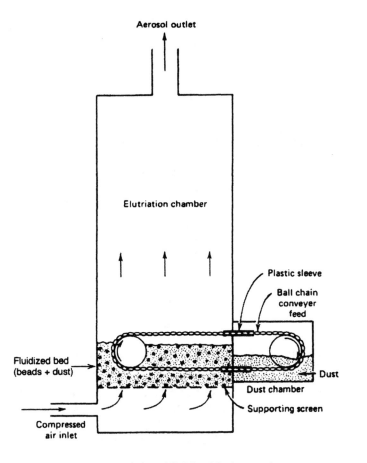

FIGURE 21.5 Diagram of fluidized-bed aerosol generator.

The low velocity at the exit of the fluidized bed prevents the output of particles with aerodynamic diameters greater than 50 μm. The dust is metered into the fluidized bed by a conveyor feed system using a continuous loop of number-3 ball chain. The space between each ball in the chain picks up a fixed volume of dust and transports it to the fluidized bed through a close-fitting tube. Dust flow into the bed, and the dust generation rate can be varied by a factor of 30 by varying the speed of the chain. Because of the long residence time in the fluidized bed, it takes several hours for the generator to achieve constant output following the initial introduction of powder. For a change in the dust-metering rate, equilibrium is reached in about 1 h. This slow response time reduces variability in the output concentration due to variations in the feed rate. A 1.2-cm-diameter respirable mass cyclone can be attached at the outlet to limit the output to respirable dust.

A special type of dry-dispersion aerosol generator is the dry-powder metered-dose inhaler (MDI). Pharmaceutical powder in premeasured capsules or packets is aerosolized and inhaled without the use of a propellant. In one version a powder-filled gelatin capsule is inserted into a rotating element in the inhalation tube of an MDI. Holes are pierced in the ends of the capsule. Vanes on the elements cause it to rotate during inhalation and disperse the powder into the turbulent inhalation airstream.

Corn and Esmen (1976) give the properties and applications of many different powders used in aerosol generation. Pollens from various plants and trees are monodisperse and can be used as test aerosols. Depending on the species, these pollens range in size from about 15 to greater than 70 μm and have densities from 0.45 to 1.05 g/cm^3. A standard material that is frequently used as a polydisperse test aerosol is AC fine air-cleaner test dust, sold by the AC Spark Plug Division of the General Motors Corporation in Flint, Michigan. This dust is a graded, naturally occurring dust that is often referred to by its place of origin as Arizona road dust, or ARD. It is composed of 58% SiO_2, 16% Al_2O_3, and 4.6% Fe_2O_3. The fine grade has 39% of its mass in particles less than 5 μm and 73% less than 20 μm.

21.4 CONDENSATION METHODS

Condensation of organic vapors is a method for generating high concentrations of submicrometer-sized aerosols for aerosol research. While condensation generators take many forms, the basic approach is to control the concentration of nuclei and vapor and allow condensation to occur under slow, controlled conditions. Under these conditions, condensation and growth are the same for each nucleus as it passes through the condensation region, and each droplet forms and grows to the same final size. Neglecting losses, such as condensation on the walls of the apparatus, the diameter of the final droplets is given by

$$d_d = \left(\frac{6C_m}{\pi \rho_L N} \right)^{1/3}$$

(21.10)

where C_m is the mass concentration of the vapor, ρ_L is the density of the liquid, and N is the number concentration of nuclei. The particle size can be controlled by controlling the concentration of either the vapor or the nuclei. Variations in the size of the nuclei do not affect the final droplet size, because nuclei are so much smaller than the droplets. Materials that are liquid or solid at room temperature and that have boiling points between 300 and 500°C are suitable for this type of generator. Appendix A.13 lists properties of some low-vapor-pressure liquids used for condensation aerosol generation. Condensation occurs through cooling by heat exchange, mixing, or adiabatic expansion. Vapor concentration can be established by blowing air over or through a heated reservoir of liquid or by nebulizing the liquid and heating the aerosol stream. Nuclei may be generated by a heated wire or a nebulizer. Condensation aerosol generators require a long period for thermal equilibration.

A simpler version of the aerosol generator just described, called the evaporation-condensation monodisperse aerosol generator and shown in Fig. 21.6, uses a nebulizer to establish both the vapor and nuclei concentrations. DOP with a small amount ($< 1\%$) of nonvolatile impurity is nebulized and heated to produce the desired vapor concentration. As each droplet evaporates, it leaves behind a residue particle, which

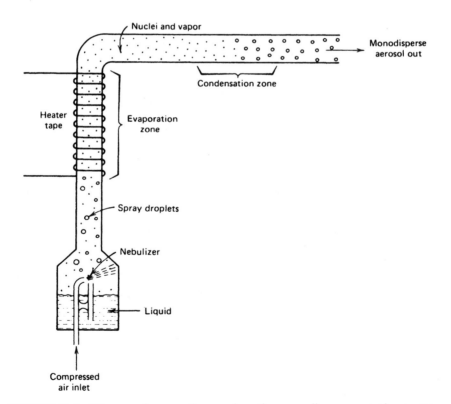

FIGURE 21.6   Diagram of evaporation–condensation monodisperse aerosol generator.

forms a nucleus for subsequent condensation. The mixture of vapor and nuclei is cooled slowly to produce a condensation aerosol. The aerosol number concentration is the same for nebulizer droplets, nuclei, and condensation-formed droplets. The latter are all grown under nearly identical conditions to produce aerosols with GSDs between 1.2 and 1.4. In the absence of a volatile solvent or loss of vapor to the walls, the diameter of average mass is the same for the spray droplets and the condensation-formed droplets, because the total number and total mass of particles remains constant. The particle size is controlled by adding to the nebulizer liquid a volatile solvent, such as alcohol, which does not condense in the condensation section. Particle sizes from 0.003 to greater than 1.0 μm can be produced with this technique. This types of generator is available from TSI, Inc., of St. Paul, Minnesota, and BGI, Inc., of Waltham, Massachusetts. Commercial polydisperse condensation generators (Air Techniques, Inc., Baltimore, MD) produce large volumes of aerosol for filter testing. These generators are simple devices in which a low-vapor-pressure liquid such as DOP (di(2-ethylhexyl) phthalate) is forced through an orifice by an inert gas such as N_2 or CO_2 to a heater block, where it is vaporized. The vapor condenses when it exits to the atmosphere, and a dense aerosol is formed

Test aerosols can also be formed by chemical reactions. Titanium tetrachloride liquid evaporates readily and forms TiO_2 and HCl by reaction with moisture in the air. The vapor reacts almost instantly to form a dense solid and liquid aerosol. This generation method is used for flow visualization studies and for smoke tubes for checking ventilation airflow.

PROBLEMS

21.1 A vibrating-orifice aerosol generator is operated at a liquid flow rate of 0.2 cm^3/min. The orifice is 15 μm in diameter and is oscillated at 100,000 Hz. The liquid that is used is prepared by dissolving 8 g of NaCl in 1 L of distilled water. What is the equivalent volume diameter of the NaCl particles obtained after drying this aerosol stream?

ANSWER: 6.1 μm.

21.2 A 0.4% suspension of 0.65-μm PSL spheres in water is nebulized and the resulting aerosol dried. A sample of the aerosol is collected on a membrane filter and examined in an optical microscope. Six fields are counted and found to contain 136 doublets and 18 triplets. Estimate the diameter of average mass of the original droplets produced by the nebulizer.

ANSWER: 3.0 μm.

21.3 A monodisperse aerosol generator of the evaporation–condensation type has an initial diameter of average mass of 2.8 μm before evaporation. If 40% of the vapor is lost to the heat exchanger walls and no nuclei are lost, what is the final diameter of average mass?

ANSWER: 2.4 μm.

21.4   A Retec X70/N nebulizer is operated at 140 kPa (20 psig) to generate a
0.42-μm PSL aerosol. If the suspension concentration is 10^{-5}, what fraction
of the droplets contains particles, and what is the ratio of singlets to multiplets in the final aerosol?

ANSWER: 8%, 3.3.

REFERENCES

Bayvel, L. P., and Orzechowski, Z., *Liquid Atomization*, Taylor Francis, Washington DC, 1993.

Chen, B. T., "Instrument Calibration," in Willeke, K., and Baron, P. A. (Eds.), *Aerosol Measurement*, Van Nostrand Reinhold, New York, 1993.

Cheng, Y.-S., and Chen, B. T., "Aerosol Sampler Calibration," in *Air Sampling Instruments*, 5th ed., ACGIH, Cincinnati, 1995.

Corn, M., and Esmen, N. A., "Aerosol Generation," in Dennis, R. (Ed.), *Handbook on Aerosols*, TID-26608, USERDA, Oak Ridge, TN, 1976.

Fuchs, N. A., and Sutugin, H. G., "Generation and Use of Monodisperse Aerosols," in Davies, C. N. (Ed.), *Aerosol Science*, Academic Press, London, 1966.

Grace, J. M., and Marijnissen, J. C. M. "A Review of Liquid Atomization by Electrical Means," *J. Aerosol Sci.*, **25** 1005–1019 (1994).

Hinds, W. C., "Dry-Dispersion Aerosol Generators," in Willeke, K. (Ed.), *Generation of Aerosols*, Ann Arbor Science, Ann Arbor, MI, 1980.

Lefebvre, A. H., *Atomization and Sprays*, Taylor Francis, New York, 1989.

Marple, V. A., Liu, B. Y. H., and Rubow, K. L., "A Dust Generator for Laboratory Use," *Am. Ind. Hyg. Assoc. J.*, **39**, 26–32 (1978).

Marple, V. A., and Rubow, K. L., "Generation of Aerosols: Basic Concepts," in Willeke, K. (Ed.), *Generation of Aerosols*, Ann Arbor Science, Ann Arbor, MI, 1980.

Mercer, T. T., *Aerosol Technology in Hazard Evaluation*, Academic Press, New York, 1973.

Pilacinski, W., Ruuskanen, J., Chen, C. C., Pan, M. I., and Willeke, K., "Size-Fractionating Aerosol Generator," *Aerosol Sci. Tech.*, **13**, 450–458 (1990).

Plinke, M. A. E., Leith, D., Boundy, M. G., and Loeffler, F., "Dust Generation from Handling Powders in Industry," *Am. Ind. Hyg. Assoc. J.*, **56**, 251–257, 1995.

Raabe, O. G., "The Generation of Aerosols of Fine Particles," in Liu, B. Y. H. (Ed.), *Fine Particles*, Academic Press, New York, 1976.

Simpkins, P. G., "Aerosols Produced by Spinning Disks: A Reappraisal," *Aerosol Sci. Tech.*, **26**, 51–54 (1997).

Appendices

Appendix A1.   Useful Constants and Conversion Factors

Fundamental units

1 micrometer (μm) = 1 micron (μ) = 10^{-6} m = 10^{-4} cm = 10^{-3} mm = 10^3 nm
   = 10^4 Å = 3.94×10^{-5} in
1 nm = 10^{-3} μm = 10^{-6} mm = 10^{-7} cm = 10^{-9} m
1 m = 39.4 in = 3.28 ft
1 km = 1000 m = 3280 ft = 0.621 mi
1 in. = 0.0254 m = 2.54 cm = 25.4 mm
1 ft = 0.305 m = 30.5 cm = 305 mm
1 newton (N) = 1 kg $\cdot$ m/s^2 = 10^5 dyn = 10^5 g $\cdot$ cm/s^2
1 lb = 0.454 kg = 454 g
K = °C + 273
°F = 1.8(°C) + 32

Volume

1 m^3 = 1000 L = 10^6 cm^3 = 10^9 mm^3 = 35.3 ft^3
1 L = 0.001 m^3 = 1000 cm^3 = 10^6 mm^3 = 0.0353 ft^3 = 61.0 in^3
1 ft^3 = 28,300 cm^3 = 28.3 L = 0.0283 m^3 = 1728 in^3
1 in.^{3} = 16.4 cm^3 = 0.0164 L = 1.64×10^{-5} m^3

Flow rate

1 m^3/s = 2120 ft^3/min = 1000 L/s
1 m^3/h = 16.7 L/min = 35.3 ft^3/h = 0.588 ft^3/min
1 ft^3/min (cfm) = 28.3 L/min (Lpm) = 1.70 m^3/h
1 ft^3/h = 0.47 L/min

Concentration

1 μg/m^3 = 1 ng/L = 1 pg/cm^3
1 mg/m^3 = 1 μg/L = 1 ng/cm^3
1 g/m^3 = 1 mg/L = 1 μg/cm^3
1 mppcf = 35.3 particles/cm^3

Velocity

1 ft/min = 0.305 m/min = 0.508 cm/s

1 ft/s = 0.305 m/s = 30.5 cm/s = 0.68 mph

Velocity of sound in air at 293 K [20°C] = 343 m/s = [34,300 cm/s]

Velocity of light in a vacuum = 3.00×10^8 m/s = [3.00×10^{10} cm/s]

Acceleration

Acceleration of gravity at sea level = 9.81 m/s^2 = [981 cm/s^2]

Pressure

1 atm = 1.01×10^6 dyn/cm^2 = 101 kPa= 14.7 lb/in^2 (psia) = 76 cm Hg
  = 760 mm Hg = 1030 cm H$_2$O = 407 in. H$_2$O

1 Pa = 1 N/m^2 = 10 dyn/cm^2 = 0.0075 mm Hg = 0.0040 in. H$_2$O

1 lb/in^2 (psi) = 6.89 kPa = 51.7 mm Hg = 27.7 in. H$_2$O

1 in. H$_2$O = 249 Pa = 2490 dyn/cm^2

Vapor pressure of water at 293 K [20°C] = 2.34 kPa = [17.5 mm Hg]

Viscosity

1 Pa · s = 1 N · s/m^2 = 1 kg/m · s = 10 Poise (P) = 10 dyn · s/cm^2 = 10 g/cm · s

Viscosity of air at 293 K [20°C] = 1.81×10^{-5} Pa · s = [1.81×10^{-4} Poise]

Energy

1 J = 1 N · m = 10^7 erg = 0.239 cal = 9.47×10^{-4} Btu

1 cal = 4.19 J = 4.19×10^7 erg = 0.00397 Btu

Ideal Gas Law $Pv = n_m RT$

R = 8.31 J/K · mol (N · m/K · mol) for P in Pa and v in m^3.

R = 82.1 atm · cm^3/K · mol for P in atm and v in cm^3

R = 8.31×10^7 dyn · cm/K · mol for P in dyn/cm^2 and v in cm^3

R = 62400 mm Hg · cm^3/K · mol for P in mm Hg and v in cm^3

Volume of 1 mol of ideal gas at 293 K [20°C] = 0.0241 m^3 = 24.1 L

Avogadro's number = N_a = 6.02×10^{23} molecules/mol

Boltzmann's constant $k = R/N_a$ = 1.38×10^{-23} J/K (N · m/K)
  = [1.38×10^{-16} erg/K (dyn · cm/K)]

Dry Air at 293 K [20°C] and 101 kPa (1 atm)

Density = 1.20 kg/m^3 = 1.20 g/L = [1.20×10^{-3} g/cm^3] = 0.074 lb/ft^3

Viscosity = 1.81×10^{-5} Pa · s = [1.81×10^{-4} Poise]

Mean free path = 0.066 μm

Molecular weight = 0.029 kg/mol = [29.0 g/mol]

Specific heat ratio = $\kappa = c_p/c_v$ = 1.40
Diffusion coefficient = 2.0×10^{-5} m²/s = [0.20 cm²/s]
Composition by volume (dry air)

N_2	78.1%
O_2	20.9%
Ar	0.93%
CO_2	0.031%
Other	0.003%

Water at 293 K [20°C] and 101 kPa (1 atm)
Viscosity = 0.00100 Pa · s [0.0100 dyn · s/cm²]
Surface tension = 0.0727 N/m = [72.7 dyn/cm]
Saturation vapor pressure = 2.34 kPa = [17.5 mm Hg]

Water vapor at 293 K [20°C] and 101 kPa (1 atm)
Diffusion coefficient = 2.4×10^{-5} m²/s [0.24 cm²/s]
Density = 0.75 kg/m³ = 0.75×10^{-3} g/cm³

Appendix A2.   Some Basic Physical Laws

The inertia of a body is that property of the body which tends to resist change in its state of rest or of motion. Mass is a quantitative measure of inertia.

$$\text{Momentum} = \text{mass} \times \text{velocity} = mV$$

Newton's second law: The rate of change of momentum of a body is proportional to the net force on the body, $\Sigma F = d(mV)/dt$. For a body with constant mass,

$$F = m\left(\frac{dV}{dt}\right) = ma$$

where force is in N (newtons) for m in kg and a in m/s², and in dyn for m in g and a in cm/s².

Work = force × distance; 1 N · m = 1 J; [1 dyn · cm = 1 erg]
Kinetic energy = $\frac{1}{2}mV^2$; 1 kg · m²/s² = 1 J; [1 g · cm²/s² = 1 erg]
Power = work per unit time; 1 W = 1 J/s = 10^7 erg/s = 0.00134 Hp

$$\text{Centripetal force} = \frac{m(V_T)^2}{R} = mR\omega^2$$

where V_T = tangential velocity, m/s [cm/s]; ω = angular velocity, rad/s; and R = radius of motion.

$V_T = \omega R$

1 radian (rad) = $(\frac{1}{2\pi})$ of a circle = $\frac{360}{2\pi}$ = 57.3°

Buoyant force = weight of displaced fluid

Flow in a Duct

Flow rate = velocity × cross-sectional area = $Q = VA$

Residence time = duct volume/flow rate

Properties of a Sphere

Circumference = $2\pi r = \pi d = 3.14d$

Projected area = $\pi r^2 = \frac{\pi d^2}{4} = 0.785d^2$

Surface area = $4\pi r^2 = \pi d^2 = 3.14d^2$

Volume = $\frac{4\pi r^3}{3} = \frac{\pi d^3}{6} = 0.524d^3$

Total solid angle = 4π steradian (sr)

APPENDIX A3. Relative Density of Common Aerosol Materials
(Multiply Values by 1000 for Density in kg/m³ and by 1.0 for Density in g/cm³)

Solids

Aluminum	2.7	Natural fibers	1–1.6
Aluminum oxide	4.0	Paraffin	0.9
Ammonium sulfate	1.8	Plastics	1–1.6
Asbestos	2.0–2.8	Pollens	0.45–1.05
Asbestos, chrysotile	2.4–2.6	Polystyrene	1.05
Carnauba wax	1.0	Polyvinyl toluene	1.03
Coal	1.2–1.8	Portland cement	3.2
Fly ash	0.7–2.6	Quartz	2.6
Fly ash cenospheres	0.7–1.0	Sodium chloride	2.2
Glass, common	2.4–2.8	Sulfur	2.1
Granite	2.6–2.8	Starch	1.5
Ice	0.92	Talc	2.6–2.8
Iron	7.9	Titanium dioxide	4.3
Iron oxide	5.2	Uranine dye	1.53
Limestone	2.7	Wood (dry)	0.4–1.0
Lead	11.3	Zinc	6.9
Marble	2.6–2.8	Zinc oxide	5.6
Methylene blue dye	1.26		

Liquids

Alcohol	0.79	Mercury	13.6
Dibutyl phthalate	1.045	Oils	0.88–0.94
Dioctyl phthalate [DOP; di-2		Oleic acid	0.894
(ethylhexyl) phthalate]	0.983	Polyethylene glycol	1.13
Dioctyl sebacate	0.915	Sulfuric acid	1.84
Hydrochloric acid	1.19	Water	1.00

APPENDIX A4. Standard Sieve Sizes

Designation		Nominal	Designation		Nominal
ISO Std.[a] (μm)	Alternate (No.)	Wire Size (μm)	ISO Std.[a] (μm)	Alternate (No.)	Wire Size (μm)
250	60	180	63	230	44
180	80	131	53	270	37
150	100	110	45	325	30
125	120	91	38	400	25
106	140	76	32	450	28
90	170	64	25	500	25
75	200	53	20	635	20

[a]Sieve opening.

APPENDIX A5.   Properties of Gases and Vapors at 293 K [20°C] and 101 kPa [1 atm]

Gas or Vapor	Molecular Weight	Relative Density[a]	Viscosity[b]	Diffusion Coefficient[c]
Air (dry)	29.0	1.00	1.81	2.0
Air (saturated)	28.7	0.99	1.79	2.0
CO	28.0	0.967	1.75	
CO_2	44.0	1.52	1.46	1.6
CH_4	16.0	0.554	1.09	
H_2	2.0	0.070	0.88	6.8
H_2O	18.0	0.62	0.96	2.4
N_2	28.0	0.967	1.75	2.0
O_2	32.0	1.10	2.03	2.0

[a]Ratio of density of gas to that of dry air. Multiply by 1.20 to get density in kg/m³ and by 0.00120 to get density in g/cm³.
[b]Multiply by 10^{-5} for viscosity in Pa·s and by 10^{-4} for viscosity in P (dyn · s/cm³).
[c]Diffusion coefficient of gas or vapor in air. Multiply by 10^{-5} for diffusion coefficient in m²/s and by 0.1 for diffusion coefficient in cm²/s.

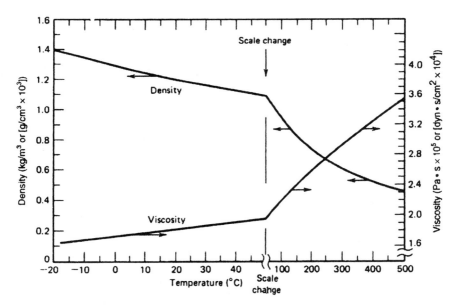

APPENDIX A6.   Viscosity and Density of Air versus Temperature

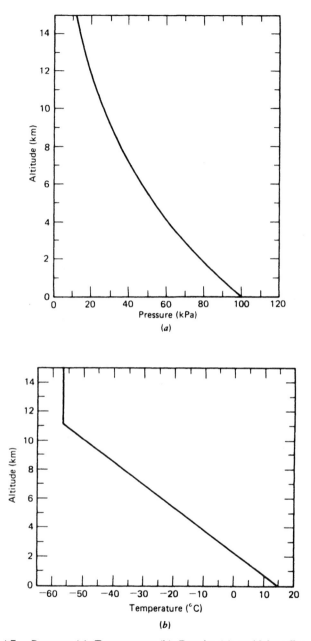

APPENDIX A7. Pressure (a), Temperature (b), Density (c), and Mean Free Path (d) of Air versus Altitude

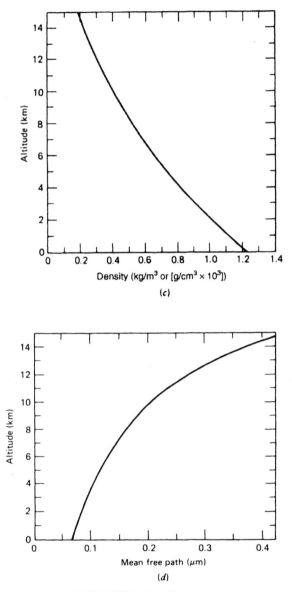

APPENDIX A7.   Continued.

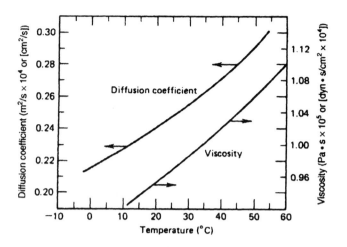

APPENDIX A8. Properties of Water Vapor

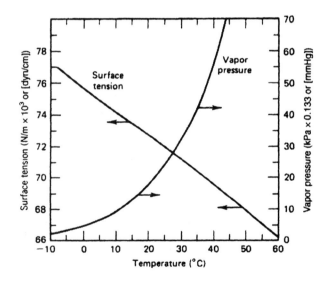

APPENDIX A9. Properties of Water

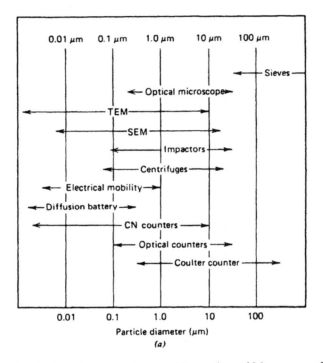

APPENDIX A10. Particle Size Range of Aerosol Properties and Measurement Instruments: (a) Application Range for Aerosol Size Measuring Instruments and (b) Size Range of Aerosol Properties (See Also Fig. 1.6)

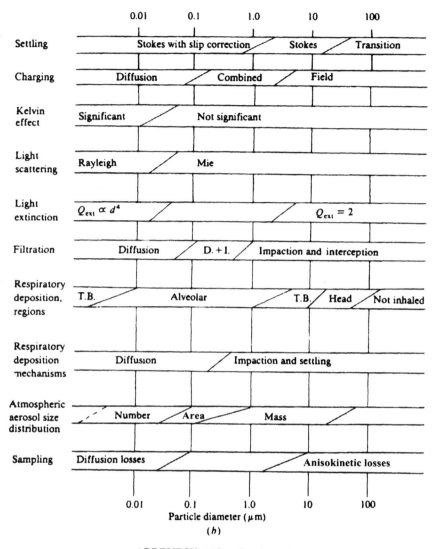

APPENDIX A10.   Continued.

APPENDIX A11 (a). **Properties of Airborne Particles at Standard Conditions (SI Units)[a]**

Particle Diameter d (μm)	Slip Correction Factor C_c	Settling Velocity V_{TS} (m/s)	Relaxation Time τ (s)	Mobility B (m/N · s)	Diffusion Coefficient D (m²/s)	Coagulation Coefficient K (m³/s)
0.001	224.332	6.75E-09	6.89E-10	1.32E+15	5.32E-06	3.11E-16
0.0015	149.752	1.01E-08	1.03E-09	5.85E+14	2.37E-06	3.81E-16
0.002	112.463	1.35E-08	1.38E-09	3.30E+14	1.33E-06	4.40E-16
0.003	75.174	2.04E-08	2.08E-09	1.47E+14	5.94E-07	5.39E-16
0.004	56.530	2.72E-08	2.78E-09	8.28E+13	3.35E-07	6.21E-16
0.005	45.344	3.41E-08	3.48E-09	5.32E+13	2.15E-07	6.93E-16
0.006	37.888	4.11E-08	4.19E-09	3.70E+13	1.50E-07	7.56E-16
0.008	28.568	5.51E-08	5.61E-09	2.09E+13	8.46E-08	8.63E-16
0.01	22.976	6.92E-08	7.05E-09	1.35E+13	5.45E-08	9.48E-16
0.015	15.524	1.05E-07	1.07E-08	6.07E+12	2.45E-08	1.09E-15
0.02	11.801	1.42E-07	1.45E-08	3.46E+12	1.40E-08	1.15E-15
0.03	8.083	2.19E-07	2.23E-08	1.58E+12	6.39E-09	1.14E-15
0.04	6.229	3.00E-07	3.06E-08	9.13E+11	3.69E-09	1.07E-15
0.05	5.120	3.85E-07	3.93E-08	6.00E+11	2.43E-09	9.92E-16
0.06	4.384	4.75E-07	4.84E-08	4.28E+11	1.73E-09	9.20E-16
0.08	3.470	6.69E-07	6.82E-08	2.54E+11	1.03E-09	8.03E-16
0.1	2.928	8.82E-07	8.99E-08	1.72E+11	6.94E-10	7.17E-16
0.15	2.220	1.50E-06	1.53E-07	8.68E+10	3.51E-10	5.83E-16
0.2	1.878	2.26E-06	2.31E-07	5.51E+10	2.23E-10	5.09E-16
0.3	1.554	4.21E-06	4.29E-07	3.04E+10	1.23E-10	4.34E-16
0.4	1.402	6.76E-06	6.89E-07	2.06E+10	8.31E-11	3.97E-16
0.5	1.316	9.91E-06	1.01E-06	1.54E+10	6.24E-11	3.76E-16
0.6	1.261	1.37E-05	1.39E-06	1.23E+10	4.98E-11	3.62E-16
0.8	1.194	2.30E-05	2.35E-06	8.75E+09	3.54E-11	3.45E-16
1.0	1.155	3.48E-05	3.54E-06	6.77E+09	2.74E-11	3.35E-16
1.5	1.103	7.47E-05	7.62E-06	4.31E+09	1.74E-11	3.22E-16
2.0	1.077	1.30E-04	1.32E-05	3.16E+09	1.28E-11	3.15E-16
3.0	1.051	2.85E-04	2.90E-05	2.05E+09	8.31E-12	3.09E-16
4.0	1.039	5.00E-04	5.10E-05	1.52E+09	6.15E-12	3.06E-16
5.0	1.031	7.76E-04	7.91E-05	1.21E+09	4.89E-12	3.04E-16
6.0	1.026	1.11E-03	1.13E-04	1.00E+09	4.05E-12	3.03E-16
8.0	1.019	1.96E-03	2.00E-04	7.47E+08	3.02E-12	3.01E-16
10	1.015	3.06E-03	3.12E-04	5.95E+08	2.41E-12	3.00E-16
15	1.010	6.84E-03	6.98E-04	3.95E+08	1.60E-12	2.99E-16
20	1.008	1.21E-02	1.24E-03	2.95E+08	1.19E-12	2.99E-16
30	1.005	2.72E-02	2.78E-03	1.96E+08	7.94E-13	2.98E-16
40	1.004	4.84E-02	4.93E-03	1.47E+08	5.95E-13	2.98E-16
50	1.003	7.55E-02	7.70E-03	1.18E+08	4.76E-13	2.98E-16
60	1.003	1.33E-01	1.11E-02	9.80E+07	3.96E-13	2.98E-16
80	1.002	1.72E-01	1.97E-02	7.34E+07	2.97E-13	2.98E-16
100	1.002	2.49E-01	3.07E-02	5.87E+07	2.37E-13	2.98E-16

[a] Calculated for standard density spheres at 293 K [20°C] and 101 kPa (1 atm).

APPENDIX A11 (b). **Properties of Airborne Particles at Standard Conditions (cgs Units)[a]**

Particle Diameter d (μm)	Slip Correction Factor C_c	Settling Velocity V_{TS} (cm/s)	Relaxation Time τ (s)	Mobility B (cm/dyn · s)	Diffusion Coefficient D (cm²/s)	Coagulation Coefficient K (cm³/s)
0.001	224.332	6.75E-07	6.89E-10	1.32E+12	5.32E-02	3.11E-10
0.0015	149.752	1.01E-06	1.03E-09	5.85E+11	2.37E-02	3.81E-10
0.002	112.463	1.35E-06	1.38E-09	3.30E+11	1.33E-02	4.40E-10
0.003	75.174	2.04E-06	2.08E-09	1.47E+11	5.94E-03	5.39E-10
0.004	56.530	2.72E-06	2.78E-09	8.28E+10	3.35E-03	6.21E-10
0.005	45.344	3.41E-06	3.48E-09	5.32E+10	2.15E-03	6.93E-10
0.006	37.888	4.11E-06	4.19E-09	3.70E+10	1.50E-03	7.56E-10
0.008	28.568	5.51E-06	5.61E-09	2.09E+10	8.46E-04	8.63E-10
0.01	22.976	6.92E-06	7.05E-09	1.35E+10	5.45E-04	9.48E-10
0.015	15.524	1.05E-05	1.07E-08	6.07E+09	2.45E-04	1.09E-09
0.02	11.801	1.42E-05	1.45E-08	3.46E+09	1.40E-04	1.15E-09
0.03	8.083	2.19E-05	2.23E-08	1.58E+09	6.39E-05	1.14E-09
0.04	6.229	3.00E-05	3.06E-08	9.13E+08	3.69E-05	1.07E-09
0.05	5.120	3.85E-05	3.93E-08	6.00E+08	2.43E-05	9.92E-10
0.06	4.384	4.75E-05	4.84E-08	4.28E+08	1.73E-05	9.20E-10
0.08	3.470	6.69E-05	6.82E-08	2.54E+08	1.03E-05	8.03E-10
0.1	2.928	8.82E-05	8.99E-08	1.72E+08	6.94E-06	7.17E-10
0.15	2.220	1.50E-04	1.53E-07	8.68E+07	3.51E-06	5.83E-10
0.2	1.878	2.26E-04	2.31E-07	5.51E+07	2.23E-06	5.09E-10
0.3	1.554	4.21E-04	4.29E-07	3.04E+07	1.23E-06	4.34E-10
0.4	1.402	6.76E-04	6.89E-07	2.06E+07	8.31E-07	3.97E-10
0.5	1.316	9.91E-04	1.01E-06	1.54E+07	6.24E-07	3.76E-10
0.6	1.261	1.37E-03	1.39E-06	1.23E+07	4.98E-07	3.62E-10
0.8	1.194	2.30E-03	2.35E-06	8.75E+06	3.54E-07	3.45E-10
1	1.155	3.48E-03	3.54E-06	6.77E+06	2.74E-07	3.35E-10
1.5	1.103	7.47E-03	7.62E-06	4.31E+06	1.74E-07	3.22E-10
2	1.077	1.30E-02	1.32E-05	3.16E+06	1.28E-07	3.15E-10
3	1.051	2.85E-02	2.90E-05	2.05E+06	8.31E-08	3.09E-10
4	1.039	5.00E-02	5.10E-05	1.52E+06	6.15E-08	3.06E-10
5	1.031	7.76E-02	7.91E-05	1.21E+06	4.89E-08	3.04E-10
6	1.026	1.11E-01	1.13E-04	1.00E+06	4.05E-08	3.03E-10
8	1.019	1.96E-01	2.00E-04	7.47E+05	3.02E-08	3.01E-10
10	1.015	3.06E-01	3.12E-04	5.95E+05	2.41E-08	3.00E-10
15	1.010	6.84E-01	6.98E-04	3.95E+05	1.60E-08	2.99E-10
20	1.008	1.21E+00	1.24E-03	2.95E+05	1.19E-08	2.99E-10
30	1.005	2.72E+00	2.78E-03	1.96E+05	7.94E-09	2.98E-10
40	1.004	4.84E+00	4.93E-03	1.47E+05	5.95E-09	2.98E-10
50	1.003	7.55E+00	7.70E-03	1.18E+05	4.76E-09	2.98E-10
60	1.003	1.33E+01	1.11E-02	9.80E+04	3.96E-09	2.98E-10
80	1.002	1.72E+01	1.97E-02	7.34E+04	2.97E-09	2.98E-10
100	1.002	2.49E+01	3.07E-02	5.87E+04	2.37E-09	2.98E-10

[a]Calculated for standard density spheres at 293 K [20°C] and 101 kPa [1 atm].

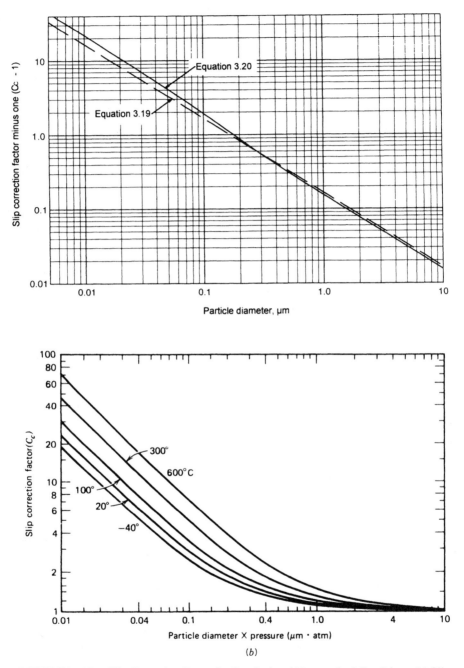

APPENDIX A12. Slip Correction Factor for Standard and Nonstandard Conditions: (a) Slip Correction Factor Minus One versus Particle Diameter at Standard Conditions; (b) Slip Correction Factor versus Particle Diameter Times Pressure for Temperatures from 233 to 893 K [–40 to 600°C][a]

[a]Calculated by Eqs. 2.1, 2.25, and 3.20.

APPENDIX A13. Properties of Selected Low-Vapor-Pressure Liquids

Liquid	Molecular Weight	Relative Density[a]	Viscosity at 293 K [20°C] (Pa · s)	Surface Tension (N/m × 10^5) or [dyn/cm]	Boiling Temperature (°C)	Vapor Pressure[b] at 293 K [20°C] Pa [mmHg]	Refractive Index
DBP[c]	278	1.05	0.020	—	340	0.012 [9 × 10^{-5}]	1.493
DOP (DEHP)[d]	391	0.984	0.082	31	350	3.5 × 10^{-6} [2.6 × 10^{-8}]	1.484
DOS (DEHS)[e]	426	0.915	0.027	32	248	—	1.448
PAO[f](3004)	—	0.819	0.027	29	401	—	1.456
Glycerol	92	1.26	1.4	63	290	0.027 [2 × 10^{-4}]	1.475
Mineral oil[g]	—	0.86–0.88	(0.6)	—	(decomposition) 95% >360	—	1.48
Oleic acid	283	0.894	—	—	360	0.012 [9 × 10^{-5}]	1.458
PEG[h]	380–420	1.13	0.11	45	(decomposition) range	—	1.465
Silicone oil[i]	—	0.97–1.1	0.1–1	3–21	—	—	—

—Indicates data not available.

[a] Density relative to water. Multiply by 1000 for density in kg/m³ or by 1.0 for density in g/cm³.

[b] Extrapolated to 293 K [20°C] from published data.

[c] Di-butyl phthalate.

[d] Di (2-ethylhexyl) phthalate.

[e] Di (2-ethylhexyl) sebacate.

[f] Poly alpha olefin, Emery 3004.

[g] White mineral petroleum oil, USP.

[h] Polyethylene glycol, PEG 400, Union Carbide.

[i] Range for DC 200, 550, and 710 cS. Dow Corning.

APPENDIX A14. Reference Values for Atmospheric Properties at Sea Level and 293.15 K [20°C][a]

Property	Symbol	Value, (SI Units)	Value, (cgs Units)
Absolute Temperature[b]	T	293.15 K	293.15 K [20°C]
Acceleration of Gravity[c]	g	9.8066 m/s^2	980.66 cm/s^2
Atmospheric Pressure[c, d]	p	1.0132×10^5 Pa	1.0132×10^6 dyn/cm^2
Avogadro Constant[c]	N_a	6.0222×10^{23} /mol	6.0222×10^{23} /mol
Boltzmann Constant[c]	k	1.3806×10^{-23} J/K	1.3806×10^{-16} dyn · cm/K
Density[e]	ρ_g	1.2041 kg/m^3	1.2041×10^{-3} g/cm^3
Diffusion Coefficient[f]	D	1.99×10^{-5} m^2/s	0.199 cm^2/s
Mean Free Path[g]	λ	0.066 µm	0.066 µm
Molar Volume[e]	v_m	0.024053 m^3/mol	24.053 L/mol
Molecular Concentration[e]	n	2.5036×10^{25} m^{-3}	2.5036×10^{19} cm^{-3}
Molecular Weight[c]	M	0.028964 kg/mol	28.964 g/mol
Molecular Collision Diameter[h]	d_m	3.7×10^{-10} m	3.7×10^{-8} cm
Molecular Velocity (mean)[e]	$\bar{c}$	462.90 m/s	46290 cm/s
Ratio of Specific Heats (c_p/c_v)[i]	κ	1.400	1.400
Speed of Sound[e]	V_s	343.23 m/s	34323 cm/s
Universal Gas Constant[c]	R	8.3143 J/mol · K	8.3143×10^7 dyn · cm/K · mol
Viscosity, Dynamic[j]	η	1.8134×10^{-5} Pa · s	1.8134×10^{-4} Poise

[a]Values rounded off to five significant figures where available.

[b]Value given is the standard temperature used in this book. It differs from the adopted sea level temperature of 288.15 K [15°C] used in USSA (1976).

[c]USSA (1976) value.

[d]Also 760 mm Hg exactly, USSA (1976).

[e]USSA (1976) value adjusted to the standard temperature of 293.15 [20°C].

[f]Bolz and Tuve (1973) value.

[g]Calculated by Eq. 2.25 using the adopted value of 0.37 nm for the molecular collision diameter. (Note, this value of λ differs slightly from the USSA (1976) value corrected to 293.15 [20°C] because of a different adopted value (0.365 nm) for molecular collision diameter.)

[h]Adopted value. This value differs slightly from the adopted value used in USSA (1976) of 0.365 nm. (Note, this affects the mean free path as calculated by Eq. 2.25.)

[i]USSA (1976) adopted value.

[j]USSA (1976) value adjusted to the standard temperature of 293.15 [20°C] by the Southerland equation (see the example problem in Section 2.4).

References

USSA, *U.S. Standard Atmosphere, 1976*, National Oceanic and Atmospheric Administration (NOAA), National Aeronautics and Space Administration (NASA), and United States Air Force (USAF), Washington, DC, 1976.

Bolz, R. E. and Tuve, G. L., *CRC Handbook of Tables for Applied Engineering Sciences*, 2d ed., CRC Press, Boca Raton, FL, 1973.

APPENDIX A15.   Greek Symbols Used in This Book

A α	Alpha	Θ θ	Theta	Σ σ	Sigma
B β	Beta	K κ	Kappa	T τ	Tau
Γ γ	Gamma	Λ λ	Lambda	Φ φ	Phi
Δ δ ∂	Delta	M μ	Mu	X χ	Chi
E ϵ ε	Epsilon	Π π	Pi	Ω ω	Omega
H η	Eta	P ρ	Rho		

APPENDIX A16.   SI Prefixes

Factor	Prefix	Symbol	Factor	Prefix	Symbol
10^{12}	tera	T	10^{-3}	milli	m
10^{9}	giga	G	10^{-6}	micro	μ
10^{6}	mega	M	10^{-9}	nano	n
10^{3}	kilo	k	10^{-12}	pico	p
10^{-2}	centi	c	10^{-15}	femto	f

Index

Absorbance, 350, 353
Absorption coefficient, 350, 353
Absorption efficiency, 353
AC fine test dust, 443
Acceleration
    particle, curvilinear, 117
    particle, straight-line, 112–116
    total, 28
Accumulation mode, 307, 311
ACGIH inhalable criterion, 246
ACGIH respirable dust criteria, 250
Acoustic coagulation, 275
Actinomycetes, 395
Activity median diameter, 88
Adhesion energy, 146
Adhesion, particle, 141–144
    electrostatic forces, 143
    surface tension forces, 143
    van der Waals forces, 141–142
Adhesive force
    effect of humidity, 143–144
    empirical equation, 144
    table, 144
Adiabatic expansion, 278, 280–281
    condensation nuclei counter, 292
Aerocolloidal system, 1
Aerodynamic cutoff, 96
Aerodynamic diameter, 53–55, 154, 406
    definition, 53
    measurement by
        aerosol centrifuge, 67
        cascade impactor, 126–128
        horizontal elutriator, 65–67
        sedimentation cell, 65
        outside Stokes region, 54
Aerodynamic equivalent sphere, 53
Aerodynamic particle sizer (APS), 136–137
Aerodynamic separation of particles,
    65–67, 249–257, 434
Aerosizer, 137–138

Aerosol, homogeneous, 8
Aerosol spray cans, 1, 434
Aerosol technology, 1–2
Aerosol therapy, 1
Aerosols
    anthropogenic, 304
    atmospheric, 304–314
    background, size distribution, 306
    centrifuge, 67
    cloud motion, 379
    definition, 1, 3–9
    generation, 428–445
    global emissions, table, 305
    man-made, 304
    measurement instruments, 456
    monodisperse, see Monodisperse
        aerosol
    natural background, 304–308
    polydisperse, see Polydisperse aerosol
    primary, 8
    properties, 457–459
    science and technology, general
        references, 13–14
    secondary, 8
    spectrometer, 66–67
    stratospheric, 305–306, 312
    tropospheric, 306–307
    urban, 307–312
    urban, size distribution, 307–312
Agar, 397
Agglomerates, 10, 51, 141, 188, 260, 411
    interparticle adhesion, 144
    porous, 10, 265
    transient molecular clusters, 284
Agglomeration, 260
Air
    pressure versus altitude, 452
    properties of, 29, 448–449
    temperature versus altitude, 453
    viscosity, 24–27, 29, 452

Air ions, 322, 331, 334
Air pollution, urban aerosol, 307–312
Airborne particles, properties of, 457–459
Albedo, earth's, 312
Altitude, effect on pressure, temperature,
    density, and mean free path of air,
    453–454
Alveolar deposition, 241
  mouth and nose breathing, 252
  respirable dust criteria, 252
Alveolar region, 233
AMD, see Activity median diameter
Analytical balance, for filter samples, 218
Anemometer, hot wire, 32
Aneroid pressure gauge, 39
Angular aperture, objective lens, 415
Anisokinetic sampling, 206–213
Anthropogenic aerosols, 304
Apparent density, particle, 10
APS, aerodynamic particle sizer, 137–138
Area mean diameter, 86
Arithmetic mean diameter, 80, 89
Arizona road dust (ARD), as test dust, 443
Asbestos counting, 422–425
Asbestos fibers, 5
  alveolar deposition, 422
  identification, 423
Asbestosis, 422
Aspect ratio, fiber, 422–423
Asperities, particle surface, 142
Aspiration efficiency, human head, 245–247
Atmospheric aerosol, 304–314
  background, 304–307
  growth, 305
  size distribution, 307–312
  sources, 305
Atmospheric properties, reference values,
    462
Atmospheric visibility, 364
Atomization of liquids, 428–434
Atomization of particles in liquid
    suspensions, 434–437
Atomizer
  electrostatic, 429
  jet, 428
  pneumatic, 428–434
  pressure, 428
  rotary, 429
  swirl, 428
  ultrasonic, 429

Atomizers, 428–434
Average particle size, see specific type
Avogadro's number
    determination of, by Lord Rayleigh, 369
    value, 448
Avogadro's principle, 16

Babinet's principle, 356
Bacteria, 9, 394
Bag filter, 184
Bend, tube, loss in, 217
Bernoulli effect in nebulizers, 429
Beta gauge, direct reading instrument, 224
Bimodal size distribution, 96
Bioaerosol
    background concentration, 396
    sampling, 396–400
    survivability, 396
    viability, 396
Bioaerosols, 4, 394–401
    size ranges, 9
Bipolar ions, charge neutralization, 335–337
Blow-off, 145
Boltzmann distribution of molecular
    velocity, 19, 324
Boltzmann equilibrium charge distribution,
    267, 335–337, 439
Boltzmann's constant, 19, 151, 153, 448
Bouguer's law, 352–357
Bounce, particle, 146
Boundary, layer, 145
Boundary layer, diffusion, 164–165
Bourdon tube pressure gauge, 39
Boyles law, 16
Brownian coagulation, see Coagulation,
    thermal
Brownian displacement, 156–160
    distribution of, 158
    probability, 158
Brownian motion, 150–154, 375
    and diffusion, 157
    as a fractal, 412
    displacement, 157
    intensify of, 153
    of ions, 324
    particle kinetic energy, 151
    particle mean free path, 154–156
    particles in filter, 194–195
    respiratory deposition, 237
    rms displacement, 157
    rotational, 160, 173

Brownian rotation, effect on coagulation of
  nonspheres, 267
Brush, rotating, dust disperser, 441
Bubble flowmeter, 36
Bulk motion of aerosols, 379–385
Bulk properties of aerosols, 1, 379
Buoyancy, effect on settling velocity, 46
Buoyant force, cloud, 383–384
Byssinosis from cotton dust, 256

Capillary pore membrane filter, 183–184,
  203, 422, 425
Capture efficiency, kinematic coagulation,
  273
Carina, impaction on, 237
Carrier gas stream, 66
Cascade impactor, 128–131
  commercial, 130, 132
  data reduction, 129–132
  high volume, 132
  inter stage losses, 133–134
  stage, 128
CCN (cloud condensation nuclei), 292
CEN inhalable criterion, 246
Centrifugal
  acceleration, 47
  force, 449
  sampler for bioaerosols, 397–398
  velocity, 47, 123
cgs system of units, 2
Characteristics of dry-dispersion aerosol
  generators, 440
Charge
  build-up during powder dispersion, 439
  build-up in electron microscope, 422
  distribution, equilibrium, 335–337
  electron, 317–318, 321–322
  equilibrium with bipolar ions, 335
  limits, 333–334
  maximum on particle, 333–335
  neutralization, 338, 434, 439
  neutralizer, 338
  saturation, 325, 327, 338
Charging
  combined mechanisms, 327–328
  diffusion, 324–325
  field, 325–327
  mechanisms, 323–331
  static electrification, 324
Charles Law, 16

Chemical composition, fine and coarse
  particles, 312, 314
Chi-square test for lognormal distribution,
  104
Cigarette smoke, 12, 276
Classifier, size-selective, with dust
  generators, 439
Clean Air Act of 1977, 364
Clearance mechanisms, lung, 235
Cloud, 4
  condensation nuclei (CCN), 292
  density, net, 381
  formation, atmospheric, 1, 289
  Reynolds number, 381
  settling, 379–385
CMAD (count median aerodynamic
  diameter), 88
CMD (count median diameter), 84,
  100–102
CN counter (condensation nuclei counter),
  292–294
CNC condensation nuclei counter, 292–294
Coagulation
  kinematic, 260, 272–275
    acoustic, 275
    differential settling, 272–273
    gradient, 274–275
    gravitational, 272–273
    orthokinetic, 272–275
    shear, 274–275
    stirred, 275
    turbulent flow, 275
    ultrasonic, 275
  thermal, 260–272, 311
    atmospheric aerosol, 311, 313
    change in diameter of average mass,
      270
    coefficient, 262, 268–271
    effect of electrical forces, 267
    effect on cloud settling, 383
    lognormal distribution, 269–270
    nonspherical particles, 267
    particle size increase, 263–267
    polydisperse, 268–272
    rate, 262
    simple monodisperse, 260–268
    Smoluchowski, 260
    stop by dilution, 267
    time to double particle size, 266
Coal fly ash, 3

Coarse fraction, separation of, 131–136
Coarse particles, 311–312
Coarse-particle mode, 311–312
Coefficient of drag, 43–44, 55–62
Coefficient of restitution, 147
Coefficient of viscosity, 24
Coincidence error, optical particle counters, 374–375
Coincidence, in time-of-flight instruments, 138
Collection efficiency
    capillary pore membrane filter, 203
    electrostatic precipitator, 339
    fabric filter, 184–185
    fibrous filter, 182, 186, 191, 196–200
    filter
        minimum, 187–199
        single fiber, 190–196
    horizontal elutriator, 66
    impactor, 122, 125, 226
    membrane filter, 202–203
    point-to-plane electrostatic precipitator, 340
    respirable mass cyclone, 253
    thermal precipitator, 176
Collision diameter, air molecule, 21
Collision efficiency, kinematic coagulation, 273
Collisions, rate of, particle, 262
Colony counting, 397
Combustion of dust, 388
Compression wave, causing dust explosion, 386
Concentration, aerosol, 10
    by piezoelectric crystal, 222
    decay (settling), 64
    direct reading instruments, 222–225
    gradient, 150, 157
    measurement of, 217–225
    range, 10
Concentration ratio, anisokinetic sampling, 209–213
Condensation coefficient, 285
Condensation mode, 311
Condensation
    atmospheric aerosol, 312–313
    growth, 285–286, 306, 309, 312
    nucleated, 286, 443–445
    nuclei, 278, 288–294, 443–445

nuclei counter, 292–294
    of droplets, 278–281
    on insoluble nuclei, 288
    on soluble nuclei, 289–291
    test aerosols, 443–445
Condenser lens, 413
Confidence limits, size distribution, 103
Constants, numerical, 447–449
Contact charging, 324
Continuum region, 51
Contrast
    apparent, 366–367
    inherent, 364–365
Contrast limen (threshold of brightness contrast), 368
Contrast reduction, 364, 366–367
Conversion factors, 447–449
Corona discharge, 331–333, 336, 338
Cotton dust standard, vertical elutriator, 256–257
Coulomb's Law, 316
Coulombic force, 318
Coulter counter, 345
Count mean diameter, 89, 101
Count median aerodynamic diameter, 88
Count median diameter (CMD), 84, 101
    confidence limits, 103
    use of instead of geometric mean, 92
Count, standard deviation of, 225–226
CPC condensation particle counter, 292–294
Critical orifice, 34
Critical saturation ratio, 284
Culture medium, 397
Culture plates, 397
Cumulative deposition
    diffusion, 162
    thermophoresis, 175
Cumulative mass distribution, 84–86
Cumulative particle size distribution, 79–80, 92, 405
Cunningham correction factor, 49
Curvilinear motion, 117
    impaction, 121–128
    Stokes number, 119–121
Cut point (effective cutoff diameter), 125
Cutoff diameter, 125
Cutoff, respirable, 250
Cutoff size, inhalable particles, 246–247

Cutsize, 125
Cyclone, respirable mass, 252–253, 439
Cylinder, Stokes number for, 121

Dalton's law of partial pressure, 278
Data inversion, size distribution, 104
Davies criteria, still air sampling, 213–215
Deconvolution, size distribution, 104
Defense mechanisms, respiratory, 233
Deflagration, 388
Density
    aerosol cloud, 379–381
    aerosol materials, 451
    air, 29
    apparent, 10
    bulk, 53
    particle, 10
    standard, 10
Denudation of dust layer, 145
Denuder, 168
Deposition
    diffusion, from turbulent flow, 164–165
    diffusive, 160–165
    electrostatic, 338
    fraction
        alveolar, 244–245
        head airways, 244
        total, 245
        tracheobronchial, 244
    from turbulent flow, 164–165
    inertial, from turbulent flow, 165
    mechanisms
        filter, 191–192
        respiratory, 235–242
        transport losses, 216
    on surface, cumulative, 162
    parameter, diffusive losses, 163, 166
    respiratory, 233–259
    thermophoretic, 174–175
    velocity, 162
Detachment force, 144
Detonation, 388
Deutsch-Anderson equation, 339–340
DeVilbiss nebulizer, 429
Di(2-ethylhexyl)phthalate, 196, 409, 431,
    444
Diameter
    activity median, 88
    aerodynamic, 53–55, 406

area mean, 86
arithmetic mean, 80
automatic determination of, 424–425
average diffusion coefficient, 167
average mass, 83, 86, 102, 270, 445
average projected area, 354
average property, 83
average surface, 83, 89, 102, 354, 408
average volume, 89, 102
collision, air molecule, 21
count mean, 86, 102
count median (CMD), 84, 101–102
count median aerodynamic, 88
effective fiber, for filters, 197
equivalent, 10
equivalent volume, 51, 53, 406
Feret's, 402
fiber, for porous membrane filter, 202–203
for 50% collection efficiency, 125–126
geometric mean, 80–82
Kelvin, 282–283
length mean, 102
length median, 102
Martin's, 403
mass mean, 86, 100. *See also* Diameter,
    average mass
mass median (MMD), 84, 100
mass median aerodynamic, 88
mean volume-surface, 86, 102
median, 80–81
modal, 80–81
moment average, 82–84
of settling sphere, high Reynolds
    number, 60–62
projected area, 403, 406
Sauter, 86, 102
second moment average, 84
statistical, 402
Stokes, 53
surface mean, 86–87, 102
surface median, 102
third moment average, 83
various, 102
volume mean, 102
volume median, 102
weighted mean, 86–88. *See also specific*
    diameter name
    from moment average diameters, 87
Diaphragm vacuum pump, 230

Dichotomous sampler, 135, 255
Dielectric constant, 325–327
Differential mobility analyzer, 342–345
Diffracted light, 356, 363
Diffusion
  and Brownian motion, 157
  and growth by condensation, 286
  battery, 165–168
  boundary layer, 164–165
  charging, 324–325, 327–330, 342
  coefficient
    air, 29
    measured by diffusion battery,
      165–167
    measurement by light scattering, 375
    particle, 26, 150, 155
    water vapor versus temperature, 455
  convective, 150, 158, 160
  dryer, 431
  eddy, 150, 158
  Fick's law, 26, 150
  force, 151–152
  gas exchange in respiratory system, 235,
    241
  in filter, 191, 203
  particle, 150–154
  respiratory deposition, 237–239
  single fiber efficiency, 194–195
  thermal, 150, 158
  to surfaces, 160–165
  to walls of tube, 162–164
Diffusiophoresis, 179
Digital image, automatic sizing, 425
Dipole force, 141
Direct-reading instruments, aerosol
    concentration, 222–225
Discharging rate, 336
Disperse system, 1
Dispersibility of powder, 438
Dispersion of powder, 438–443
Displacement equation, accelerating
    particle, 114–115
Distribution function, particle size, *see*
    Frequency function
Distribution, particle size
  exponential, 105
  lognormal, 90–93
  mass, 84–86
  moment, 84–88

normal, 90
Nukiyama-Tanasawa, 104
power function, 104
properties of, 75–82
Rosin-Rammler, 104
see also specific distribution
DMA and CNC, 343
DMA, differential mobility analyzer,
    342–345
DOP, *see* Di(2-ethylhexyl)phthalate
Doppler effect, 376
Doublets, 435–437
Drag coefficient, 43–44, 55–62
Drag force, 43, 46
  cloud, 380
  effect of particle shape, 51–52
  effect of slip correction, 49
  irregular particles, 51–52
  Newton's, 43–44
  on filter fibers, 202
  Stokes, 45–46
Droplet, 6
  effect on drag, 46
  evaporation, 294–301
  heating, condensation, 286
  lifetime, 294–301
  mode, 311
  temperature, equilibrium, 286
Dry dispersion aerosol generation, 438–443
Dry gas meter, 37
Drying time, droplet, 294–301
Dust, 4
Dust counting cell, 227
Dust dispersion, methods, 438–443
Dust explosions, 368–392
  control, 391
  dispersibility of dust, 391
  maximum pressure, 389–390
  minimum concentration, 386–388
  minimum energy for spark ignition, 391
  minimum ignition temperature, 389
  rate of pressure build-up, 387, 389
  requirements, 386
  severity, 386, 390–391
Dust free layer
  evaporating surface, 179
  thermophoresis, 171
Dust generators, characteristics of, 440
Dust loading, 200

Dust metering, 438
Dustiness of powder, 438
Dynamic light scattering (DLS), 375
Dynamic shape factor, 51–53, 406
  use at high Reynolds number, 62

EAA (electrical aerosol analyzer), 342
Eddy diffusion, 150, 160
Effective fiber diameter, filter, 197, 202
Efficiency, electrostatic precipitator, 339
Efficiency, *see* Collection efficiency
Electret fibers, 196
Electric field, 318–320, 338
  earth's atmosphere, 320
  intensity, 318–320
  nonuniform, 331
  strength, 318–320
Electrical aerosol analyzer (EAA), 342
Electrical force, effect on coagulation, 267.
    See also Electrostatic force
Electrical mobility, 320–322, 327, 338
Electrical mobility analyzer, 342
Electrical properties, 316–38
Electrical resistance, size measurement by,
    345
Electrobalance for filter samples, 218
Electrolytic charging, 324
Electromagnetic radiation, 350
Electron microscope
  calibration, 419
  microprobe analysis, 422
  scanning, 419–422
  sizing procedure, 405
  transmission, 419–420
Electron microscopy, 416–422
Electronic charge, 317–318
Electrospraying, 434
Electrostatic atomizer, 434
Electrostatic collection in filters, 192,
    195–196
Electrostatic deposition, respiratory, 239
Electrostatic dispersion, 268
Electrostatic field, *see* Electric field
Electrostatic force, 316–317
  adhesion, 141, 143
  Millikan cell, 321
Electrostatic precipitation, 338–341
Electrostatic precipitator
  collection efficiency, 339–340

for electron microscopy sampling,
    340–341, 419
  penetration, 339
  point-to-plane, 340–341
  scaling equation, 339–340
Electrostatic system of units, 316–317
Electrostatic units, conversion factors, 317
Electrostatic velocity, 320–322
  high Reynolds number, 321
Electrostatics, basic principles, 316–320
Elemental analysis, electron microscope,
    422
Elemental analysis, particle, 425
Elutriation spectrometer, 66, 384
Elutriator
  horizontal, 65–66, 252
    cotton dust, 256–257
    respirable mass sampling, 252
  vertical, 65, 439
Endospores, 395
Endotoxins, 394
Equation of motion, 114, 120
  general form, 114
Equivalent diameters, 10, 51–53, 402–408
Equivalent volume diameter, 51, 53, 406
Equivalent volume sphere, 51
Erosion of dust layer, 145
Evaporation, atmospheric aerosol, 312
Evaporation, droplet, 294–301
  correction for Fuchs effect, 296
  correction for Kelvin effect, 296
  correction for settling velocity, 297–298
  correction for volatile liquids, 296
  correction for wind effect, 297–298
  effect of corrections, 300
  effect of material, 301
  effect of relative humidity, 299
  rate, 294–301
  self-cooling, 296
  surface temperature, 296
Evaporation region, saturation ratio, 290
Evaporation-condensation monodisperse
    aerosol generator, 444
Expansion ratio, adiabatic cooling,
    280–281
Explosions, dust, 386–392
Explosive concentration, 386
Explosive dusts, characteristics of, 389
Explosive gas, 386, 388

Explosive limit, dust, 386
Exponential distribution, 105
External force, 111
Extinction, 352–358
    and apparent contrast, 367–368
    and diameter of average surface area,
        354
    and visual range, 368
    coefficient, 352–353, 368
    efficiency, 353–357
    for gases, 354
    for polydisperse aerosols, 353–354
    maximum, particle size for, 357
    measured by transmissometer, 371
    paradox, 356
    per unit aerosol mass, 357
    use in condensation nuclei counter, 292
    wavelength dependent, 370
Eyepiece lens, 413–414

Fabric filtration, 184
Face velocity, filter, 186
FAM, fibrous aerosol monitor, 424
Feret's diameter, 402–403, 411, 416, 425
Fiber
    alignment in flow stream, 53
    asbestos counting, 422–425
    fibrous aerosol monitor, 424
    filters, 182
    respiratory deposition, 239
    Reynolds number, 190
    slip correction, 53
Fibrogenic dust, 235
Fibrosis, alveolar region, 235, 422
Fibrous filters, 146, 182–202, 190
Fick's law of diffusion, 26, 150, 157, 160,
        261
Field charging, 325–330, 338
    particle size range for, 327
    rate, 325–327
Field lines, field charging, 325
Field strength (electric), 318–320, 338
    around charged particle, 325
    around point charge, 319
    between parallel plates, 319
    in cylindrical tube, 319, 331–332,
        338–339
    tropospheric, 320
Filar micrometer, 416

Filled site, colony counting, 399–400
Filters
    analysis by single fibers, 190
    artifact formation, 188
    capillary pore membrane, 183–184
    characteristics, 189
    clogging, 188
    collection efficiency, 186
    dust loading, 188, 200
    effect
        of humidity, 218–219
        of loading, 200
        of static charge, 220
        of storage, 219
    equilibration before weighing, 218
    fabric, 184
    face velocity, 186
    fiber diameter, 187
    fibrous, 182–202
    gravimetric analysts, 187–188
    holders, 220
    isolated fiber theory, 190
    macroscopic properties of, 182–190
    mass collection efficiency, 185
    membrane, 182, 202–203
    overall efficiency, 196–200
    packing density, 186
    particle size for minimum collection
        efficiency, 187
    penetration, 186–187, 191
    pressure drop, 187–188, 200–202
    quality, 188, 200
    resistance, see Filters, pressure drop
    samples
        chemical analysis, 188
        gravimetric analysts, 217–222
    sampling, 185
    single fiber efficiency, 193–196
    single fiber theory, 190
    solidity, 186, 202
    tare weighs, 218
    velocity for minimum collection
        efficiency, 187, 199
    weight stability, 188, 218–219
Filtration, 182–204
    fabric, 184
    granular bed, 185
Fine fraction, separation of, 135
Fine particle mode, effect on visual range,
        370

Fine particles, 311–312
Flame charging, 323
Flame propagation, 388
Flame speed, dust explosion, 390
Flow around obstacle, particle motion, 120
Flow controller, 34
Flow rate, gas, measurement of, 33–36
Flow rate in duct, 33–36
Flow regime, 27
Flow similarity, 27
Fluidized-bed aerosol generator, 442–443
Fly ash, coal, 3
Fog, 5
Force
    adhesive, 141–144
    buoyancy, 46
    centrifugal, 47
    coulombic, 316–317
    diffusion, 151–152
    diffusiophoretic, 179
    drag
        Newton's, 43–44
        Stokes, 45–46
    electrostatic, 143, 316–317
    external, 112
    gravity, 46
    osmotic, 151–152
    photophoretic, 178
    radiation pressure, 178
    surface tension, 143
    thermophoretic, 172–175
    van der Waals, 141–142
    see also specific force
Forest fires, global emission, 305
Formation, particle
    with nuclei, 288–292
    without nuclei, 283–284
Fractal, 408–413
Fractal dimension, 10, 408–413
Fractal morphology, 409
Free molecule region, 51
Frequency function, 78–79
    lognormal distribution, 92
    normal distribution, 90
    relation to cumulative distribution
        function, 80
Frequency histogram, 75–77
Friction factor, *see* Drag coefficient
Frictional force, fluid element, 27

Fuchs correction factor, 286, 288
Fuchs effect, evaporating droplets, 296, 300
Fume, 5
Fungal spores, 9, 394–395
Fungi, 394–395

Gas
    constant, ideal gas law, 16, 448
    exchange region, respiratory system,
        233
    flow rate, measurement, 33–36
    kinetic theory of, 15–18
    molecules, light extinction by, 354
    properties, 15–41, 452
    resistance force, 43
    volume measurement, 36
Gas-to-particle conversion, 284, 305,
    312–313
Gas-to-particle conversion, atmospheric
    aerosol, 312–313
Geometric mean diameter, 80–82, 92, 102
Geometric standard deviation, 92, 95
    confidence limits, 103
    from log probability plot, 95
Global emissions, aerosol, 305
Global warming, 312
Gradient coagulation, 274–275
Gradient, concentration, 25–26
Granular bed, filtration, 185
Graticule, eyepiece, 403
Gravimetric analysis, filters, 188, 217–222
Gravitational settling, 46–48
    in filters, 192, 195
Greek symbols, 464
Greenberg-Smith impinger, 227
Greenhouse effect, 312, 314
Grouped data, size distribution, 75–76
Growth by condensation, 285–286
Growth region, saturation ratio, 290
GSD, *see* Geometric standard deviation

Hamaker constant, 147
Hansen filter, 196
Hartmann apparatus, 387
Hatch-Choate equations, 97–100
    coefficient, 98–99, 102
    derivation, 105–108
    general form, 98
Hazard evaluation, aerosols, 233

Haze, 5
Haze, summertime, salt nuclei, 291
Head airways region, respiratory
    deposition, 233, 239–240
Heterogeneous nucleation, 288
Hi-vol sampler, 230
High-volume sampler, 230
Higher order Tyndall spectra (HOTS),
    371
Histogram
    frequency, 76–78
    standardized for sample size, 77
Homogeneous nucleation, 283–284, 305
Horizontal range, see Stopping distance
Hot wire anemometer, 32
Hydrological cycle, 1
Hydrosol, 1
Hygroscopic nuclei, 289–291
Hygroscopic particles
    and visibility, 370
    deposition in respiratory system, 237
    hysteresis with humidity change, 291
    in accumulation mode, 311
Hyphae, 395

Ice nuclei, 288
ICRP respiratory deposition model,
    242–245
Ideal gas law, 16
Ignition temperature, minimum for dust
    explosion, 389, 391
Illumination, effect on threshold of bright-
    ness contrast, 368
Illumination for microscopy, 413
Image analysis, automatic, 425
Image charge, 195, 239
Image-splitting eyepiece, 416
Immersion oil, optical microscopy, 415
Impaction, 121–128, 311
    atmospheric aerosols, 311
    in nebulizers, 429
    jet, 121
    nozzle, 121
    parameter, 121
    plate, 121, 125, 128, 133
        oil coating of, 133
    respiratory deposition, 235
    on moving rod, 397

Impactors
    cascade, 128–131. See also Cascade
        impactor
    collection efficiency, 122
    cutoff curve, 125, 130
    downstream pressure, 126
    jet diameter, 122
    jet to plate spacing, 125
    low pressure, 127
    multijet, 397–399
    nozzle, 125
    oil coating of plates, 133
    overloading, 134
    partial, 439
    particle bounce, 133, 146
    precollector, 134
    separation distance, 125
    similarity, 123
    slit, 397
    Stokes number, 122–123
    to control size distribution, 431
    virtual, 134–136, 312
Impinger(s), 226–227, 398
Inclined manometer, 38
Incompressible flow, 45
Index of refraction, 350–351
Inertia, gas, 43–44
Inertial force on fluid element, 27
Inertial impaction, see Impaction
Inertial range, see Stopping distance
Inhalability, 246
Inhalable fraction, 244–249, 251
Inhalable particle inlet, 247–249
Inhalable particle sampling, 246, 398
Inhalable particles, 250
Inhalable samplers, 247–248
Inhalation hazard, silica dust, 249
Inhalation of dust, 237
Integrating nephelometer, 372
Intensity, light, see Light, intensity
Interception
    in filters, 191–192
    respiratory deposition, 239
Interception parameter, 192
IOM personal inhalable sampler, 248–249
Ion mobility, 323, 325–327
Ion wind, 332

Ions
  air, 322
  as nuclei, 278, 289
  atmospheric, 307
  bipolar, 336
  charged molecular clusters, 288
  creation by corona discharge, 331
  field charging, 325
  mean thermal speed, 325
Irregular particles, equivalent sizes, 10,
      51–53, 402–408
ISO inhalable criterion, 246
Isokinetic sampling, 206–213
Isolated fiber theory, filtration, 190

Junge Layer, 306
Junge size distribution, 308

Kelvin diameter, 281–283, 292
Kelvin effect, 281–282, 289–300
Kelvin equation, 281–282, 290
Kelvin ratio, 281–282
  droplet size for soluble nuclei, 289
Khrgian-Mazin distribution, 105
Kinematic coagulation, *see* Coagulation,
      kinematic
Kinetic energy
  gas, 19
  particle, Brownian motion, 151, 154
Kinetic theory of gases, 15–18, 285
Knudsen number, 22
Kohler curves, 290–291
Koschmieder's equation, 367
Krypton-85 charge neutralizer, 338
Kuwabara hydrodynamic factor, 193

Lambert-Beer law, light extinction, 352
Laminar flow, 27, 29–30
Laminar-flow element, 36
Laminar sublayer, 145
Laskin nebulizer, 431
Latex particles, *see* Polystyrene latex
      spheres
Law of proportionate effect, 105
Least squares regression for log-probability
      plot, 103
Length mean diameter, 89, 102
Length median diameter, 102

Lifetime of particles in atmosphere, 306
Light
  absorption, 350
  angular scattering, 358
  attenuation along axis, 352
  back-scattering, 359
  forward scattering, 359
  frequency, 350
  intensity, 351
  multiple-scattered, 358, 371
  polarized, 352
  pressure, 178
  scattering, 349, 351, 358–364
    angle, 358
    effect of wavelength, 359
    effect on visual range, 364
    elastic, 349
    geometric optics, 349
    measurement with, 370
    Mie equations, 361–363
    molecular, 359
    photometer, 371
    plane, 358
    polarized components, 359
    Rayleigh region, 359
    size parameter, 351, 358–359
  single-scattered, 358
  transmission, use with condensation
      nuclei counter, 292
  unpolarized, 352
  velocity in vacuum, 350
  wavelength, 350
Limit of resolution, 415, 423
Liquids, low-vapor-pressure, properties of,
      461
Loading of filters, 188
Log-probability graph, 94–97
  bimodal distribution, 96
  effect of cutoffs, 86
  errors, 103
  fitting procedure, 103
  moment distributions, 97
  nonlinear, 96
  plotting procedure, 94
  truncation error, 103
  urban aerosol, 309
Logarithmic transformation, 90

Lognormal distribution, 90–93
  chi-square goodness-of-fit test, 104
  frequency function, 92
  Hatch-Choate equations, 97–100
  theoretical basis, 105
London-van der Waals forces, 141
Low-pressure impactor, 127
Lower explosive limit, dust, 386–387
Luminance, 365–366
Lung
  characteristics of, 235–236
  deposition in, 235–245

Magnification, maximum useful, 416
Manometer, 37–39
Martin's diameter, 402, 425
Mass concentration, 10–11
  aerosol cloud, 379
  and nephelometer measurements, 370,
    372
  measurement by photometer, 371
  relation to number concentration, 83
Mass distribution, 84–86
Mass flow meter, 35
Mass median aerodynamic diameter, 88
Mass median diameter (MMD), 84, 100,
    102
Mass median diameter obtained by
    cascade impactor, 129
Mass subsidence, 379
Mass transfer
  molecular, 26
  particle, 157
Maxwell-Boltzmann distribution of
    molecular velocities, 19, 324
Mean free path, 29
  gas, 21–23, 26, 49, 199, 285, 454
  particle, 154–156
Mean mass diameter, see Diameter, average
    mass
Mean, particle size, definition, 80
Mean volume-surface diameter, 86, 102
Mechanical mobility, 47–48, 111, 117, 153,
    320, 322
Median diameter, definition, 80–81
Membrane filter, 202–203, 225–226, 398,
    423
Meteorological range, 369
Metered dose inhaler, 434

Metered-dose inhaler, dry powder, 443
Micromanometer, 39
Micrometer, 8
Microscope
  objectives, characteristics of, 417
  optical, 413–417
  scanning electron, 419–422
  transmission electron, 416–420
Microscopic measurement of particle size,
    402–408
Microscopic particle counting, 249
Microscopic point of view, need for, 1, 8
Microscopy
  dark field, 416
  electron, 416–422
  optical, 413–417
  phase contrast, 416, 423
  polarizing, 416
Midget impinger, 227
Mie intensity parameters, 361–363
Mie scattering, 361–363
Mie scattering region, 349
Migration velocity, 47
Millikan oil drop experiment, 321–322
Minute volume, respiratory deposition, 245
Mist, 5
MMAD (mass median aerodynamic
    diameter), 88
MMAD, see Mass median aerodynamic
    diameter
MMD, see Mass median diameter
Mobility
  electrical, 320–322, 327, 338
  electrical analyzer, 342
  mechanical, 47–48, 111, 117, 153, 320,
    322
Mobility differential mobility analyzer,
    342–345
Mode, 80–81
Molecular velocity, mean, 21
Molecular-kinetic region, 50–51
Molecules
  adhesive forces, 141
  clusters, formation, 283
  collisions, 16–17, 21, 285
  concentration, 18, 285
  diameter, 23
  light-scattering, 349
  mass transfer, 25–26

mean free path, 15, 26, 49, 285, 454
momentum transfer, 24, 172, 179
    spacing, 21, 23
    velocity, 18–19
Moment average diameters, 82–84, 88–89, 102
Moment distribution, 84–88
    averages of, 86–88, 102
    log probability plot, 97
    lognormal distribution, 92
Moment sums, 87
Monodisperse aerosol, 8
    generation, 432–444
    settling, 62–64
    sizing, 370
    use of, 123, 428
Monolayer of particles, 146
Motion at high Reynolds number, 321
mppcf, 11
Mucociliary escalator, lung, 235
Multiplets, 435–437
Mycotoxins, 394

Nanoparticles, 9
Natural background aerosol, 308
Navier-Stokes equations, 44, 67, 122
NBS dust generator, 442
NCRP deposition model, 242–243
Nebulizer, compressed air, 429–434
Nebulizers, characteristics of, 430
Nephelometer, integrating, 369–370, 372
Neutralization rate, 336–338
Newton's drag equation, 43–44, 381
Newton's law of viscosity, 23
Newton's region, 43, 55
Newton's resistance law, 42–45
Newton's second law of motion, 112, 449
"NIOSH method" asbestos counting, 422
Nonrespirable particles, 250, 253
Nonspherical particles, 51–53
    coagulation of, 267
    settling at high Reynolds number, 62
Nonspherical particles at high Re, 138
Nonviable particles, 394, 399
Normal distribution, 90–91, 103, 158
Nucleated condensation, 288–292, 443–445
Nucleation
    atmospheric aerosols, 305
    homogeneous, 283–284

Nuclei
    condensation, 278, 289–291
    effect on evaporating droplets, 301
    soluble, effect on equilibrium particle
        size, 289–291
Nuclei mode, 311
Nuclepore filter, see Capillary pore
    membrane filter
Number concentration, 11
    measurement of, 225–226, 292–294, 374
    relation to mass concentration, 83
    standard deviation of count, 226
Numerical aperture, objective lens, 415

Objective lens, 413–414
Oil coating for impaction plates, 133
Oleic acid, 409
OPC, see Optical particle counter
Optical cutoff size, 96
Optical microscope, 413–417
    for measurement of number concentra-
        tion, 225–226
    maximum useful magnification, 416
    objectives, characteristics of, 417
Optical microscopy, for asbestos counting,
    423–424
Optical particle counter, 372–375
    characteristics of, 373
    coincidence error, 374–375
    effect of refractive index, 374
    scattering volume, 374
    truncation error, 375
    use with condensation nuclei counter,
        292
Optical properties of aerosols, 349–370
Orifice meter, 33
Orthokinetic coagulation, 275
Osmotic pressure, 151–152
Owl instrument, 370
Ozone depletion, 312, 314
Ozone layer, 305

Packing density, filter, 186
Partial pressure, vapor, 278–279, 296
Particle density, 10
Particle mean free path, 154–156
Particle Reynolds number, 29
Particle shape, 10, 402

Particle size, 8
    measurement
        automatic methods, 424–425
        direct methods, 402–407, 456
        indirect methods, 65–67, 128–131,
            165–168, 341–345, 370, 456
        range of aerosol measuring instruments,
            456
        range of aerosol properties, 457
Particle size distribution, see Distribution,
    particle size
Particle size-selective sampling, 245–257
Particle velocity
    centrifugal, 47
    gravitational, 47
Particles
Particulate matter, 6–8
Particulate suspensions, types, 1
Peclet number, 194–195
Penetration, electrostatic precipitator, 339
Penetration, see Collection efficiency
    diffusion battery, 165–168
    electrostatic precipitator, 339
    horizontal elutriator, 66
    tube, 163–164
Permittivity, 316–317
Permittivity, relative, 325, 327
Persistence, particle, 121, 193
Personal sampling pump, 228–230, 254
Personal sampling, type of filter holder,
    220
Phagocytic cells, pulmonary region, 235
Photochemical smog, 311
Photochemical smog, formation, 278, 284,
    304, 311
Photometer, 371
Photon correlation spectroscopy (PCS),
    375–376
Photophoresis, 178
Photophoretic force, 178
Piezoelectric mass measurement, 222, 340
Piston-meter, 36
Pitot tube, 31–32, 212
Plateout, 160–165
PM-10 sampling criteria, 255
PM-2.5 criteria, 255
Point-to-plane electrostatic precipitator,
    340–341
Poisson distribution of count, 226

Pollak counter, 293–294
Pollen, 9, 394, 396–397, 443
Polydisperse aerosol
    coagulation coefficient, 268–271
    definition, 8
    test, 438–443
Polystyrene latex spheres, 435
Polystyrene-divinylbenzene spheres, 435
Pore size, porous membrane filter, 183
Porosity, filter, 182
Porous membrane filter, 182–183, 202
Portacount, 294
Porton graticule, 403–405, 419
Positive hole, colony counting, 399–400
Potential gradient, 319
Powder, dispersion of, 438–443
Powder technology, 2
Powders for aerosol generation, 443
Power function size distribution, 104
Pressure
    conversion factors, 448
    gas, measurement, 37–39
Pressure drop, filter, 187, 200–202
Pressure rise index, 387, 389
Primary aerosol, 8
Probability density function, see Frequency
    function
Probability plot, 94
Probe size, still air sampling, 213–214
Probit, 94
Projected area diameter, 403, 406, 425
PSL, see Polystyrene latex spheres
Pulmonary region, 233
Pumps, sampling, 228–232

Quartz, 4, 52, 249
Quartz crystal microbalance (QCM), 221
Quasi-elastic light scattering (QELS),
    375–376

Radiation pressure, 178
Radiometric force, 178
Rainout, 305, 311–312
Rayleigh charge limit, droplets, 334
Rayleigh scattering, 359–360, 364
Rayleigh-Taylor instability, 384
Rebound energy, 146
Red bands, scattered light, 371
Reentrainment, 145

References, general, on aerosol science and technology, 13–14
Refractive index, 350, 370, 415
Refractive index, low-vapor-pressure liquids, 461
Regional deposition, respiratory system, 239–242
Relative humidity
 and saturation ratio, 279–280
 and visual range, 370
 effect on adhesive force, 143–144
 effect on droplet lifetime, 299
Relaxation time, 111–112, 117, 156
 and settling velocity, 112
 and stopping distance, 117
Reserve air, lung, 234
Residual air, lung, 235
Residual charge distribution, 335
Resistance, filter, 187, 200–202
Resolution, microscope, 415–416, 419–420
Respirable dust criteria, 250
Respirable fraction, 251
Respirable fraction samplers, characteristics of, 250–251
Respirable mass
 of dust sample, 252
 sampling, 249–257
 standards, dusts, 251
Respirable particle sampling, 249–254, 398
Respirable samplers, 250
Respiratory defense mechanisms, 233
Respiratory deposition, 233–259
 effect of breathing rate, 239
 mass per unit time, 245
 mechanisms, 233–259
  alveolar, 243
  head airways, 239
  regional, 239, 245
  regional, 239–242
  tracheobronchial, 244
 models, ICRP, 242–245
 models, NCRP, 242–243
 total, experimental measurements, 239
Respiratory regions, 233
Respiratory system, 233–235
Resuspension, 145–146
Resuspension, dust explosion, 386, 390

Reticule, eyepiece, 403
Reverse photophoresis, 178
Reynolds number, 27–31
 cloud, 381
 fiber, 190
 flow in pipes, 30–31
 impactor jet, 127
 particle, 29–31, 44, 48, 55–62, 68, 121
 settling of large particles, 55, 59, 61
Richardson plot, 409
rms
 average diameter, 84
 forward velocity, particle, 154
 molecular velocity, 18–19
Rosin-Rammler distribution, 104
Rotameter, 34–35
 pressure correction, 35
Rotary vane pump, 230
Rotorod sampler, 397–398

Sample volume, requirement for filter samples, 220
Samplers, inhalable, 247–248
Sampling, 182, 206–217
 anisokinetic
  misalignment, 208–209
  velocity mismatch, 210–212
 area, 221
 bioaerosols, 396–400
 breathing zone, 254, 423
 criteria, PM-10, 255
 definitions, 9
 filters, 185
 from ducts, 212–213
 isokinetic, 206–213
 personal, 220
 probe, Stokes number, 209
 pumps, 228–232
 respirable mass, 249–257
  criteria, 249–257
 stack, 220
 subisokinetic, 206, 210
 superisokinetic, 206, 210
 total mass, 222
 tube, losses in, 216–217
Satellite droplets, spinning disk aerosol generator, 434
Saturation charge, 327, 338

Saturation ratio, 279
and Kelvin diameter, 281–283
critical, 284
equilibrium, for droplets, 281–283
nucleated condensation, 288
required for homogeneous nucleation, 283
Saturation vapor pressure, 279
Sauter diameter, 86, 102
Scanning DMA, 344
Scanning electron microscope, 419–422
Scattering coefficient, 353
measurement by nephelometer, 372
Scattering efficiency, 353
Scattering of light, see Light, scattering
Scavenging of aerosol particles by raindrops, 273
Secondary aerosol, 8
Sedimentation cell, 65
Sedimentation, see Settling
Self-cooling of evaporating droplets, 296
Self-nucleation, 283–284
Self-preserving size distribution, 272
Self-similarity, 409
Sensitive volume, optical particle counter, 375
Separation distance, 127, 142, 316
Settling, 46–48, 55–62
atmospheric aerosol, 312–313
chamber, 65
in filters, 192, 195
in respiratory system, 237
of atmospheric aerosol, 306
polydisperse aerosol, 64
stirred, 63–64
tranquil, 62–63
Settling tube, 65
Settling velocity, 46–48
all conditions, summary, 62
at high Reynolds number, 55–62
cloud, 379, 382
effect of buoyancy, 46
effect of particle shape, 51–53
effect of pressure, 51
effect of slip correction, 49
effect on droplet evaporation, 297–298
in terms of mobility, 111
in terms of relaxation time, 112
terminal, 46–48
time to reach, 114

Shape, effect on settling velocity, 51–53
Shape factor
dynamic, 51–53
volume, 406
Shear coagulation, 274–275
Shock wave, 390
SI system of units, 1
SI units, prefixes, 464
Sieves, 10, 451
Silica dust, inhalation hazard, 249
Silver iodide, 288
Similarity
flow, 27
particle motion, 121
Single fiber deposition mechanisms, 190–196
Single fiber efficiency, 193–196
diffusion, 194–195
diffusion-interception interaction, 195
electrostatic attraction, 195–196
impaction, 193–194
interception, 192–193
settling, 195
total, 196
Single fiber theory, filtration, 190
Single particle counter, see Optical particle counter
Singlet ratio, 435–437
Singlets, 435–437
Size distribution, see Distribution, particle size
Size parameter, light scattering, 351
Size selective sampling, 245–257
Skewed distribution, 81, 90
Slip correction, 48–51
filter fibers, 202
impactor cutoff, 126
Slip correction factor, 48–51
effect of pressure, 51
effect of temperature, 460
fibers, 53
for aerodynamic diameter, 54
irregular particles, 52
Smog, 5
Smoke, 6
Smoke tubes, 445
Smoluchowski coagulation, 260. See also Coagulation, thermal
Soap bubble spirometer, 36

Sodium chloride nuclei, 289–291
Soil dust, 304
Soil dust, global emissions, 305
Soil particles, in urban air, 311
Solidity, 202
Solidity, filter, 186, 193
Sound, velocity of, 40
Specific surface, 109–110
Sphere, properties of, 450
Spheres, standard, 435
Spinning disk aerosol generator, 432–434
Spirometer, 36
Spray, 6
Spray electrification, 324
Spreading factor, liquid droplets, 408–409
Sputter coating apparatus, 422
Stabilizer, in polystyrene latex suspension,
    437
Stack sampling, 221
Stack-sampling, EPA method 5, 221
Stage micrometer, 404, 416
Stagnant settling, 63
Standard conditions, 8
Standard density sphere, 53
Standard deviation, 90
Standards
    chrysotile asbestos, 423
    cotton dust sampling, 256
    mineral dust, 226
    nuisance dust, 11
    particulate matter, 11, 221
    respirable dust, 251
    US PM-10, 11
"Stat" electrical units, 316–317
Static charge on sampling filters, 220
Static detachment, 145
Static electrification, charging, 324
Static pressure, 31
Statistics
    descriptive, 75
    particle size, 75–110
Stefan flow, 171, 179
Still air, sampling, 213–216
Stirred coagulation, 275
Stirred settling, 63–64
Stokes diameter, 53
Stokes drag, 45, 151, 179. *See also* Drag
    force

Stokes number
    cylinder, 121
    50% efficiency, 125–126
    filter fiber, 193–194
    general form, 121
    gravitational coagulation, 274
    impactor, 122
    impactor, 125
    sampling tube, 209
Stokes region, 44–45, 112, 117, 320–322
Stokes's law, 42, 44–46, 49, 113
    derivation, 67
    dynamic shape factor, 51–53
Stokes-Einstein equation, 153
Stopping distance, 117–119
    in respiratory impaction, 237
    outside Stokes region, 117
    relation to mean free path, 155
    use of, in Stokes number, 121
Stratified counting, 406
Stratospheric aerosol, 305–306, 312
Streamlines, flow around an obstacle, 120,
    192, 206, 237–239
Stripper, 168
Submicrometer particles, 8–9
Supersaturated vapor, 279–280
Supersaturation, 279–280
    condensation nuclei counter, 292
    in atmosphere, 288–289
    required for condensation on insoluble
        nuclei, 288
    required for condensation on ions, 288
    required to prevent droplet evaporation,
        283
Surface deposition, 161
Surface field for spontaneous charge
    emission, 333–334
Surface mean diameter, 86, 89, 102
Surface median diameter, 102
Surface tension
    adhesion, 141, 143
    charge limit, 334
    low-vapor-pressure liquids, 461
    of water versus temperature, 455
Survey instruments, 222–225
Survivability of bioaerosols, 396
Suspended particulate matter, 1
Suspensions, particulate, types, 1
Sutherland equation, 25

Tapered element oscillating microbalance (TEOM), 223
TDMA, tandem differential mobility analyzer, 344–345
Temperature depression correction, 296
Temperature gradient
    gas, 171–173, 176
    particle, 173
Terminal centrifugal velocity, 47
Terminal electrostatic velocity, 320–322
Terminal settling velocity, 46, 49
Terminal velocity, time to reach, 114
Test aerosol, DOP, 431, 444–445
Test aerosols, production of, 428–445
Thermal coagulation, see Coagulation, thermal
Thermal conductivity of air, 173–174
Thermal conductivity, various materials, 174
Thermal deposition, 174–175
Thermal force, 172–173
Thermal precipitation, use in air cleaning, 176–178
Thermal precipitator, 176–178
Thermal velocity, mean
    gas, 21
    particle, 154
Thermodynamic diameter, 167
Thermophoresis, 171–178
Thermophoretic velocity, 172–173, 217
Thomson-Gibbs equation, 281–282
Thoracic fraction, 251, 255
Threshold limit value, 11, 226
Threshold of brightness contrast, 367–368
Tidal volume, 234
Time to double particle size by coagulation, 266
Time to reach terminal velocity, 114
Time to stop, particle, 117
Time-of-flight instruments, 136–138
Titanium tetrachloride, 445
Total mass, sampling, 222
Toxicity of aerosol particles, 1
Tracheobronchial region, 233, 240–241
Tranquil settling, 62–63
Transition region, 51
Transmission electron microscope, 422
Transmissometers, 371

Transmittance, light, 352
Transport losses, 216–217
    in tube bend, 216
    laminar flow, 217
    turbulent flow, 217
Tropospheric aerosol, 306–307, 312
Tropospheric background aerosol, 306
Tropospheric electric field, 320
Tube, diffusion to walls, 162–164
Turbulent air flow, 145
Turbulent eddies, 145
Turbulent flow, 27
    coagulation, 275

U-tube manometer, 37
Ultrafine particles, 8–9
Ultramicroscope, see Sedimentation cell
Ultrasonic coagulation, 275
Ultrasonic nebulizer, 429, 431–432
Ultraviolet fluorescence of bioaerosols, 400
Ultraviolet light and homogeneous nucleation, 284
Unipolar ions, 324, 331
Units
    cgs system, 6
    electrostatic, 316–317
    SI system, 1
Unsaturated, vapor, 279
Urban aerosol, 304, 307–312
US PM-10, 11
UV-APS bioaerosol monitor, 400

van der Waals force, 141–142, 283
van't Hoff's law, 151
Vapor pressure, 279
    at droplet surface, 296
    effect of dissolved salt, 289–291
    equilibrium, 279
    low-vapor-pressure liquids, 461
    of water versus temperature, 455
    saturation, 279
Vapors, properties of, 452
Variable area meter, 34–35
Variable head meter, 33–34
Velocity
    boundary layer, turbulent flow, 277
    changing external force, 116
    deposition, 162

gas
  for still air sampling criteria, 215
  measurement of, 31–32
gradient, coagulation, 274–275
head, 32
particle, adjustment of, 115
pressure, 31–32
terminal
  centrifugal, 47, 123–124
  electrostatic, 320
  for constant force, 112
  settling, 46–48. *See also* Settling
    velocity
  thermophoretic, 172–173
Venturi meter, 33
Venturi scrubbers, Stefan flow in, 180
Viability of bioaerosols, 396
Viable particles, 394, 396, 398
Vibrating orifice aerosol generator,
  432–434
Virtual impactor, 134–136, 255, 431
Viruses, 9, 395–396
Viscosity
  aerosol, 379
  air, 24–27, 29, 448, 452
  coefficient of, 24
  effect of temperature, 24–25
  Newton's law of, 23
  temperature dependence, 24–25
Viscous force, gas, 23, 44
Visibility, 311, 364–370
Visual range, 364–370
  and extinction coefficient, 368
  effect of contrast, 364
Volcanic aerosol, 6

Volcano, aerosol emissions, 304–305, 314
Volume fraction
  of fibers, filter, *see* Solidity, filter
  suspensions of spheres, 435–437
Volume mean diameter, 89, 102
Volume median diameter, 102
Volume ratio, adiabatic cooling, 280
Volume shape factor, 406, 408

Wall losses in tubes, diffusive, 162–164
Walton-Beckett graticule, 423
Washout, 305, 311–312
Water, properties of, 455
Water vapor, properties of, 455
Wavelength of light, 350
Wavelength of light and microscope
  resolution, 415
Weibull distribution, 104
Weight stability, filters, 188, 218–219
Weighted distribution, *see* moment
  distribution
Weighted mean diameters, 86–87
  from moment average diameters, 87
Wet test meter, 37
Wilson cloud chamber, 289
Wind, effect on droplet evaporation,
  297–298
Wind, effect on aspiration efficiency, 246
Winnowing gas stream, 66
Wire-in-tube geometry, field strength, 319,
  331–332, 338–339
"Whitehouse" effect, 314
Wright dust feed, 439–441

Yeasts, 395